THE NEW NATURALIST

A SURVEY OF BRITISH NATURAL HISTORY

THE OPEN SEA—ITS NATURAL HISTORY

THE WORLD OF PLANKTON

The aim of this series is to interest the general reader in the wild life of Britain by recapturing the inquiring spirit of the old naturalists. The Editors believe that the natural pride of the British public in the native fauna and flora, to which must be added concern for their conservation, is best fostered by maintaining a high standard of accuracy combined with clarity of exposition in presenting the results of modern scientific research. The plants and animals are described in relation to their homes and habitats and are portrayed in the full beauty of their natural colours.

THE OPEN SEA: ITS NATURAL HISTORY

Part I : The World of Plankton

by

SIR ALISTER HARDY

M.A., D.Sc., F.R.S

Fellow of Merton College,
Hon. Fellow of Exeter College,
and Linacre Professor of Zoology
and Comparative Anatomy
in the University of Oxford

WITH 142 WATERCOLOUR DRAWINGS
BY THE AUTHOR
67 PHOTOGRAPHS IN BLACK AND WHITE
BY DOUGLAS WILSON AND OTHERS
AND 300 LINE DRAWINGS AND MAPS

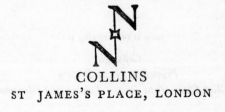

COLLINS
ST JAMES'S PLACE, LONDON

First edition 1956
Reprinted, with revisions, 1958
Reprinted 1962
Reprinted 1964
Second edition, 1970
Reprinted 1971

ISBN 0 00 213164 1

Copyright
Printed in Great Britain
Collins Clear-Type Press: London and Glasgow

CONTENTS

PLATES IN COLOUR

PLATES IN BLACK AND WHITE

*It should be noted that throughout this book Plate numbers in Arabic
figures refer to Colour Plates, while Roman numerals are used for
Black-and-White Plates*

EDITORS' PREFACE

PROFESSOR HARDY began his marine biologist's life over a third of a century ago on his return from service in the first world war. After Oxford and a scholarship to the Stazione Zoologica at Naples, he soon became a member of the Fisheries Department in the Ministry of Agriculture and Fisheries; and, in the middle 'twenties, served as Chief Zoologist to the *R.R.S. Discovery* expedition, to the Antarctic seas, making a special study of plankton. His subsequent professorships— first at University College (now the University of) Kingston-upon-Hull; next at Aberdeen; and since 1945 at the University of Oxford— have brought him the highest academic status and honours, but have not kept him away from his beloved sea. In the closing stages of the writing of this volume, as the editors well remember, he was correcting the typescript, and completing his unique and wonderful colour illustrations, on the deck of the latest Royal Research Ship, *Discovery II*, scanning the contents of each netting or dredging, sketching new or rare creatures of the sea before their colour faded, applying himself to his research with an enthusiasm excelling that of most naturalists of half his age.

If the editorial board were asked to select from Professor Hardy's many scientific qualities that which has contributed most to the creation of this extraordinary book, they would perhaps settle for enthusiasm. Throughout *The Open Sea* it is quite apparent that he is devotedly obsessed by, and interested in, animals; he is eternally curious about the nature of their adaptations and lives, brilliantly critical in the examination of their mysteries, acutely lucid and at the same time highly artistic in his depiction of them in his remarkable plates. It was a welcome burst of enthusiasm that caused Professor Hardy to write so much and so well of the life of the sea that he has written us two books instead of one. It is the first of these, concerned with the general natural history of the open sea and the world of its plankton, that we here welcome. The second part of *The Open Sea* concerns the open sea's fish and fisheries, and will be published some time in 1957 or early 1958; like the present book, it will be illustrated by Professor Hardy's own colour paintings, which represent what no colour-camera has yet been able to catch, and by black and white photographs by that most distinguished marine biologist and skilful photographer, Douglas P. Wilson.

To most readers the subject of this first of Professor Hardy's two contributions to our series—the world of Plankton—will be relatively unknown and mysterious; but here the enlightened amateur naturalist is shown how, with modest equipment he may investigate it himself. The world of plankton is a world of complex anatomy, much of which can be understood only with the lens of the microscope. The life-histories of the animals are also complicated; some of them are extraordinary. To describe the plankton of our seas, and to set it in its pattern of community, climate, sea-scene and season is a major task. Professor Hardy has brought vast knowledge and experience and scholarship to a synthesis never before attempted.

THE EDITORS

AUTHOR'S PREFACE

ORIGINALLY it had been intended that the whole natural history of the sea, apart from that of the sea-shore and of the sea-birds already dealt with in the *New Naturalist* series, should be treated in one general volume. As the writing proceeded, however, it became clear that to do justice to the subject it would be impossible to include all its different elements within a single cover. There is the life of the plankton in almost endless variety; there are the many kinds of fish, both surface and bottom living; there are the hosts of different invertebrate creatures on the sea-floor; and there are those almost grotesque forms of pelagic life in the oceans depths. Then there are the squids and cuttlefish, and the porpoises, dolphins and great whales. In addition man's fisheries now play such an important part in the ecology of our waters that they also must form a part of any general natural history of the sea.

Certainly there is too much material to go into one volume. There occurs, however, a fairly natural division between the teeming planktonic world and the other categories of life it supports: the fish, the whales and the animals of the sea-bed. This first book on the open sea deals mainly with the plankton; it aims at giving the general reader a non-technical account, save for the necessary scientific names, of its many remarkable animals and showing how, with only a little trouble, quite a lot of them may be seen and studied alive. Perhaps to some it may introduce a new world of life—a world so unusual that few of its inhabitants have homely English names at all. It is hoped, too, that it may be a guide to the plankton for university students who are beginning their studies in marine biology. The book also deals with the water-movements and the seasons in the sea; and it contains an account of the squids and cuttlefish, and of those queer creatures, including the deep-water (and often luminous) fish, swimming in the great depths only a little way beyond our western coasts. It will conclude by showing how a study of the plankton is helping us to have a better understanding of the lives of our commercially important fish. Later, and before very long, will come the sequel: a separate volume devoted mainly to fish and fisheries, but also including whales, turtles and other marine animals which are likewise, directly or ultimately, dependent on the plankton for food.

Before going any further I must thank the publishers and editors, not only for all the trouble they have taken over the production of this book, but also for the patience they have kindly shown over my delay in its completion. I accepted their invitation to write it in August 1943, some twelve years ago; it has, however, meant more than the writing. My excuse for its late arrival will be offered after I have made my main acknowledgment.

The value to the book of the remarkable collection of photographs by Dr. D. P. Wilson of the Plymouth Laboratory will be clear to all, but just how wonderful they are and consequently how lucky I am to have them as illustrations, may not at once be fully appreciated by those who are not yet familiar with the living plankton of the sea. Douglas Wilson has long been recognised as the leading photographer of marine life and his beautiful pictures in black-and-white and in colour which graced Professor C. M. Yonge's *The Sea Shore* in this series of volumes will, I am sure, have been seen and admired by most of my readers. I, too, am showing some of his studies of the larger forms of life, such as those of cuttlefish or his unusual view of that strange jelly-fish, the Portuguese-man-of-war, taking a meal; it is, however, his photographs of the tiny plankton animals to which I particularly wish to draw attention here. Though they are taken through a microscope, they are photographs of creatures swimming naturally, very much alive and certainly kicking. Never before has such a series of photomicrographs of living members of the plankton been published; they are unique and will, I believe, be of immense value not only to marine naturalists but to all students of invertebrate zoology. They are the fortunate result of a remarkable combination; Dr. Wilson has brought his skill and artistry to work with that very modern invention the electronic flash. For the first time this device has made possible such instantaneous pictures at a very high magnification. It is not only that invention, however, which makes these pictures unique; while others will follow him, Dr. Wilson's photographs will always have a quality of their own, because he is an artist as well as a scientist. He is not satisfied until he has produced a photograph that has an appeal on the score of composition as well as on that of scientific value. All his photographs except two (the stranded jelly-fish and squid) are of living animals. A few excellent black-and-white pictures by other photographers are included in some of the plates and these are acknowledged in the captions or the text.

It was my hope, and that of the editors, that in addition to his black-and-whites Dr. Wilson would have been able to contribute a series of colour photographs of the living plankton and especially of the richly pigmented animals from the ocean depths. At that time the electronic flash was only just being developed and he felt unable to attempt them. The movement of the ship at sea, he said, would prohibit the use of a long enough exposure to enable the deep-water animals to be photographed in colour by ordinary means; they quickly die and fade, and so must be taken as soon as they are brought to the surface. I had already had some experience in making water-colour drawings of living plankton animals on the old *Discovery* during the Antarctic expedition of 1925-27; the editors kindly allowed me to undertake a series of such studies to form the accompanying twenty-four colour plates. To obtain and make drawings of the full range of animals which I felt to be desirable, meant a considerable delay and this was added to an earlier postponement of my start on the book caused by my being appointed to the chair of Zoology at Oxford soon after I had accepted the invitation to write it. For several years the work of my new department and research to which I was already committed took all my attention.

All save seven of the 142 drawings in the plates were made from living examples or, in a few cases, from those taken freshly from the net when some deep-water fish and plankton animals were dead on reaching the surface. The seven exceptions, which are noted where they occur, were drawn from preserved specimens but with memories or colour-notes from having seen them alive; I should like to have drawn these too from life, but I could delay the book no longer. It may be of interest to record how the drawings were made. All the animals, except the larger squids and jelly-fish, were drawn either swimming in flat glass dishes placed on a background of millimetre squared paper where they were viewed with a simple dissecting lens, or on a slide under a compound microscope provided with a squared micrometer eyepiece; in either case the drawings were first made in outline on paper which had been ruled with faint pencil lines into squares which corresponded to those against which the specimen was viewed. In this way the shape and relative proportions of the parts could be drawn in pencil and checked and rechecked with the animal until it was quite certain that they were correct. The outline was then gone over with the finest brush to replace the pencil by a permanent and more expressive water-colour line; next all the pencil lines, including the background squares, were rubbed out and the full colouring of the drawing proceeded with. If rough weather at sea made such a course impossible, the living animal would be sketched in pencil, and painted, in perhaps one or two different positions, to give life-like attitudes and colouration without attempting to get the detailed proportions exactly right; it would then be preserved in formalin for accurate redrawing by the squared-background system when calmer conditions returned. The animals I have selected for illustration are mainly either those which are not included in the black-and-white photographs or those for which colour can add supplementary information. I have, for example, drawn some of the transparent but iridescent comb-jellies, but not the transparent and colourless arrow-worms or salps. I am most grateful to the Sun Engraving Company, who made the blocks for the colour plates, for the great care they have taken in making such excellent reproductions.

I must now make special acknowledgments in regard to these drawings. First I must record my thanks to Dr. N. A. Mackintosh, the Deputy Director in charge of the biological research of the National Institute of Oceanography, for kindly allowing me to accompany the R.R.S. *Discovery II* on two of her biological cruises in the Atlantic in the summers of 1952 and 1954. It is to the *Discovery*, with all her equipment of deep-water nets, powerful winches and laboratory accommodation, that 71 of these drawings are due, including all those representing the remarkable animals which live in the great depths over the edge of the continental shelf to the southwest of Britain. Without such facilities they could never have been made; actually three of them date back to earlier days when I had the honour of sailing south in the old *Discovery* in 1925. Next I must thank a number of kind helpers who have sent me living specimens of plankton in specially protected Thermos flasks from many parts of the coast: Mr. J. Bossanyi from Cullercoats, Dr. E. W. Knight-Jones from Bangor, Dr. Richard Pike from Millport, Professor J. E. G. Raymont from Southampton and Mr. R. S. Wimpenny from Lowestoft. Although I made many visits to different places to draw my specimens, there were still a number I could not get myself in the time available; these were supplied by these kind friends who were on the lookout for what I wanted at widely separated points. I am most grateful to Dr. Marie Lebour and to the Council of the Marine Biological Association of the United Kingdom for kind permission to reproduce some of her beautiful drawings of living plankton animals capturing their prey; these, which form my text-figures 26, 27, 35, 40 and 41, were originally published in her papers in the Association's *Journal* in the years 1922 and '23. Then I must thank Sir Gavin de Beer and members of his staff at the British Museum (Natural History), particularly Dr. W. J. Rees and Mr. N. B. Marshall, for kindly allowing me to make many of the black and white drawings in the text from specimens in the museum collections. I am similarly indebted to Dr. J. H. Fraser of the Marine (Fisheries) Laboratory at Aberdeen who has let me draw some of the beautiful plankton animals he has caught to the north and west of Scotland; and to Dr. Helene Bargmann and Mr. Peter David who have also kindly given me much help on looking out specimens from the *Discovery* collections for me.

With no less gratitude, I must make acknowledgments regarding the text. Apart from the more normal editorial comments and suggestions I particularly want to say how much I am indebted to my old friend—and once Oxford tutor—Dr. Julian Huxley, who read the whole book with the greatest care and made many valuable suggestions for its improvement. My typescript—how reminiscent of my undergraduate essays of 1919 and '20!—was covered with his pencilled scribblings in the margin: "Surely you should refer to—, this might be made more emphatic" and the like; not all were adopted, but certainly a great many. The chapter on water movements was read by Dr. G. E. R. Deacon, and that on squids and cuttlefish by Dr. W. J. Rees; I am indeed grateful to them for a number of helpful suggestions they kindly made.

Finally I wish to draw the attention of those who are not scientists to a glossary at the end of the book giving a simple explanation of the few technical terms which have been unavoidably used; and for the zoologist I would point out that the authorities for the different specific names will be found quoted after these names in the index and not in the text where they are left out to avoid undue elaboration.

A.C.H.

CHAPTER I

INTRODUCTION

THERE IS a very simple fact about the sea which makes its inhabitants seem even more remote from us than can entirely be accounted for by their being largely out of sight. To make my point allow me to imagine a world just a little different from our own.

Suppose for a moment that we live in a country which is bounded on one side by a permanent bank of fog. It is a grey-green vapour, denser even than that often known as a London particular, and it has a boundary as definite as the surface of a cloud so that it is like a curtain hanging from the sky to meet the ground; we cannot enter it without special aids except for a momentary plunge and as quickly out again for breath. We can see into it for only a very little way, but what we do see is all the more tantalizing because we know it must be just a glimpse—a tiny fraction—of all that lies beyond. We find it has life in it as abundant as that of our own country-side, but so different that it might be life from another world. No insects dwell beyond the barrier, but other jointed-legged creatures take their place. Unfamiliar floating forms, like living parachutes with trailing tentacles, show their beauty and all too quickly fade from view; then sometimes at night the darkness may be spangled with moving points of light— living sparks that dart and dance before our eyes. Occasionally gigantic monsters, equal in size to several elephants rolled into one, blunder through the curtain and lie dying on our land.

To make a reality of this little flight of fancy all we have to do is to swing this barrier through a right-angle so that it becomes the surface of the sea. How much more curious about its unfamiliar creatures many of us might be if the sea were in fact separated from us by a vertical screen—over the garden wall as it were—instead of lying beneath us under a watery floor. Who as a child has not envied the Israelites as they passed through the Red Sea as if marching through a continuous aquarium: "and the waters were a wall unto them on their right hand, and on their left." ? What might they not have seen ? Because normally our line of vision stretches out across the sea to the skyline and carries our thoughts to other lands beyond, many of us tend to overlook this perhaps more wonderful realm beneath us, or we seem to think it must be too difficult of access ever to become a field for our exploration or delight.

The aim of this book is to give the general reader an account of the natural history of the open sea around our islands and at the same time show how he may, with only modest equipment, see something of this strange world for himself. The amateur naturalist afloat—whether on a yachting cruise, on a fishing vessel, or just out in a rowing boat—may see much if he has the right kind of quite simple gear and knows how to use it; he may perhaps also be lucky and make original observations which will be a contribution towards finding an answer to one of the many unsolved riddles of the sea. The book will also give a sketch of some of the factors upon which the success of our great sea fisheries depend. The lives of the different fish are like threads woven in a web of life—a network of inter-relationships between many various creatures large and small, as complex as any on the land. The story of fishery research, which belongs mainly to our subsequent volume, is so closely linked with this unseen web, that it is hoped an account of these less familiar animals may be as interesting to the fishermen as to the naturalist; indeed many fishermen *are* naturalists and have much of importance to tell the scientist.

As our title indicates, the book will deal with the open sea—the sea beyond the coastal waters. The life of the intertidal zone has already been beautifully treated in this series of volumes by my friend Maurice (C. M.) Yonge (1949). The sea-shore can be studied by direct observation as the tide recedes and has long been a happy hunting-ground for the naturalist; he can lift up the fronds of seaweed, turn over stones, probe into rock-pools and dig into the sand and mud. Our methods of studying the life of the open sea must be very different; it is far from 'open' to the investigator, being in fact a hidden world, but this makes its exploration all the more exciting. Deep-sea photographic and television cameras are important new developments which promise much for the future; they, however, as also submarine observation chambers like the bathysphere, must for some time to come be regarded as very costly and specialist equipment giving us here and there direct confirmation of what we usually have to find out by other means. The diving helmet and the aqualung may help us to see something of this enchanting world in shallow water, but for the discovery of what is happening over wide stretches of the underwaters of the open sea we must devise more indirect methods.

The fact that we can see only a very little way below the surface indicates a property of water, and particularly of the sea, which is of fundamental importance to the life it contains. Held up in a glass, water appears so very transparent that we are at first surprised to find how quickly light is absorbed in the sea itself and what a little distance its rays will penetrate. Measurements made in the English Channel off Plymouth show that at a depth of five metres (just over 16 feet) the intensity of light is less than half that just below the surface, while at 25 metres it is only an insignificant fraction, varying between 1½ and 3 per cent. This at once tells us that the green plants, which must have sunlight in order to live, will only be found in the upper layers of the water.

The one real difference, of course, between animals and plants is a matter of their mode of feeding. We know that an animal of any kind, whether mammal, fish, shrimp, or worm, must have what we call organic food: proteins, carbohydrates (sugars, starches and the like) and fats, which have been built up in the bodies of other animals or plants. One animal may feed upon another kind of animal which in turn may have lived upon other kinds, and perhaps these upon yet others, but always these food-chains, long or short, must begin with animals feeding upon plants. Only the green plants, with that remarkable substance chlorophyll acting as an agent, can build themselves up from the simple inorganic substances by their power of using the energy of sunlight (photosynthesis); they split up the molecules of carbon dioxide, liberate the oxygen, and combine the carbon with the oxygen and hydrogen of water to form simple carbohydrates, which are then elaborated into more complex compounds by being combined with various minerals in solution. On the land we are all familiar with this elementary fact of natural history; my reason for recalling it is to emphasise that it is of universal application. The plants are the producers and the animals the consumers as much in the sea as on the land. Indeed 'all flesh is grass'.

Where then in the sea, we may ask ourselves, are all the plants upon which the hordes of animals must depend? They cannot grow in the darkness or dim light of the sea-floor, and the seaweeds, forming but a shallow fringe along the coasts, are of no real importance in the economy of the open sea. From the deck of a ship, or even from a rowing boat, we can see no plant-life floating near the surface; yet we know it must be there. Another little flight of imagination will, I think, help us to get some idea of the extent of this elusive vegetation.

Let us suppose for a moment that the herring is not a fish, but a land animal. We know that some three thousand million herring are landed every year at ports in the British Isles; these, together with all those landed in other countries, must be only a small fraction of their total number, for we also know that herring are the food of so many other abundant animals of the sea. For simplicity let us consider them to be feeding directly upon plants—and let us imagine them in their unnumbered millions sweeping across the continent. If we do this it needs no imagination to see that the countryside would be stripped of vegetation as if by locusts. Now let us think of the other fish in the sea besides the herring: the cod, haddock, plaice, skate and such that fill our trawlers (as distinct from the herring-drifters) to the extent of more than a million tons a year; then also think of the crowded invertebrate life of the sea-bottom. If all these animals were on the land as well, what an immense crop of plants it would take to keep them supplied with food! There are indeed such luxuriant pastures in the sea but they are not obvious because the individual plants composing them are so small as to be invisible to the unaided eye; we can only see them through a microscope. Their vast numbers make up for their small size.

As an introduction to all that follows let us consider the natural economy of the sea in its simplest terms. We have the sun shining down, its rays penetrating the upper layers of the water; we have the gases, oxygen and carbon dioxide, dissolved in it from the atmosphere; we have also the various mineral salts—notably phosphates and nitrates and iron compounds—continually being brought in by the erosion of the land, and there are minute traces of some essential vitamin-like substances. These are ideal conditions for plant growth. Just as these are spread through the water, so is the plant life itself scattered as a fine aquatic 'dust' of living microscopic specks in untold billions. In a shaft of sunlight slanting into a shaded room we have all watched the usually invisible motes floating in the air, floating because they are so small and light; these tiny plants remain suspended in the water in just the same way. Many of them are provided with fine projections like those of thistledown to assist in their suspension.

Feeding upon these tiny floating plants, and also like them scattered through the sea in teeming millions, are little animals. Crustacea, little shrimp-like creatures of many different kinds, predominate; mostly they range in size from a pin's head to a grain of rice, but some are larger. There are hosts of other animals as well: small worm-like forms, miniature snails with flapping fins to keep them up, little jelly-fish, and many other kinds which surprise us with their unexpected shapes and delicate beauty when first we see them through the microscope.

All these creatures, both animals and plants, which float and drift with the flow of tides and ocean currents are called by the general name of *plankton*. It is one of the most expressive technical terms used in science and is taken directly from the Greek πλανκτον. It is often translated as if it meant just 'wandering', but really the Greek is more

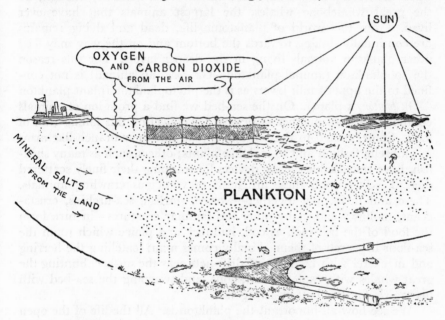

FIG. 1

A diagrammatic sketch illustrating the general economy of the sea.

subtle than this and tells us in one word what we in English have to say in several; it has a distinctly passive sense meaning *'that which is made to wander or drift'*, *i.e.* drifting beyond its own control—unable to stop if it wanted to. It is most useful to have one word to distinguish all this *passively* drifting life from the creatures such as fish and whales which are strong enough to swim and migrate at will through the moving waters: these in contrast are spoken of as the *nekton* (Gk *nektos*, swimming). Actually when they are very young, the baby fish are strictly speaking part of the plankton too, for they are also carried along at the mercy of the currents until they are strong enough to swim against them. Photographs taken through a microscope of some typical planktonic plants and animals are shown in Plates I and II; they have been caught by drawing a net of fine silk gauze through the water. Their natural history will be dealt with in later chapters.

The simple sketch in Fig. 1 (p. 5) shows this general economy of the sea in diagram form. A number of fish, including the herring, pilchard, sprat, mackerel and the huge basking shark, feed directly upon the little plankton animals; and so also, curiously enough, do the great whalebone whales, the largest animals that have ever lived. From this world of planktonic life, dead and dying remains are continually sinking towards the bottom and on the way may feed other plankton animals living in the deeper layers. For this reason the zoöplankton (animal plankton—Gk *zoön*, an animal) is not confined to the upper sunlit layers as is the phytoplankton (plant plankton —Gk *phuton*, a plant). On the sea-bed we find a profusion of animals equipped with all manner of devices for collecting this falling rain of food. Some, rooted to the bottom, spread out their branch-like arms in umbrella fashion and so look like plants; others, such as many shell-fish, have remarkable sieving devices for trapping their finely scattered diet. Feeding upon these are hosts of voracious crawling animals. These and their prey together—worms, starfish, sea-urchins, crustaceans, molluscs and many other less familiar creatures—in turn form the food of the fish such as cod, haddock and plaice which roam the sea-floor in search of them. Finally comes man: catching the herring and mackerel with his fleets of drift-nets near the surface, hunting the great whales with explosive harpoons, and sweeping the sea-bed with his trawls for the bottom-living fish.

We see how all-important the plankton is. All the life of the open sea depends for its basic supply of food upon the sunlit 'pastures' of floating microscopic plants.

Our knowledge of life in the sea has been built up step by step by many pioneer naturalists. Oceanography is still quite a young science; its beginnings were made only a little over a hundred years ago. It is worth while looking back.

The vast community of planktonic animals and plants was unsuspected till it was discovered by the use of a very simple device, the tow-net: a small conical bag of fine silk gauze or muslin, usually with a little collecting jar at its end, towed on a line behind a boat. In nearly all the text-books of oceanography it is stated that the tow-net was first used in 1844 by the German naturalist Johannes Müller, and I have myself been guilty of repeating this error. It is certain that

Müller's researches excited the scientific world and led many others to follow him; but our own great amateur naturalist J. Vaughan Thompson, when serving as an army surgeon in Ireland, was using a tow-net to collect plankton from the sea off Cork as early as 1828. It was there that he first described the zoëa, the young planktonic stage of the crab. A little later, 1833, he discovered the true nature of the barnacles and so solved an age-long puzzle. These enigmatic creatures, fixed to rocks or the bottoms of ships, had been thought to be aberrant molluscs. Thompson caught little undoubted crustaceans in his tow-net and found that they settled down to be transformed into barnacles. His classical discoveries were described in privately printed memoirs which he published in Cork; they are among the rarest items of biological literature. He showed that the plankton consisted not only of little creatures permanently afloat, but also of the young stages—larvae, as the scientist calls them—of many bottom-living animals; these latter more sedentary forms throw up their young in clouds to be distributed far and wide by the ocean currents, just as many plants scatter their seeds in the wind for the same purpose. Charles Darwin also used a tow-net before Müller, on his famous voyage in the *Beagle*; in his *Journal of Researches* (1845) under the date of 6 December 1833 he writes: "During our different passages south of the Plata I often towed astern a net made of bunting and thus caught many curious animals." Today many forget that our famous T. H. Huxley, champion of Darwinism, began his career as did Darwin before him, as a great field naturalist; in 1846 he sailed for the South Seas as surgeon in H.M.S. *Rattlesnake* and by his use of the tow-net laid the foundations of our knowledge of those remarkable composite jellyfish-like animals, the siphonophores, which we will later discuss (p. 111).

Another simple device, the naturalists' dredge—a coarse netting bag on a rectangular iron frame—dropped and dragged along the bottom of the sea revealed another new world of life. It was first used by two Italian zoologists, Marsigli and Donati, in the middle of the eighteenth century, but it was another of our own great marine naturalists, Edward Forbes, who became the leading pioneer in this work; he began his dredging in 1840, both in British waters and in the Aegean Sea.

How deep in the sea can life exist? This became the subject of much controversy among scientists following the discoveries made by the use of an ingenious device invented by just a boy—a brilliant young midshipman in the U.S. Navy—J. M. Brook. He hit on the idea of attaching a quill to the sounding lead used in plumbing the ocean depths and so bringing to light a sample of the ooze from the bottom into which it had penetrated. It gave only a tiny sample— but how exciting! That was in 1854, and soon from all over the Atlantic basin, from any depth over 1000 fathoms, came samples of oozy sediment containing minute calcareous shells. These were shells of animals belonging to the group of the Protozoa (single-celled animals) known as Foraminifera and nearly all belonging to one genus, called *Globigerina* on account of the spherical form of their shells. This form of deposit has consequently become known as Globigerina ooze. Did the creatures which made the shells actually live at these great depths, or did the shells fall from near the surface when their floating owners died? That was the problem. It is amusing for us now to recall that most of those who held the latter and correct view did so on quite false grounds: they believed that it would be quite impossible for life to exist at these great depths and that therefore the shells must have fallen from above. A drawing of a living *Globigerina* is shown in Plate 2 (p. 81).

Edward Forbes had considered there was what he called a *zero of life* at about 300 fathoms—a boundary below which no life could stand the great pressure of the depths. This fallacy was soon to be exposed. The laying of submarine cables was just beginning. In the Mediterranean one of these after a little use had parted and was hauled up for repair in 1858; it came up encrusted with bottom-living animals, some of them at points on the cable which must have lain at a depth of over 1000 fathoms. Once it is pointed out, the truth of the matter seems obvious: an aquatic animal should feel no ill effects of pressure provided it has no spaces or bubbles filled with air or gas inside it. All liquids are only very slightly compressible. A body made up of fluid or semi-fluid protoplasm, and covered with a flexible or elastic skin, will contract only very slightly even under the greatest pressure; its contents too will be of course at the *same pressure* as the surrounding water. With the stresses inside and outside the body perfectly balanced in this way, the animal can have a most delicate structure and make the finest movements just as well in the great depths as can one living nearer the surface. Even the seemingly rigid armour-platings of such animals as crabs are in fact made up of a number of parts separated from one another by thinner flexible joints, so that changes of pressure are equalised inside and out; the same applies to the starfish and sea-urchins, whose armour is actually not strictly on the outside of the body, but just below the skin.

Simple as the explanation seems to us now, the discovery of these animals living under great pressure came as a real surprise to most people. This was all the more extraordinary, for actually there was in existence a thoroughly attested instance of a remarkable starfish-like animal (one of the Gorgonocephalidae with branching arms) being brought to the surface from a depth of 800 fathoms; it came up entangled round a sounding line on Sir John Ross's expedition to Baffin Bay in 1818. It had been forgotten or overlooked by the naturalists of a later generation, who also did not appreciate the significance of the dredgings reported by his even more famous nephew Sir James Clark Ross. Accompanied by the young Joseph Hooker, he had made a number of rich hauls from depths down to 400 fathoms during those great south polar voyages in the *Erebus* and *Terror* from 1839 to 1843. Unfortunately these important deep-sea collections, which contained marine invertebrate animals in great variety, were subsequently lost to science.[1]

The waters immediately to the north and west of the British Isles may perhaps be regarded as the cradle of oceanography; they became the scene of the pioneer deep-sea dredging expeditions in the naval surveying ships *Porcupine* and *Lightning* led by Dr. W. B. Carpenter and Professor (later Sir) Wyville Thomson. During the summers of 1868-70 they made nearly 200 dredge hauls over a wide area and reached a depth of 2,435 fathoms; as far down as they went they revealed a wealth of life and opened up a new world to the naturalist. Thomson's great book *The Depths of the Sea* (1873) is still fascinating reading. It was their remarkable discoveries, together with the interest taken in the new venture of laying transoceanic cables and the consequent need for a more accurate knowledge of the ocean floor, that led in 1872 to the dispatch by the British Government of H.M.S. *Challenger* on her famous expedition; under the leadership of Sir Wyville Thomson she sailed on a three and a half years' voyage to explore all the oceans of the world. The results of this magnificent venture filled more than 50 large volumes with a wealth of information not only of the life of the ocean and of the nature of the sea-floor as revealed by tow-net and dredge, but also about the physics and chemistry of the sea at

[1] In the "Summary of the Scientific Results of the *Challenger*" Part 1, p. 79, Sir John Murray refers to the deep dredging of the *Erebus* and *Terror*. He says: "Sir James Ross was an indefatigable zoological collector, but it is to be regretted that the large collections of deep-sea animals, which he retained in his own possession after the return of the expedition, were found to be totally destroyed at the time of his death. Had they been carefully described during the cruise or on the return of the expedition to England, the gain to Science would have been immense, for not only would many new species and genera have been discovered, but the facts would have been recorded in journals usually consulted by zoologists instead of being lost sight of as was the case."

different depths. Oceanography as an organised branch of science had come into being. Other nations followed the example of the *Challenger* and sent out similar expeditions.

Having mentioned *The Depths of the Sea* I must also refer to another great book of similar title which I believe will always be a classic in the literature of Oceanography: *The Depths of the Ocean* by Sir John Murray and Professor Johan Hjort, published in 1912. Murray was on the *Challenger* with Wyville Thomson and later, when Thomson's health failed, directed the Challenger Office, seeing to the completion of all the work and the editing of the great series of Reports; Hjort, who died as recently as 1948, was the great Norwegian marine biologist and Director of his country's fisheries research. I shall be referring to this book again, particularly in Chapter 12, for it contains the results of a very successful expedition which the two authors made in 1910 in the Norwegian research ship *Michael Sars* to study the deepwater life of the North Atlantic. I draw attention to it here, however, because it is also a splendid introduction to our science in general, with a valuable chapter on the early history.[1]

Like the correction of the idea of a zero of life, the destruction of another early myth, the mystery of the 'Bathybius', is also amusing history. In 1857 H.M.S. *Cyclops* had made a line of soundings across the Atlantic with an improved modification of Brook's apparatus for collecting samples of the sea-bed. The deposits of ooze were found to contain a strange gelatinous substance which was supposed to be a primitive form of life. Carpenter and Wyville Thomson again found it in their deep-sea dredgings. To give an idea of the universal interest it aroused at the time I will quote from '*The Depths of the Sea*'; Wyville Thomson is here referring to his deepest haul from 2,435 fathoms.

"In this dredging, as in most others in the bed of the Atlantic, there was evidence of a considerable quantity of soft gelatinous organic matter, enough to give a slight viscosity to the mud of the surface layer. If the mud be shaken with weak spirit of wine, fine flakes separate like coagulated mucus; and if a little of the mud in which this viscid condition is most marked be placed in a drop of sea-water under the microscope, we can usually see, after a time, an irregular network of matter resembling white of egg, distinguishable by its maintaining its

[1] Sir John Murray also wrote an excellent little introduction to oceanography: *The Ocean* (1913). For those who wish to make a fuller study of the development of our science, and particularly of the early plankton investigations by the German Hensen school, there is J. Johnstone's important book on *Conditions of Life in the Sea* (1908). Two other valuable introductions should be mentioned: Fowler and Allen's *Science of the Sea* (2nd edition, 1928) which gives much practical advice on collecting specimens and the working of gear, and Russell and Yonge's charming general natural history of *The Seas* (1928) which deals with the life of tropical waters as well as of our own.

outline and not mixing with the water. This network may be seen gradually altering in form, and entangled granules and foreign bodies change their relative positions. The gelatinous matter is therefore capable of a certain amount of movement, and there can be no doubt that it manifests the phenomena of a very simple form of life."

"To this organism, if a being can be so called which shows no trace of differentiation of organs, consisting apparently of an amorphous sheet of a protein compound, irritable to a low degree and capable of assimilating food, Professor Huxley has given the name of *Bathybius haeckelii*.   If this has a claim to be recognised as a distinct living entity, exhibiting its mature and final form, it must be referred to the simplest division of the shell-less rhizopoda, or if we adopt the class proposed by Professor Haeckel, to the monera. The circumstance which gives its special interest to Bathybius is its enormous extent: whether it be continuous in one vast sheet, or broken up into circumscribed individual particles, it appears to extend over a large part of the bed of the ocean. . . ."

The 'Bathybius' however came to an inglorious end. It was shown by the naturalists of the *Challenger* to be a precipitate thrown down from the sea-water associated with the deposits by the alcohol used in their preservation, and T. H. Huxley made a public retraction of his earlier ideas.

Towards the end of the century came the founding of the famous marine stations, the first at Naples in 1872, and then those at Plymouth and Millport in this country and Woods Hole in America; in these laboratories and many more to be founded later, researches into the structure, development, physiology, life and habits of marine creatures of all kinds have been continued to the present day.

It then began to be realised that progress in oceanography was essential to a better understanding of fishery problems and to the development of a more rational exploitation of the sea. The rapid development of trawling, with the introduction of steam power and the replacement of the old beam-trawl by the much larger and more efficient otter-trawl, gave rise to some concern as to the possible depletion of the stocks of fish; this led a number of nations, our own included, to set up fishery investigations. As we shall see, those fears were indeed well founded. In 1899 King Oscar II of Sweden invited all the nations of Europe interested in sea fishing to send representatives to a conference in Stockholm; the discussions which took place led to the foundation in 1901 of the International Council for the Exploration of the Sea. The different nations began a series of investigations to form part of one great plan. In spite of the temporary suspension of activities in two world wars the work of the Council still goes on.

The scientists of the various fishery departments are not only enquiring into the natural history of the fish themselves, their life-histories, food and feeding habits, migrations, growth, birth-rates and so forth; but with continually improved equipment they are studying the distribution of the different planktonic forms upon which they depend, the conditions under which they live, the flow of the ocean currents, the physics and chemistry of the sea and the varying nature of the sea-bottom and its life. It was wisely realised from the start that in order to provide answers to such questions as: 'Why are fish sometimes plentiful and sometimes scarce?'; 'Can the future of a fishery be forecast?'; 'Is this or that area being overfished?' and so on, the natural history of the sea must be investigated in all its different aspects.

Everything the naturalist wants to find out about the conditions under which fish live he must grope for in this unseen world, often below a storm-tossed surface; he must always remember too that the sea is a very big place—he must work with a sense of proportion and perspective. He stops his research ship at intervals to let down his instruments on wires and ropes: some to take the sea's varying temperature and to collect samples of water from different depths for analysis, others to measure the amount of light reaching different levels, and others again to record the speed and direction of ocean currents. He samples its life with all kinds of nets to estimate not only the varying quantities of the plankton but of the eggs and fry of the fish themselves. Building up a picture of life in the sea is like putting together a huge jig-saw puzzle made up of tiny pieces, but much more difficult. Not only have we a very imperfect idea of what kind of picture will emerge, but all the pieces to be fitted together are not on the table before us; they are lying about somewhere underneath it and we must feel about for them in the darkness. It is certainly a fascinating pursuit, but full of disappointments. Some bits of the puzzle—perhaps a stage in a life-history or some evidence of a migtation—can only be picked up during a short period of the year; before the missing pieces can be found, stormy weather may intervene and we must wait a whole year before we can try again.

In spite of these obstacles the picture of life in the sea is continually growing: the chapters which follow will endeavour to sketch an outline of what has been achieved. The amateur naturalist should not be discouraged by these difficulties for it is because of them that there are so many gaps in the story yet to be filled in. There are still many original discoveries to be made. And the difficulties add spice to the game; a golf-course would indeed be a dull one if there were no bunkers on it.

CHAPTER 2

THE MOVEMENT OF THE WATERS

O NE OF THE MOST important features in the world of a marine animal is the movement of the sea itself. Apart from the wave-action near the surface and the to-and-fro tidal streams in shallow coastal areas, the whole water-mass is in continual flow as part of a greater system of oceanic circulation. Carried with the moving waters go the floating animals and plants. The main surface currents in the North Atlantic are shown in Fig. 2, p. 14.

In some places the sea may be richer in plankton and in others poorer—like contrasting regions of luxuriance and barrenness on the land; unlike them, however, these areas in the sea are day by day changing their positions in relation to the coasts and the sea-bed. Not only is the basic food-supply on the move, but the delicate and helpless young of many fish and bottom-living animals are carried in one direction or another. Clearly these water movements are profound in their effect. To understand the lives of the inhabitants of our seas we must first have a knowledge of the main average pattern of the ocean currents flowing round our islands. They are by no means fixed in an unalterable course: variations are continually occurring. Sometimes these changes have striking effects; occasionally some exotic creatures from warmer seas, such as the beautiful Portuguese Man-of-war (*Physalia*) with its iridescent float and long trailing tentacles, or the smaller blue *Velella* riding the surface with a little 'sail', is driven ashore upon the coasts of Devon and Cornwall. (See p. 118 and Plate 5, p. 112). Normally we find typical Atlantic water outside the western entrance to the English Channel, but at times there may be a marked incursion of water from the Bay of Biscay carrying with it these and other planktonic visitors from further south. It is probable that long periods of strong south-westerly winds may contribute to this movement.

The general systems of the surface currents of the world have been worked out largely from the innumerable observations of navigators. Suppose a ship steers from point A on a particular course by compass

Fig. 2
The surface currents of the North Atlantic Ocean. Drawn, with
kind permission, from Admiralty Chart No. 5310 (1949),
omitting some of the detail.

which should carry it to a point B in, say, a day's time; if at the end of
the 24 hours the navigator finds by reference to the sun or stars that
he has instead reached a point five miles to the south-east of it, then
if there is no wind to account for his drift he knows that the surface
water must be flowing in a south-easterly direction at a speed of five
miles a day. The great pioneer in organising a systematic recording
of winds and ocean currents throughout the world was Lieut. M. F.
Maury of the U.S. Navy who wrote the first *Physical Geography of the Sea*
in 1855, followed by many editions.[1] The charts he produced of the
wind and current systems of the world rendered immense service to com-
merce; before their production the average time of sailing between England and Australia was 124 days, but with their use it was reduced to 97,

[1] See Addendum note on p. 315.

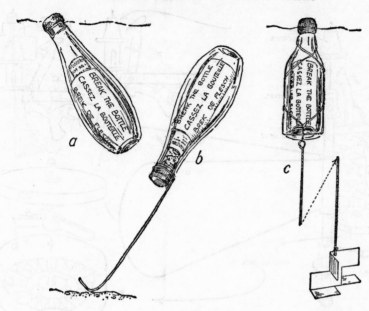

FIG. 3

Types of drift-bottle for investigating ocean currents. *a*, surface drifter, old
type; *b*, bottom drifter; *c*, surface drifter, modern type; for further explanation
see text.

or the average passage to California (presumably from New York)
was 183 days and this was reduced to 135 (from Maury, loc. cit.).

Observations on the movements of ships are suitable for indicating
the surface drift only in the wide open oceans. For working out the
current systems in more detail in more confined areas science has made
use of that romantic object, the shipwrecked sailor's bottle with a mes-
sage to his home. Thousands and thousands of these drift bottles have
been cast into the sea at many different points. Each bottle displays
through its glass sides, a notice in several different languages asking
the finder to break it and read the instructions within; these tell him
that if he will fill in and send off an enclosed postcard giving particulars
of the place and date of recovery, he will receive a small reward. Many
hundreds of these cards have been returned from various points
along the coasts of different lands. In this way Dr. T. W. Fulton of the
Scottish Fishery Board first charted the main surface movements of
the North Sea at the beginning of the century. Later Dr. G. P. Bidder
introduced bottom-drifting bottles, each provided with a trailing wire

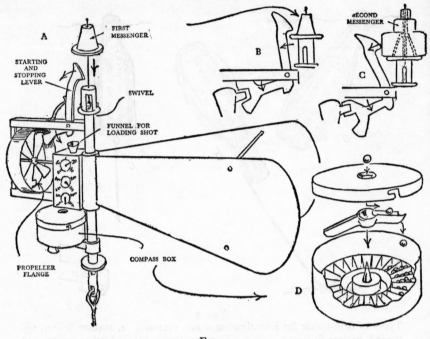

FIG. 4

Diagrammatic sketches showing the working of the Ekman current-meter.

and weighted so as to drift along with this just touching the sea-bed; these are designed to be recovered by the many trawlers dragging their nets along the bottom of the North Sea and English Channel. Thus knowledge has been gained not only of the surface currents but also of the movements of the lower layers of water, which are often very different. Objects floating on the very surface of the sea may be driven by the wind much faster than will the true upper layer of water; surface drift bottles have now been improved by Dr. J. N. Carruthers (1928) who has provided them with little weighted 'sea anchors' to keep them drifting with the main body of water. Sketches of these different bottles are shown in Fig. 3 (p. 15) and an example of their use in studying the drift of plaice eggs is shown in Fig. 5 (p. 18). Quite recently oceanographers have started using water-proof plastic envelopes instead of bottles.

For the more exact measurement of the speed and direction of current flow at different depths special instruments have been invented

to be used from anchored ships. One of the most successful is that of
Dr. Ekmann. It is such a beautiful and ingenious device that it is
worth examining in some detail to see how it works. It is suspended
in the sea on a wire, as shown in Fig. 4 (p. 16), and, being provided with
a vane like a weather-cock, always points head on into the current.
It also has a little 'propeller' which will be turned by the water flowing
past it; this however is held by a catch until the measurement of flow
is about to begin. On the underside of the apparatus, and held horizontally, is a circular box, like a large and rather deep pill-box. The
lower half of this is divided into a number of compartments by partitions radiating from the centre like the spokes of a wheel; swinging
above these compartments, and pivotting upon a spike in the centre,
is a stout compass needle having a slightly sloping groove on its upper
surface running from its mid-point to the end pointing to the north.
In addition the compass box has a small hole in the centre of the lid
and immediately above it is another little box containing a supply of
bronze shot. To take a current measurement the instrument is lowered
to the desired depth—perhaps 10, 50 or 100 metres; then a small
brass weight, called a messenger, is threaded on to the suspending
wire and sent sliding down to release the catch and set the propeller
free to turn. After one, two or more hours as may be desired, the recording is ended by the dispatch of a second 'messenger', which moves the
catch further over to stop the propeller once more, and the machine is
hauled up to the surface. The number of turns the propeller has made
in the interval of time is recorded on a series of little dials and so enables
the speed of the current to be estimated; also after every so many
revolutions a little valve opens and allows a shot to drop into the box
below and be guided by the grooved compass-needle into one of its
compartments. The compass-box, being fixed to the instrument, is
always orientated in relation to the direction of the current; thus as
the instrument swings in response to changes in current flow, each
shot is dropped into whatever compartment happens to lie below the
north-pointing end of its compass needle. By counting the shot found
in each of the compartments at the end of a recording we know to
what extent the current has varied during the time of observation and
also the average direction of its flow. The direction is, of course, given
by the angle between the mid-line of the box and that of the compartment which most often pointed north, as revealed by the shot in it.
Such observations are made at several depths to show how the speed
and direction of the water flow varies at different levels.

The upper layers of sea are nearly always travelling faster than the lower ones. The surface layer is affected by the wind; apart from this, however, the lower layers will be retarded by the frictional resistance of the sea-bed, just as in a river the water in the middle travels faster than that near the bank. We shall later see (on p. 213) that these differences in speed may have a marked effect on the distribution of plankton animals which make extensive day and night migrations

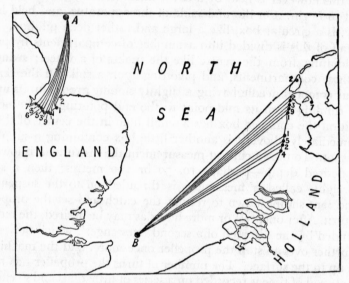

FIG. 5

Showing where drift-bottles liberated at A and B were recovered. A map reproduced from one in the museum of the Fisheries Laboratory at Lowestoft showing the results of one of many experiments made by the late Mr. J. O. Borley when investigating the drift of plaice eggs and larvae from the spawning areas to the coastal nursery grounds. The numerals show the number of bottles picked up at each point.

between the upper and lower layers. In deep ocean basins we may even find currents at different levels flowing in opposite directions.

There are several different kinds of current-meter but most of them are similar in general principle to that described. Dr. Carruthers (1926) has devised a much more robust machine which is used on a number of lightships anchored round our coasts to record automatically the main drift of water for periods of a month at a time. It records the to-and-fro tidal stream as well as the residual current-drift. These instruments have given much valuable information on the variations in the flow of water through the English Channel into the southern North Sea (Carruthers, 1930). It is in the Flemish Bight that so many plaice congregate in winter to lay their floating eggs at the point where the Channel water enters; these eggs, and later the hatched-out fry, are carried by the current and so bring the new generation to settle down as small flat-fish on the nursery feeding grounds in the shallow waters off the Dutch and Danish coasts. Herring fry in vast numbers are also carried by the same current into the North Sea from one of the largest spawning grounds off Cape Gris Nez. The spawning migrations of many fish have been evolved in relation to the prevailing current systems; selection has naturally acted to preserve the offspring of those parents who migrate to lay their eggs at the point up-stream most favourable for their survival; eggs spawned elsewhere are less likely to be successful because they are carried to less suitable nursery grounds. Fig. 5 (p. 18) shows the results of two of many experiments made with drift-bottles during the English fishery investigations into the life-history of the plaice. Those liberated at a point in the main spawning area in the Southern Bight have all been picked up on the Dutch coast, whereas those from the lesser spawning area off the Yorkshire coast are carried into the Wash, which is another smaller 'nursery' ground; the number against each point on the coast indicates the number of bottles found there. In some years a marked variation from a normal current-flow may well have a considerable effect on the relative success of a particular brood of fry; if the flow is weaker than usual they may not reach the best ground, if it is too strong they may be carried beyond it. The speed of flow of Channel water into the North Sea may be affected by the wind at a critical time apart from any variation in the fundamental current system; a prolonged westerly wind may accelerate the flow, and an easterly one may have the reverse effect. These are just some of the many factors we must take into account in puzzling out the possible causes for this or that unusual event in the fisheries; it is so often that such factors have an effect which is most marked in the fishery several years later, i.e. when the young fish of that brood will have grown to maturity—or failed to.

Without the use of drift bottles or current-meters some indication of the direction of current-flow may also be got by mapping the contours of varying salinity (measurements of salt-content) obtained by the analysis of water-samples collected at a number of different points in the area; for instance, a tongue of very salt water projecting into a less saline area might indicate a flow of ocean water into a more coastal region where the salt water has been diluted by drainage from

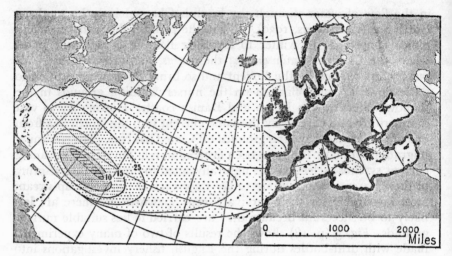

FIG. 6

The distribution of the common European Eel (*Anguilla vulgaris*) during its various stages of development. The contoured areas represent those in which the larvae of various sizes, 10, 15, 25 and 45 mm. are found; the line *ul* represents the limit of occurrence of unmetamorphosed larvae; the black bands along the coasts indicate the countries where the adult is found in fresh water (after Schmidt).

the land. Indeed modern oceanography has developed an elaborate mathematical system for estimating the direction and relative speeds of water movements from a knowledge of the varying densities of the water at a number of different points. We shall later see how at times certain planktonic animals and plants characteristic of one particular type of water may be used as indicators of the incursion of such water into other areas: in fact one such example, that of the tropical *Physalia* and *Velella* reaching our coasts, has already been mentioned, and we shall discuss others at the end of the chapter. We are apt to think of the use of plankton animals as current indicators as rather a modern idea; I was interested to find that Alexander Agassiz in 1883 was emphasising the importance in this respect of the animals just mentioned. "This group of Hydrozoa," he wrote "is eminently characteristic of the Gulf Stream, and wherever its influence extends these Velellae and Physaliae have been found. In fact these surface animals are excellent guides to the course of the current of the Gulf Stream— natural current bottles, as it were."

In the Department of Natural History in the University of Aberdeen there is a cabinet containing a remarkable collection of South American and West Indian seeds picked up on the shores of the Outer Hebrides. They were gathered from 1908 to 1919 by William L. MacGillivray who was a nephew of a former Regius Professor; most of them he found on the West Sand of Eoligarry, Barra, where he lived, but some came from Lewis and the Island of Fudag. There are Brazil nuts, the seeds of a leguminous liana *Diodea*, the Virgin Mary nut, palm seed of different kinds, the pecan nut, the Calabar bean, nutmeg, the seeds of the Central American soapberry tree and many others. Altogether seventeen tropical species are represented. Just as these seeds are drifted to our shores, so also are the baby eels carried round by the ocean circulation from where they were spawned—from a small area situated between Bermuda and the Leeward Islands—and eventually scattered by the Gulf Stream to enter the rivers along the whole seaboard of Europe. This was the amazing discovery made by the famous Danish oceanographer Professor Johannes Schmidt. On many special voyages he plotted the distribution of the tiny eel fry all over the Atlantic until at last he could show that there is only one limited area where the very smallest and newly hatched young are to be found—a breeding ground some 3,000 miles from the rivers in which they grow to maturity. In Fig. 6, opposite, I reproduce his map showing the spread of the fry of different sizes. Their drift round the ocean to Europe takes from 2 to 2½ years, and during this phase they are little flat and quite transparent creatures having the shape of a willow leaf. They used to be thought to be a separate species of fish, called *Leptocephalus brevirostris*, until the Italian naturalists Grassi and Calandruccio kept some in an aquarium and were surprised to find they turned into the common elvers, as the young freshwater eels are called when they ascend the rivers from the sea. The story is now very well known and I only recall it here because it is, to use Agassiz's simile, nature's greatest drift-bottle experiment and demonstrates so clearly the constancy of this vast current system; year after year, for many millions of years, the eel-fry must have been transported in this way with never a break in the sequence. How the adult eels navigate back to breed in this one place is one of the most profound mysteries of the sea; a discussion of this, however, belongs to a chapter on fish and must await the subsequent volume.

The causes of the great circulations of water—as distinct from mere tidal streams—are of three kinds; oceanographers, however, are still not fully agreed as to which is the most important: indeed all three play their part together. Primarily there are the effects of the prevailing winds over wide stretches of ocean, particularly the north-east and south-east trade winds blowing obliquely towards the equator from north and south respectively. These certainly take a great part in driving the equatorial water towards Central America, so that from the Gulf of Florida emerges the powerful warm Gulf Stream to flow across the North Atlantic and give to our islands and the north of Europe so temperate a climate compared with that of the corresponding latitudes of North America. The latter are cooled by the Arctic Stream of the Labrador Current. The Gulf Stream, or the North Atlantic Drift as it is more correctly called on this side of the ocean, has a profound effect upon our waters.

Although marine physicists are beginning to believe that the stress of the wind on the sea surface, together with the effect of the earth's rotation, can give rise to slow movements of water in the deep layers of the ocean as well as near the surface, they have to consider a second kind of cause: the action of what are often termed Archimedian forces. These are the forces due to internal changes in the water-mass causing alterations in its density. Such changes may be due to the expansion or contraction of the water on being warmed or cooled; they may also be due to an increase in the salt-content caused by excessive evaporation of water at the surface, as in the tropics, or to a decrease in saltness caused by large additions of fresh water from melting ice or excessive rainfall. Whether these causes actually produce the deep water movements, or do no more than make the water take the path of least resistance, they have far-reaching effects, particularly in giving rise to vertical as well as horizontal differences in the great ocean basins. One of the most surprising discoveries of world-wide oceanography was made between the two world wars—indeed through the work of our own *Discovery* Expeditions and the German *Meteor* Expedition, in the Antarctic and Atlantic Oceans; it is the fact that the heavy snow or rainfall and the melting ice in the Antarctic seas have an immense influence that extends across the equator into the northern hemisphere. It is so striking an example of these Archimedian forces that I cannot resist using it as an illustration.

The great ice-cap at the south pole dominates the oceans of the world. The Antarctic continent rising in a plateau to elevations of some 8,000 feet is covered with a sheet of ice many hundreds of feet thick; this is continually being added to by the frequent heavy falls of snow, and is constantly and slowly moving as a vast glacier to the coast and beyond into the ocean where it juts out as the floating ice barrier. This is shown on the left of the accompanying diagram: Fig. 7.

At its edge this barrier from time to time breaks up into the massive tabular ice-bergs so characteristic of the south polar seas. This is freshwater ice, which on melting, helps to form a cold but *light* surface layer; in spite of being colder it is lighter than the normal sea-water because its salt content is reduced by the addition of the fresh-water. All round the pole this cold surface layer flows away to the north. Below this is water that is heavier because it is just cooled and not diluted with fresh-water; this sinks and forms a cold current, also flowing north but over the ocean floor. To take the place of these

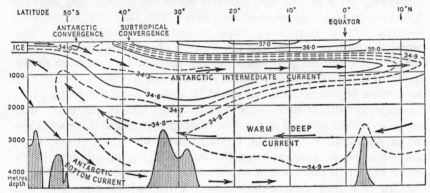

FIG. 7

A section through the Atlantic Ocean, from latitude 55°S to 15°N along the meridian 30°W, showing the water of varying saltness (34-0 o/oo to 37-0 o/oo) and the directions of the main water movements at different depths. It shows how the great Antarctic icecap extends its influence into the northern hemisphere. Redrawn in diagrammatic form from Deacon (1933).

two streams of water flowing away from the pole, a mass of warmer water flows southwards and wells up against the ice, to be itself diluted, cooled and turned north again. The surface current thus formed continues till it meets warmer water which, although more saline, is lighter because it is so much warmer; the cold current now dips below this warmer water but still travels northwards and can be traced to a point some 30° of latitude *north* of the equator; it then sinks and joins in the return flow going south again to complete the circulation.

Here we have a striking case of waters at different levels travelling in opposite directions: a layer going south flows in between two layers coming north. We shall see in a later chapter how the behaviour of some plankton animals is adapted in a most remarkable manner to take advantage of this fact.

Dr. G. E. R. Deacon, who has done so much to increase our knowledge of this remarkable system (1933 and 1937), while on one of the *Discovery* expeditions took water-samples all the way along the path of this northward-flowing current after it had dipped below the surface; when he had analysed them he found something very extraordinary. We have said that this water was of low salt-content because of the melted ice; it also has a high oxygen-content because of a great production of planktonic plants in the polar surface waters (due to the rich nutrient salts and to the long hours of daylight in high latitudes). Now as this water travels north the salt-content increases by diffusion from the surrounding layers and the oxygen-content is lowered by the respiratory requirements of animals. As he went along the path of the current Dr. Deacon obtained a clear indication that the salt and oxygen values did not increase and decrease respectively in a perfectly steady manner as one might have expected, but in a series of waves. As far as could be judged from the graph of the increase in saltness there were seven undulations in the curve from south to north; likewise in a graph of oxygen-decrease there were also seven undulations. Now the crests of the undulations of one curve corresponded in position with the *troughs* of the undulations of the other curve; in other words as we passed along the stream of water, regions of higher oxygen-content and lower salinity alternated with regions of lower oxygen-content and higher salinity. This water had come originally from the Antarctic surface layer in which *during the summer* more ice melts and also more plants are produced than in the winter. More ice melting means a lowering of salinity and more plants mean a greater production of oxygen. Clearly these regions of lower salinity and higher oxygen-content alternating with regions of higher salinity and lower oxygen-content represent the water which left the Antarctic surface layers in past summers and winters respectively. Dr. Deacon tells me that he now fears that there are not sufficient observations along the path of the current to make their number quite certain; but since the indicated rates of water movement agree very well with most other estimates, he feels that they give us a fairly reliable time scale for this great circulating system. This water takes *at least* seven years on its journey from the Antarctic to the northern hemisphere!

There is a somewhat similar system of a cold and less saline surface current flowing away from the north polar basin due to melting ice and the fall of snow and rain, but it is not so far-reaching as that from the south because this precipitation is less and in addition the Arctic

THE MOVEMENT OF THE WATERS

Ocean is almost entirely enclosed by submarine ridges; this cold stream dips below the warmer water at the northern boundary of the Gulf Stream. It is partly to replace this cold water stream that the extension of the Gulf Stream—the North Atlantic Drift—is carried so far to the northward of our islands and up the northern coasts of Scandinavia. Here we see how these Archimedian forces may contribute to the North Atlantic system.

The third factor affecting ocean-currents is a much more subtle one, due directly to the actual spin of our planet. This deflecting force of the earth's rotation is sometimes called Corioli's force, after the French physicist, though in fact it was carefully worked out by his countryman Laplace 60 years before; it applies to the atmosphere as well as the sea. It is not a cause of the initial motion of the water but a cause of its *deflection*. A body of water moving in any direction is deflected to the right in the northern hemisphere and to the left in the southern hemisphere. The effect is greater towards the poles and reduced towards the equator; on the actual equator itself there is no such effect at all. It applies not only to water but to any moving object; we can perhaps understand it best by considering the effect upon a swinging pendulum. Let us suppose we could hang a fairly heavy weight, say of some 20 lbs., on a long string from a 100 foot tall gallows-like structure at the north pole; now if we set the weight swinging to-and-fro in the same direction as a straight line drawn in the snow beneath it, we should soon observe that its line of swing would deviate from the line in the snow. Its swing would be deflected in a regular fashion in a clockwise direction; even in ten minutes its path would be deflected $2\frac{1}{2}°$. Unperceived by us the earth and the gallows would be rotating in an anti-clockwise direction, but the heavy pendulum weight is swinging free and its path is not affected although the string at the top will twist. If we watched it for a full twenty-four hours we should see the path of the pendulum complete a deflection of 360° and once more for a moment swing directly above the line in the snow. If we repeated this experiment at the equator—drawing our line in the sand—we should see no such effect; if the pendulum was set swinging say north and south it would continue to swing thus as it would also continue to swing in any other direction in which we might choose to start it; here the earth makes no turning motion *in relation* to the swing; for the line in the sand and the line of swing are carried on together by the earth's motion round its axis. At the south pole we should of course get a similar effect to that at the north pole except that the pendulum would appear to be deflected in an *anti-clockwise* direction. Swinging such a pendulum at places in different degrees of latitude will give a different amount of deflection. At a place situated in latitude 30° north the pendulum will swing through 180° in the 24 hours, for, speaking mathematically, the effect depends on the sine of the angle of latitude; in London at 51.5° latitude it will swing through 281°. The effect was shown very clearly by Foucault, the French physicist and inventor of the gyroscope, by swinging a hundred-foot pendulum at the Great Exhibition of 1851. This demonstration which is to be seen in the Science Museum in London and in a number of provincial museums is not difficult to set up in any building with a high roof or in any house that has a fairly wide staircase well above the entrance hall; it is an impressive sight to see in the matter of a few minutes the apparent change of motion of the pendulum, which really indicates the rotation of the hall itself or, indeed, the earth.

Just as the pendulum is deflected in relation to the objects in the hall, so any body of water *in motion* tends to be deflected to the right in the northern hemisphere and to the left in the south in relation to the surrounding land masses and the ocean floor; account has to be taken of it in every practical treatment of tides, wind drifts and ocean currents. Whenever a water-mass meets an obstruction, either a mass of land or an opposing water-mass, it will, other things being equal, turn to the right in the north rather than to the left, and *vice versa* in the south. There is another important effect. We have seen how through differences in temperature and saltness the water varies in density; the lighter water will naturally be on top. In a current system the water of a particular density—say the lightest water at the top—is not lying in a layer of uniform depth; owing to the earth's rotation the lighter water is pushed more to the right-hand side of the current stream than the heavier water, so that imaginary surfaces separating waters of different density are not horizontal but tilted. It is from a consideration of the deflection of waters of different densities that the speed and direction of ocean currents can be mathematically worked out as mentioned earlier in the chapter.

With this slight introduction, intended merely to give an idea of the kind of forces at work to produce the circulatory ocean-systems, we may now briefly review the main streams of water in the seas around our coasts. Fig. 8, opposite, is based on the account by Comd. J. R. Lumby (1932), hydrologist at the Fisheries Laboratory, Lowestoft, with a small revision which he has kindly made in the drawing for this figure. The water in the North Sea and English Channel is slightly less salt than the Atlantic Ocean water; it is typically coastal water diluted

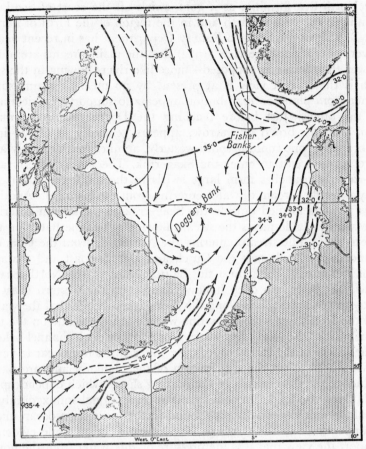

FIG. 8

The varying saltness (31.0 o/oo to 35.4 o/oo) of the surface waters and the
main circulation typical of the North Sea and English Channel in winter.
Drawn from a chart kindly provided by Commander J. R. Lumby, of the
Fisheries Laboratory, Lowestoft.

by freshwater drainage from the land. The Baltic has a much lower
salinity still. A stream of Atlantic water flows into the North Sea from
the north, mainly passing round to the east of the Shetland Islands
to flow due south and not usually entering between the Orkneys and
the Shetlands as was originally thought; a less powerful stream flows
up the English Channel and enters it from the south. The northern
influx is generally thought to flow on a broad front down the middle

of the North Sea forming, as it goes, swirls off the coast of Scotland especially in the Moray Firth and in the region of the Firth of Forth. Dr. J. B. Tait of the Scottish Fishery Department has in recent years, however, put forward the view (1952) that the main streams are much narrower than formerly supposed—more like rivers flowing in the sea. Which is the correct view is at present by no means certain; some evidence from plankton distribution appears to support one view and some the other. Just before reaching the Dogger Bank the main stream, whether broad or narrow, appears generally to divide into three branches: one running south-westerly, another south-easterly and a third turning east to enter the Skagerrak. The south-westerly and south-easterly branches form large swirls in the southern North Sea as they meet the stream of water entering from the Channel. Another smaller swirl is formed outside the Skagerrak as the stream entering on the southern side meets the stream flowing out of the Baltic on the northern side.[1] The stream entering the North Sea from the Channel flows north-eastwards past the Dutch and Danish coast and some of it joins the stream going into the Baltic. Most of the North Sea is shallow, but there is a deep hollow running up the western coast of Norway to the north; it is along this Norwegian trough that the water leaves the North Sea—the less saline water from the Baltic on the top and the bulk of the North Sea water proper in the deep channel below.

The extent of the inflow of Atlantic water varies from year to year; such variations affect the distribution of the plankton and are likely to influence the distribution of the herring shoals which depend upon the plankton for food. In some years of exceptional influx numbers of plankton animals usually only found in the more open ocean make their appearance in the northern North Sea. There is some evidence to support the view that it is the pressure of this water from the north (produced by the main wind systems) which, apart from the occasional effects of local winds already referred to, controls the inflow of Channel water into the southern North Sea. If the pressure from the north is high it seems that the Channel flow is reduced; if it is weak then a larger influx from the Channel seems to take place. It is in the study of this inflow into the northern North Sea that the charting of the relative movements of certain Atlantic plankton animals in different years can be most helpful. We shall see in a later chapter (p. 309) how, by the use of plankton-recording machines towed at monthly

---

[1] A valuable review of the changes in the southern North Sea, due to variations in the influence of the low salinity water from the Baltic and that of the higher salinity water from the Channel, has been made by Lucas and Rae (1946).

intervals by commercial steamships on regular routes, we can compare the areas of invasion of these more oceanic forms in different months and years. We shall find that not only does the extent of the Atlantic inflow vary from year to year, but the time of the advance of typical invading organisms will vary: in some years it may be a month earlier or later than in other years. There is an interesting suggestion now being investigated that the time of the appearance of the shoals of herring at different points down the east coast of Scotland and England, and consequently the time of the different fisheries, may be earlier or later in different years depending on whether this Atlantic inflow is earlier or later.[1] Whether this indication—it is no more at the moment—will be proved correct or not, there can be no doubt that the fluctuations that are found to occur in the water movements round our islands must have a profound effect upon the fish and other life inhabiting our seas.

A more definitely established connection between water changes and fisheries has been demonstrated at the western entrance to the English Channel. The water of the greater part of the Channel is like that of the southern North Sea—coastal water which is less saline and less rich in plankton than the Atlantic water that flows into it. This more coastal water can readily be distinguished from the more oceanic water by the presence of certain of these indicator plankton species—particularly two species of *Sagitta*, the slender transparent arrow worm shown in Plate IX (p. 132); *Sagitta setosa* being found in the coastal water and *Sagitta elegans* in the more oceanic water.[2] The boundary between the two waters formerly used to lie somewhere in the region of Plymouth where sometimes the plankton would have *elegans* predominating in it and sometimes *setosa*; during the investigation up to 1929 it was more usually *elegans*, indicating Atlantic water richer in phosphates and other nutrient salts. The importance of these salts in the economy of the sea will be discussed in Chapter 4. Since 1929 the boundary between *elegans* and *setosa* water has lain much further to the west so that the water off Plymouth has been of the coastal type and much poorer in plankton. Since this date there has also been a marked reduction in the number of young fish of many kinds present in the plankton as well as a change in the herring fishing; since that time the herring which used to visit the Plymouth area around Christmas have not turned up in their usual numbers so that this winter fishery, once quite a prosperous one, now no longer takes place. An

---

[1] This will be referred to again in chapter 15, p. 313.
[2] The difference between the two is shown in Fig. 42, p. 143.

excellent account of this trend was given by the late Dr. Stanley Kemp in his presidential address to the Zoology Section of the British Association in 1938. More recently some other interesting differences between the *elegans* and *setosa* water have been discovered; these will be referred to later when the various chemical constituents of the water are being considered (p. 65).

Mr. F. S. Russell, the present Director of the Plymouth Laboratory, who carried out these studies on *Sagitta* (1935, 1936) and young fish (1940), made cruises to trace the boundaries between the different kinds of plankton. He has told me how very abruptly one type of

Map showing the distribution of three kinds of water round Great Britain each characterised by a different species of the arrow-worm *Sagitta: serratodentata* in open ocean water, *setosa* in coastal water and *elegans* in oceanic water mixing with the coastal water. The conditions are those which might be expected in the autumn of a year with a strong Atlantic influx into the North Sea from the north. From Russell (1939), but modified in the north in the light of more recent surveys and with some other details omitted.

water may give place to another. Maury (1861) has also commented on these sharp dividing lines in the sea; he writes "Often one half of the vessel may be perceived floating in Gulf Stream water while the other half is in common water of the sea—so sharp is the line. . . ."[1] Fig. 9, above, shows the general distribution of the *elegans* and *setosa* water round the British Isles as it might be expected in the autumn of a year in which there is a strong influx of Atlantic water into the North Sea from the north; it is taken from another of Dr.

[1] In the first edition I erroneously attributed such a statement to Dr. Russell thinking he had told it me concerning the *elegans* and *setosa* water.

Russell's papers (1939). These different waters may also be some-times discernible by a difference in their colour, a contrast of shades of blue and green making a line across the sea. In 1923, when

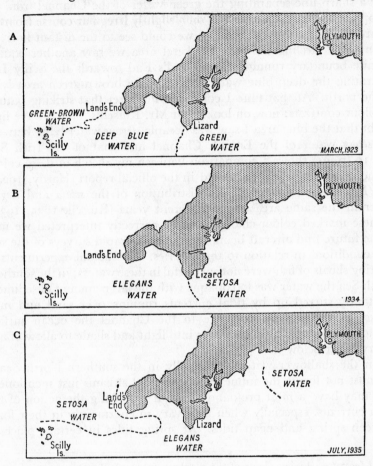

FIG. 10

Well defined areas of blue and green water (A) seen from the air during mackerel spotting tests off Cornwall in 1923 drawn from the chart by Hardy (1924) and compared (B and C) with the distribution of western and Channel water as indicated by the arrow-worms *Sagitta, elegans* and *setosa*, charted by Russell (1935 and 1936).

on the staff of the Fisheries Laboratory at Lowestoft, I acted as observer in some attempts to locate shoals of herring and mackerel from the air. In flying from Plymouth to the western mackerel grounds we passed over a sharp line separating the green water of the Channel from the deep blue of the Atlantic; it ran on a slightly irregular course from the Lizard to the south-west as far as we could see to the distant horizon. Then while circling over the mackerel area we saw another equally definite boundary running from Land's End towards the Scilly Isles separating the deep blue water from a more brown-green area lying to the north. At that time I could not interpret that striking pattern of colour contrasts; now on looking at Mr. Russell's maps I have little doubt that the blue area I saw was oceanic *elegans* water lying between the *setosa* water of the English Channel and that of the Irish Sea. Fig. 10 (p. 31) shows a comparison between my sketch of these colour boundaries, which was published in the official report (Hardy, 1924*a*) and Mr. Russell's maps of the distribution of the *setosa* and *elegans* water in the same area but in different years (Russell, 1935, 1936). If these marked colour-changes can be correctly interpreted we may in the future find aircraft being used to make rapid surveys of the surface conditions in relation to the fisheries. The actual experiments in spotting shoals of fish were not successful in these waters; in the southern North Sea the water was too opaque with the large amount of sediment constantly stirred up by tidal currents running over sand and mud banks; at the western entrance to the Channel the ocean surface was too much broken up by waves into light and shade to allow of any observations below it.

In the shallower waters—especially in the southern North Sea— we must not forget the influence of the tidal streams just mentioned; they may have a most profound effect in modifying the action of the main currents, especially when they vary so enormously in their force between spring and neap tides. At spring tides in certain places a

---

Plate I (above) a. Living plants of the plankton (phytoplankton), × 110.
Chains of cells of several species of *Chaetoceros* (those with spines), a chain of *Thalassiosira condensata* (at and pointing to bottom right corner), and a chain of *Lauderia borealis* (above the last named).
(below) b. More living phytoplankton from the English Channel, × 60. Diatoms: *Biddulphia sinensis* (4 large cells linked together), *Coscinodiscus conicinus* (single large cell), *Melosira borreri* (chain of 8 small cells, pointing 'N.N.W.') and *Rhizosolenia faeröense* (chain of 6 small cells, pointing 'N.N.E.'), also the dinoflagellate *Ceratium tripos* and closely related species (anchor-shaped cells). Both by electronic flash.   *(Douglas Wilson)*

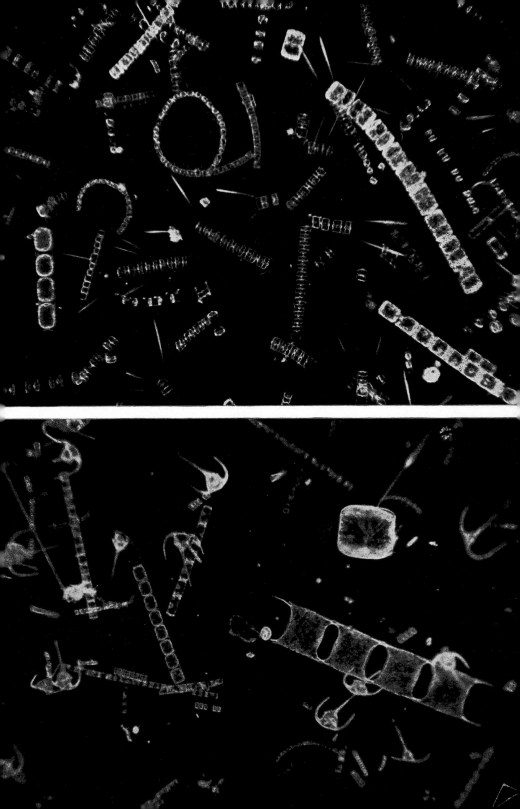

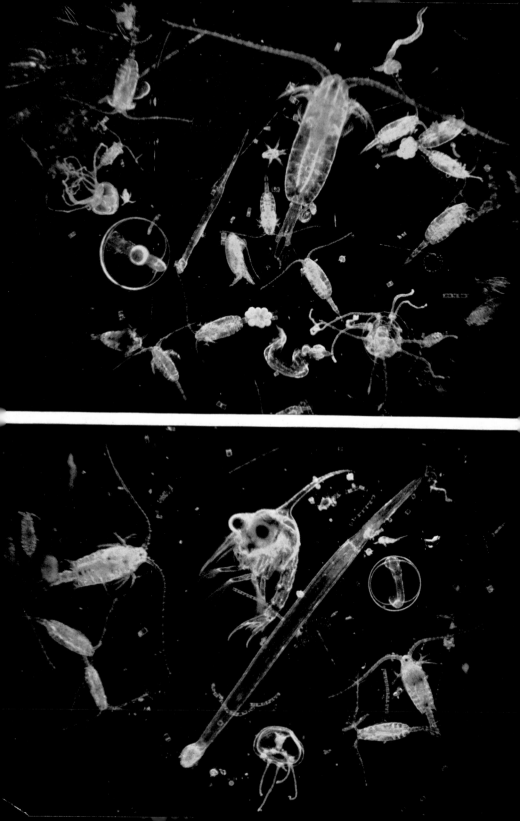

mass of water may be moved for some thirty or forty miles in each direction.

The movements of water in the Irish Sea are also dominated by tidal currents; these flow into it from both ends and follow the general direction of the coast lines. Professor K. F. Bowden, who has given us such an excellent account of these tidal streams (1953), writes, "Knowledge of the non-tidal drift, however, is much less certain and is based on indirect evidence. It was recognised at an early date that the distribution of salinity indicated a north-going drift and in 1907 Knudsen estimated that the rate of flow was such that the water in the Irish Sea would be completely renewed in a year." After saying that "this implies a flow through the Dublin-Holyhead channel at an average rate of just over a kilometre a day," he later stresses that, although there seems little doubt about this average northward movement, "its magnitude, its variations and the degree of its dependence on the wind are still uncertain."

We have now dealt with the main water movements round our islands; later in the book we shall see instances of more local effects and how upwellings and the mixing of waters may be important in producing a richer plankton. I will end the chapter by referring to some surprising and significant plankton records being made by Dr. J. H. Fraser (1952*b*, 1955) of the Scottish Fishery Laboratory at Aberdeen. In some years, over a wide area to the north of Scotland, he finds plankton animals which we should more usually associate with the latitudes of the Mediterranean; they indeed indicate a very unexpected movement of water. It now appears that some of them may in fact actually have come from the Mediterranean Sea itself.

It has long been known that a surface stream of Atlantic water

---

Plate II (*above*) *a.* Living animals of the plankton (zoo-plankton), × 16.
The copepods *Calanus finmarchicus* (largest animal) and *Pseudocalanus elongatus* (similar in shape, but much smaller than *Calanus*, and one with cluster of eggs); two small anthomedusae with long tentacles; a fish egg (circular object); a young arrow-worm *Sagitta* (to right of fish egg); small nauplius (larval stage) of copepod (close to left side of *Calanus*) and the planktonic tunicate *Oikopleura* (curly objects top right and middle bottom).
(*Below*) *b.* More living zooplankton, × 16. The transparent arrow-worm *Sagitta setosa* (diagonal); zoea, young stage of crab (above arrow-worm); fish egg, with developing fish, pilchard (spherical); anthomedusan (bottom centre); and copepods; 2 specimens of *Centropages* (left of zoea and below fish egg) and the slightly smaller ones are *Acartia*.
Both by electronic flash, but partially narcotised. (*Douglas Wilson*)
TOS—D

flows eastwards through the Straits of Gibraltar and that this influx
is balanced by an outpouring (at a lower level) of Mediterranean
water of very high salinity; this spreads out from the Gulf of Gibraltar
underneath the North Atlantic water and some of it is carried north up
the edge of the European continental shelf. This movement is well
summarised in Sverdrup, Johnson and Fleming's important book *The
Oceans* (1942, pp. 646, 685–6) which, for the serious student, gives

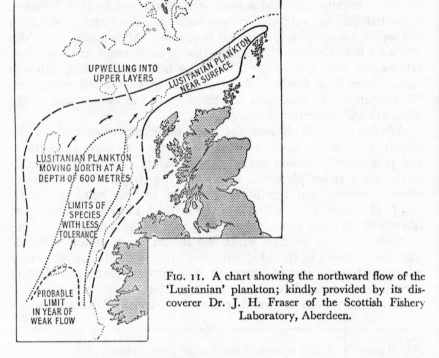

FIG. 11. A chart showing the northward flow of the
'Lusitanian' plankton; kindly provided by its dis-
coverer Dr. J. H. Fraser of the Scottish Fishery
Laboratory, Aberdeen.

such an excellent account of the main results of modern oceanography.
Dr. L. H. N. Cooper of the Plymouth Laboratory has recently (1952)
made a study of the distribution of this water to the west of the British
Isles as it continues northward below the Atlantic water at a depth of
some 600 to 1,200 metres. How far north it goes seems to vary greatly
in different years; in some it appears to go no further than the west
of Ireland, but in other years it flows onwards to upwell and spread
over the continental shelf. Its course has been followed by Dr. Fraser
by finding its typical but exotic fauna in his plankton nets; to the west
he finds it deep down—but let me quote from his recent paper.

"It apparently follows the edge of the Hebridean Continental shelf, mixing on its western edge with open oceanic water, and upwells somewhat on the east side to overflow and mix with coastal water on the shelf. In some years this current may not reach Scotland or is too weak to be recognised, but on occasions it is sufficiently strong to continue into the South side of the Faroe Channel, though it only rarely penetrates beyond the north of Shetland. Frequently, however, it mixes with the coastal water on the shelf and the resulting mixture floods the area to the west of Orkney and often passes through the Fair Isle-Orkney Passage into the Moray Firth area."

How lucky we are to have such a remarkable current carrying its rich and southern life far below the surface and then spreading it out, as it were, on our northern doorstep for our examination. Fraser calls this a planktonic 'Lusitanian fauna' and lists no fewer than 43 species characteristic of it; he has kindly prepared for me a chart of its typical distribution which is reproduced in Fig. 11 opposite. Through his recent publications I have been able to add to my account some very interesting animals which I shall be describing in Chapters 7 and 8 and which hitherto I should never have dared to include as inhabitants of British waters. He defines his Lusitanian fauna as that which "originating in the outflow from the Mediterranean, has become modified by admixture with fauna from the area between the Azores and Bay of Biscay." This work is an outstanding example of the importance of natural history in helping us to have a better understanding of the physics of the sea. I will give a final quotation from his work:

"The whole of this oceanic system to the north and west of Scotland overlies a south tending mass of artic or boreal water. The main flow of this water mass is to the west of Faroe from whence it thrusts southwards in deep water (below about 1,000m.), but part also penetrates the Faroe Channel where it is checked by the Wyville Thomson Ridge.[1] Although this water affects the inflowing system where it mixes at its interface it is not of such importance as are the more massive cold water currents on the other side of the Atlantic.

"Each of the above water masses has a typical plankton fauna (see Russell 1939, and earlier works), which varies within certain limits, in the abundance and in the proportions of its constituent species from year to year. As these organisms are transported further from their natural habitat they gradually die as their limit of tolerance is reached, and they are replaced by other species through mixing either with other oceanic streams or with coastal water. The fauna of an incoming water mass thus gradually changes along its length; for example, few of the oceanic species noted off Scotland normally reach north-western Norway (Wiborg 1954). The degree of survival of the original fauna gives a measure of the purity of the inflow, and the relative life of the species less tolerant to various factors may give an indication of the type of dilution or change involved."

[1] See p. 220.

CHAPTER 3

PLANTS OF THE PLANKTON

HAVING DISCUSSED the movement of the waters it might perhaps seem more logical to pass on at once to consider other physical characters of the sea and something of its chemistry before proceeding to deal with any of the life within it. On the other hand, since the plants of the open sea are so intimately dependent upon their physical and chemical background it will be more interesting if we know what kind of plants we are dealing with before we actually discuss the conditions which are most favourable for their growth.

The vegetation of the open sea must be floating freely in the water in order to be sufficiently near the surface to get enough light; the great difference between it and that of the coasts or land, is that it consists entirely of plants of microscopic size. They are, as we have already seen, part of the plankton: the phytoplankton. Each is composed of just one unit of life—a single cell—instead of being made up, as are the larger plants, of a vast number of such units. Instead of having various kinds of cells specialised to perform different functions in organs such as roots, stems, leaves and reproductive bodies, all these activities of life are carried out by just one highly organised unit. It does at first sight seem strange that there should not be even a few larger plants adapted for such a floating existence. There is the famous Gulf-weed *Sargassum* which, buoyed up by the little floats upon its fronds, is found in masses drifting round that great eddy of the tropical Atlantic—the Sargasso Sea; this however is an accidental open-ocean plant derived from the coasts of Central America and the West Indies.[1] In our own waters we may sometimes meet with patches of bladder-wrack, *Fucus vesiculosus*, torn from the sea-shore by storms and floating in the same way.

The microscopic plants must have some great advantage over

[1] In my first edition I followed Gran (1912) in regarding the *Sargassum* weed as the same as the coastal form; Professor Randolph Taylor of Michigan, however, informs me that two distinct pelagic species, *S. natans* and *S. fluitans*, are now recognised.

larger plants in this floating drifting life. The smaller an object is the larger is its surface *in relation* to its volume. If we increase the size of an object—keeping its shape in the same form—the volume increases by the cube of linear measurement but the surface does so only by the square. This elementary fact is so important in the present discussion that it may be well to emphasise it by a simple concrete example. If we have eight small cubes of soap of the same size and press them together to form one big cube, the volume of this new cube will then be eight times that of one of the smaller ones, whereas its surface will be only four times as large. We have of course lost all the surfaces that were pressed and fused together. Inversely, the more we cut up our soap into smaller and smaller cubes, the more surface will each new cube have in proportion to its volume. A cube the size of one of our little plants will have a surface-volume ratio *many hundreds of times* as large as that of a cube with a side no more than an inch or two. A large surface-volume ratio is a great advantage to our little plants in at least two important respects. Firstly, the larger the surface in relation to mass the greater will be the frictional resistance to the water which will retard its sinking and so enable it to remain more easily in the upper sunlit layers. Secondly, since absorption must take place through the surface, the larger its surface in proportion to its volume the more readily will it be able to take up for its needs enough of the necessary mineral salts which may be present in the water in only very small amounts. This indeed may be the cardinal factor which has prohibited the development of larger plants in the plankton; but for this they might well have evolved bladder-like floats to support their larger mass, as some animals have done. Each tiny plant, as a single cell, can also take better advantage of the scattered sunlight than can a number of such cells massed together.

It is at first difficult to believe that these finely scattered and microscopic plants can really form a vegetation which has sufficient bulk to support all the teeming animal life of the sea: the dense populations of planktonic crustaceans, the vast shoals of fish and all the invertebrate animals on the sea-bed. Yet we know this must be so. Some estimates of the actual quantities of plants present in a cubic metre of sea-water will be given later in the chapter; here, in passing, we will only note that, given suitable conditions, the amount of plant life produced under a given area of sea may well exceed that produced for the same area in a tropical forest. Just as our coal supplies are giving us the energy stored up in the great primaeval forests of some hundred million years ago—so is the energy in the petrol, which drives our motor and flying

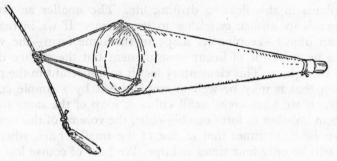

FIG. 12
A simple form of the plankton collecting tow-net.

age, derived from that originally trapped from the sunlight by the tiny planktonic plants in the seas of long ago. According to current geological theory, the great supplies of mineral oil have been formed, in the course of ages, from the remains of marine organisms buried in sedementation under specially favourable conditions which are not yet fully understood. It is most likely that the planktonic crustaceans, whose modern representatives are so rich in oil, would in the past be the main contributors to the supplies of petrol we are burning up today; those crustaceans, of course, derived their energy either directly or indirectly from such tiny plants as we are now considering.

A microscope of sufficient power to enable us to see a great deal of this world of planktonic plants and animals need not be an elaborate one, nor need it cost much more than a good pair of field-glasses. We shall want some glass slides and coverslips, small dishes (such as watch-glasses), pipettes, i.e. old-fashioned fountain-pen fillers, for picking up very small plankton animals, and some glass jam jars; apart from that, all we need is a tow-net and line with which to collect the plankton from a rowing boat or any larger vessel that can be made to go slowly enough.

A tow-net can be bought from the laboratory of the Marine Biological Association at Plymouth (address: The Laboratory, Citadel Hill, Plymouth) or it can be home-made. It consists essentially of three parts: in front is a hoop made either of light galvanised iron or strong cane and provided with three bridles of cord which will come together at a small ring or shackle for attaching to the towing rope; next comes the actual net, a conical bag made of a fabric which will act as a fine sieve; lastly at the end of the net is a small collecting jar, either a glass honey-jar or one made of zinc or copper with a slight lip. Such a simple tow-net is shown in Fig. 12 opposite. For ordinary collecting purposes a hoop of 18 inches diameter will be quite sufficient. The net is best made of the silk 'bolting cloth' used by millers for sieving flour, but a good quality muslin will do if this cannot be obtained; with a mouth of 18 inches it should be almost five feet in length. If it is to be home-made great care should be taken in cutting out the material in order to ensure that a perfect cone is formed; if lop-sided it will not fish properly. It is a good plan to pin together a paper model to serve as a pattern; this will also enable one to see how best to use the material with as little waste as possible. Round its wide mouth a canvas or calico band is sewn for attachment to the hoop; it may either be provided with a series of eyes for lashing it on or it may be folded over the hoop and sewn to enclose it, leaving gaps where the towing bridles are secured. At the hind end is sewn another canvas band to form a small cylinder, say 2½ inches in diameter, which will slip closely over the mouth of the collecting jar and be firmly held in position by a tightly tied tape. It is well to be provided with two such tow-nets; one made of the very finest material for the collection of the small plants—the finest bolting-cloth has 200 threads to the linear inch—and one of coarser material, having about 60 threads to the inch, for the capture of the somewhat larger animals. The coarser net lets most of the plants go through its mesh but filters a very much larger quantity of water more quickly and so captures the larger more active animals which are only rarely taken in the finer net.

To collect the phytoplankton the fine net should be towed just a little way below the surface. A weight, say a 7lb. lead, is slung at the end of the rope and the net attached a little way above it. The essence of successful tow-netting is to tow *very slowly*, never at more than 1½ knots. If it is towed faster the water will not be filtered quick enough; the net will just push a mass of water in front of it which will prevent any more water entering it. A ten minutes' tow may give quite a large enough sample. Most of the plankton will have passed down into the jar at the end as it is towed; a number of specimens, however, may still be sticking to the inside of the net as it is taken from the water, so that it should be carefully washed down from the outside with water from a bucket, to flush them into the jar.

Our sample will contain a vast number of both plants and animals. In this chapter we will concern ourselves only with the former, which are so small that they must be looked for with the compound micro-scope. After bringing our sample home and letting it stand for a little we should take only a few drops at a time with a fine pipette from near the bottom and place them on a slide under a coverslip; now we shall hunt with the low-power lens and then turn on the high-power to examine each new specimen we find. We shall not, of course, expect to find examples of all the different kinds in one sample but there may well be representatives of several of the more important groups. The most prominent members of the phytoplankton are the diatoms. They are unicellular algae differing from all other algae in having a cell wall which forms a siliceous external skeleton enclosing the cell like a glass box. The pigment bodies, or chloroplasts, which enable the plant to make use of the energy of sunlight are not the usual bright green of chlorophyll but a brown or brownish-green pigment closely allied to it. The siliceous skeleton is in two parts which fit together like the top and bottom of a pill-box; indeed some of the diatoms are just like a pill-box in form, but many others are drawn out into all manner of fantastic shapes. When first we see a sample of plankton rich in diatoms under the high power of the microscope it is like looking at a group of crystal caskets filled with jewels as the strands of sparkling protoplasm and groups of amber chloroplasts catch the light. Every plant or animal cell consists of a mass of protoplasm with a more or less central body, the nucleus, which appears to govern its life; it is characteristic of the diatoms that, in addition, the proto-plasm usually has large cavities in it containing clear fluid. The nucleus is usually central and surrounded by a mass of protoplasm; radiating from this and forming an irregular network are protoplasmic strands stretching across the cavities like the spokes of a wheel to join up with a layer of protoplasm which lines the inner surface of the box-like covering. More rarely, in some forms, the nucleus may be in the layer of protoplasm at the side. The pigment granules usually lie more or less regularly spaced against the cell-wall, where they are exposed to the light; if, however, the light is too intense, they come close together either down the strands to the centre or to some other part of the cell where they can partly screen one another from the harmful effects of the rays.

The top, bottom and sides of the glass-like box are not made of just plain sheets of silica; their surfaces are sculptured with all manner of striations, pits and perforations forming intricate patterns peculiar to the different species. This detail of design has always made the diatoms favourite specimens with microscopists, not only on account of their beauty, but because they are such excellent objects with which to test and display the quality of their instruments in the higher ranges of magnification. They are now being put under the electron microscope which can give a micrograph with a magnification of up to 100,000 times; this has at once revealed an arrangement of structure far more elaborate than that seen with the highest powers available in the optical systems (Hendy, Cushing and Ripley, 1954). Instead of there being just one system of pits or perforations in their walls, some forms are shown to have smaller and yet smaller ones, secondary and tertiary systems, on inner layers of silica; in others the wall is more like a basket of spiral threads intricately woven together. Some of the perforations measured were less than a ten-thousandth of a millimetre in diameter and the surrounding walls were of equal thickness. These delicate lattice systems have at least two important qualities for floating plants: they give strength with lightness and at the same time provide a framework for presenting a greatly increased surface area of protoplasm to the surrounding water.

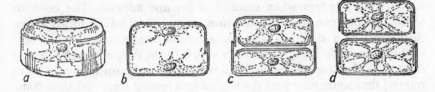

FIG. 13

Diagrams showing the division of a simple pillbox-like type of diatom. *a*, a sketch of the cell before division has begun, *b* to *d* sections through the diatom after cell division to show stages in the formation of the new skeletal cell walls. Note that the upper cell in *d* is smaller than the original cell *b*.

Diatoms normally reproduce by simply dividing in two. The nucleus divides first and then the protoplasm becomes separated into two masses, each containing a nucleus, one at either end of the box; each mass of protoplasm now forms, between it and the other mass, a new valve as the halves of the pill-box are called. These new valves each fit closely their own part of the old box; we have in fact two pill-boxes now instead of one, as is shown above in Fig. 13. They may separate entirely, or in some species they may remain attached to form long chains. It will be realised that in this process of repeated division by forming new half-boxes *within* the old, the *average* size of the diatoms so produced will tend to get smaller and smaller; at each division, as shown in the drawing, one of the new boxes will be the same size as the old one but the other *must be smaller*. Thus we find a considerable range in the size of diatoms of the same species, but there must be a limit to this reduction. After a certain number of such divisions there is formed what is called an auxospore, by which the original size is recovered; throwing off the old valves the cell becomes a bladder-like mass of protoplasm within which new valves are formed two or three times the size of the old discarded ones. Some diatoms seem to do this at definite seasons whereas others do so only at intervals of two or three years. By taking sample measurements of the diatoms forming some of the dense concentrations which are carried by currents about the North Sea, planktologists have been able to identify individual patches in their wanderings by the regular decrease in the size of their component cells (Wimpenny 1936; Lucas and Stubbings 1948). Some of these concentrations, as we shall see in Chapter 15 have been thought to have a marked effect upon the herring fisheries; their wanderings may therefore be of economic interest.

In addition to forming auxospores, diatoms may produce what are called resting spores when conditions become adverse. The contents of the cell become concentrated in a central mass which forms a new thick wall of a different but characteristic shape and the old cell-wall is discarded. They now either sink into the deeper water layers or right to the bottom where they will remain till more suitable conditions return; thus some may pass the winter in a resting state and then come up to start active life again in the spring. Planktonic diatoms are not definitely known to have any sexual phase, but in some a number of smaller spores (microspores) have occasionally been observed to be formed within the cell-wall and it has been thought that these may be gametes (sex cells), but this is not yet established.

Not all diatoms are planktonic; in the shallow coastal regions there are numbers living on the bottom where sufficient light reaches it; these have much thicker shells than the more delicate floating forms and are more uniform in character. What makes the planktonic diatoms so interesting is the variety of devices that have been evolved to assist in their flotation. It is not the purpose of this book to attempt a systematic treatment of the groups of animals and plants. Here we shall just refer to some of the more important kinds of diatoms in relation to their mode of suspension. Dr. Marie Lebour's excellent book (1930) on the planktonic forms should be studied for a full account.

Although so very small they have a considerable range in size; a few exceptionally large ones may be over one millimetre in diameter and the smallest may be but a few thousandths of this. Sketches of examples of the different genera to be mentioned are shown in Fig. 14 and Plate 1 (p. 80), and some are also included in the photographs in Plate I (p. 32). A number have the typical pill-box form, such as members of the genus *Coscinodiscus* and some of these are of comparatively large size; a few

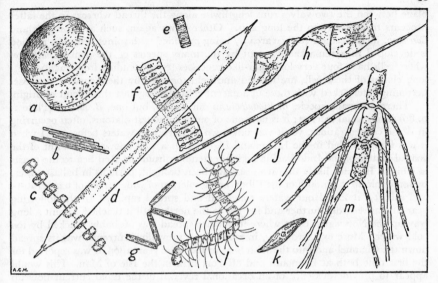

FIG. 14

Some characteristic plankton diatoms not shown in Plate I, all magnified
×90 diam. *a, Coscinodiscus concinnus; b, Bacillaria paradoxa; c, Thalassiosira
gravida; d, Rhizosolenia styliformis; e, Paralia sulcata; f, Bellarochia maleus;
g, Thalassiothrix nitzschioides; h, Streptotheca thamensis; i, Rhizosolenia hebetata*
(form *semispina*); *j, Nitzschia seriata; k, Gyrosigma sp.; l, Chaetoceros curvisetus;
m, Ch. convolutus.* The actual length represented by the longer side of this
figure is 1/20th of an inch.

like *C. concinnus* may be just visible to the naked eye. These kinds have large vacuoles
and would seem to be buoyed up by globules of oil. Members of this genus live
singly, but others of allied genera may remain after division attached together to form
long chains; the cells of *Paralia* and *Guinardia* are linked rigidly together by their
valve surfaces, those of *Thalassiosira* form flexible chains as they are strung together
by fine threads of protoplasm, and others like *Lauderia* are still more loosely held
together by irregular strands of slime. The cells of *Thalassiosira* also produce such
slime-strands but for a different purpose—around the margin of each valve (i.e.
each 'pill-box lid') are a number of small hollow spines from which can be extruded
long slender threads of slime so that they radiate on all sides like the strands of
thistle-down and indeed act in the same way to assist in parachute-like support.

The remainder of the planktonic diatoms, while essentially built on this pill-
box plan, have each valve or half-box modified into all sorts of shapes which increase
their surface area in relation to their volume and so give greater frictional resistance
to sinking. Some are flattened like thin sheets of paper and often twisted to some
extent; these usually remain attached together to form long ribbons: such are
Bellarochia, Eucampia and *Streptotheca*. Others are drawn out either into long thin
hair-like forms such as *Thalassiothrix longissima* or into the more rigid pencil or needle-
like members of the genus *Rhizosolenia*, pointed at each end; in the former the division

plane between the two valves runs lengthwise along the thread whereas in the latter it occurs transversely to the long axis. Other forms again, such as *Biddulphia* and *Corethron,* increase their surface area by being provided with spines. This last method is developed to a remarkable extent by the many species of the genus *Chaetoceros* whose cells have four very long hair-like processes (two extending from each valve); long chains of these cells are formed and held together by their curving processes becoming interlocked with those of adjacent cells close against their point of origin.

There are many species of *Biddulphia* in our waters, but one, *B. sinensis* (Plate 1, p. 80) is of special interest; it is now one of our commonest diatoms, often occurring in dense concentrations, yet it was unknown in European waters before 1903 when it was first recorded in the Heligoland Bight. It is a well known inhabitant of the coastal waters of the Indo-Pacific region, extending from the Red Sea to the coasts of China. It seems likely that it must have been brought, perhaps in ballast water, by some ship to the mouth of the Elbe; being tolerant of wide ranges of temperature and salinity, it found our waters congenial and spread rapidly. In the following years it was recorded further and further to the north until it reached a point a long way up the Norwegian coast where its further spread was probably checked by too cold water. More extraordinary, in view of the prevailing currents, was its spread down the Channel and into the Irish Sea; in the same year, 1909, it was reported for the first time both at Plymouth and off Port Erin in the Isle of Man. This would appear to provide evidence of an occasional reversal of the usual current flow up the Channel; indeed such a reversal has been suspected by some oceanographers on other grounds. Or was it transported to the western Channel and Irish Sea in the same way as it had apparently reached the Elbe? We may never know the answer to that. Some have maintained that *Biddulphia sinensis* must have been a native of our waters all the time and only just noticed at the beginning of the century; this, however, can hardly be so because of extensive collections that were made, particularly by the Kiel planktologists, throughout the eighteen-nineties.

In coastal regions we may find a number of typical bottom-living diatoms carried up into the plankton, such as the boat-shaped members of the genus *Navicula.* They are capable of a remarkable gliding movement, thought to be produced by a flow of protoplasm passed out through a slit in the wall of the cell. *Gyrosigma* is another bottom form often met with in the shallow-water plankton. Other related forms such as *Bacillaria* and *Nitzschia* are more planktonic but typical of coastal waters. *B. paradoxa* forms bands of long slender cells held together side by side like the planks in a raft yet each capable of sliding up and down along its neighbours; *N. seriata,* in contrast, forms long strings of narrow boat-shaped cells end to end with just their tips overlapping and in contact.

It is impossible to mention all the kinds of diatoms of our seas in such a general review, and I shall only refer to one other species, one which may well attract attention: *Asterionella japonica* (Plate 1, p. 80). Its cells are rod-like but thickened at one end; by these thickened ends the cells remain attached to one another to form beautiful radiating star-like clusters.

In striking contrast to the brown-green colouring of the diatoms is the brilliant green sphere of *Halosphaera viridis* which may reach a size of nearly a millimetre in diameter; it and one or two closely allied species are the only representatives in the marine plankton of the Yellow-green Algae or Heterokontae. *H. viridis* is found over the whole

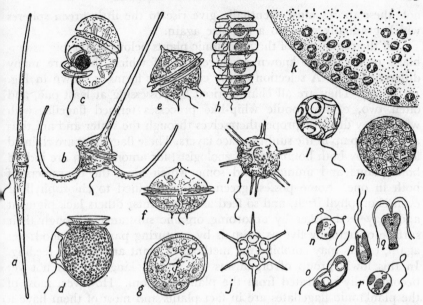

FIG. 15

Some flagellates of the plankton. *a-h,* Dinoflagellates: *a, Ceratium fusus* (× 180); *b, C.macroceros* (× 200); *c, Protoerythropsis vigilans* (× 320) (note clear spherical lens against dark eye-spot); *d, Dinophysis acuta* (× 400); *e, Peridinium granii* (× 360); *f* and *g, P. ovatum* (× 320), side and top view; *h, Polykrikos schwarzi* (× 250); *i* and *j,* the Silicoflagellate *Distephanus speculum* living and half of skeleton (× 320); *k,* a very small part of the large gelatinous capsule formed by the tiny cells of *Phaeocystis; l* and *m,* Coccolithophores (× 1000): *Coccolithus huxleyei* and *Coccosphaera leptopora; n-r,* some of the smallest flagellates (× 1500): *n, Dicrateria inornata; o, Hemiselmis rufescens; p, Isochrysis galbana; q, Pyramimonas grossii; r, Chromulina pleiades.* Original drawings except *c* from Marshall (1925), *h* from Lebour (1925), *m* from Murray and Blackman (1898) and *n* to *r* from Parke (1949).

Atlantic from the tropics to the far northern branches of the North Atlantic current off Spitsbergen. In autumn it is often brought into the northern North Sea in large numbers and is usually found floating very near the surface. It is exceptional in its mode of reproduction; it does not divide in two, but when full-grown undergoes multiple fission into a large number of small spores which break out of the surrounding envelope and swim, like the flagellates about to be described, by the use of whip-like locomotory organs. The full life-history has not yet been observed; whether after fusing with others or not, these spores must eventually give rise to the little green spheres which gradually grow to a full size again.

All the remainder of the planktonic plants belong to the big assemblage of organisms known as flagellates of which there are many different kinds. A selection of the commoner forms is shown in Fig. 15 (p. 45). They are all characterised by possessing at least one, and often two, of the motile whip-like processes termed flagella, with which they draw or propel themselves through the water and are thus able to keep up in the sunlit surface layers. These flagellates are claimed for study by both botanists and zoologists, for among them are indeed both plants and animals—and some which have the characters of both in one. Some possess green pigments allied to chlorophyll or even chorophyll itself, and so feed as true plants; others lack pigment and may feed either by absorbing organic substances through their surface or actually live as animals by capturing particulate food; yet again, others may combine the methods of plant and animal feeding. In this lowly group of organisms the animal kingdom has not yet become fully separated from the plant kingdom. However, most of the planktonic flagellates are in fact plants and most of them have a small red 'eye-spot' or *stigma* which is sensitive to light and so enables them to tell whether they are moving towards or away from the radiant energy necessary to build up their food. If you are able to obtain a plankton sample very rich in these small green flagellates you will be able to see how readily they are attracted upwards towards the light. Fill a tall narrow glass jar with the sample and cover the lower three-quarters with thick brown paper. Now if you leave it for half an hour in the full light of the window you will find on removing the paper that the top quarter of the jar is distinctly greener than the rest; the little flagellates from the whole jar have become concentrated in the sunlit zone. By standing a sample of sea water in the light you may be able to grow a more abundant culture of these little flagellates and so give a more striking demonstration of this experiment. There are some of the Dinoflagellates (members of the family Pouchetiidae) which have a much more elaborate light-sensory organ furnished with both a lens and a pigment-cup; indeed it might almost be called an eye. One of these is shown in Fig. 15c.

The Dinoflagellates are the most striking members of the phytoplankton after the diatoms and are usually present in large numbers; for a full account of them another excellent volume by Dr. Marie Lebour (1925) should be consulted. They have a cell wall made up of a number of plates of cellulose fitting together to form a mosaic and are characterised by possessing two flagella: one working transversely in a prominent groove which almost completely encircles the body like a girdle, the other projecting behind from out of a small longitudinal groove running backwards from the girdle. This latter groove is often protected by curtain-like membranes so that the flagellum may be withdrawn spirally into a sheath. Typically, as in *Peridinium*, they have a single spine pointing forward in front and two spines projecting backwards from the half of the body behind the girdle. The flagellum working in the groove sets them waltzing round as they are at the same time driven forwards by the other flagellum behind (Fig. 15, p. 45); they screw themselves through the water. There are a great many species of *Peridinium* and closely allied genera with very much the same general appearance (Plate IIIb, p. 48); one related genus, however, *Ceratium*, is most striking in having the spines drawn out into long horns, with the two posterior ones usually curving forwards to give the whole body the shape of a little anchor. Sometimes, particularly in late summer and autumn, plankton samples may be full of *Ceratium tripos* or perhaps another of the many species of the genus distinguished by only small differences (Plate Ib, p. 32). Two species which are very common in our waters stand out in contrast to the rest: *Ceratium furca* (Plate 1, p. 80) in which the posterior spines are rather short and point straight backwards, and *C. fusus* (Fig. 15, p. 45) in which there is only one posterior spine, long and only very slightly curved, just like the anterior one. There is a remarkable range of colour in *Ceratium* from a bright green to a yellow-brown.

Dinoflagellates, like other flagellates, multiply by simple fission into two and, like diatoms, each daughter-cell retains one half of the old cell-wall and forms another half anew; but unlike the diatoms the new halves are not formed within the old cell wall and so there is no gradual diminution in size. Occasionally recently-divided individuals of *Ceratium* may remain adhering together to form little chains. The cell-wall grows in thickness but when the little plates of armour become too heavy they fall off to be replaced by new and extremely thin ones. The long horns of *Ceratium* are certainly organs to assist in suspension. Species in the warmer waters generally have much longer ones than those in colder waters. Warmer water is more fluid—or less viscid—than colder water, for the molecules move over one another more freely with greater heat-motion; in the tropics an object will sink twice as fast as will one of similar density and shape in the polar seas. In general plankton organisms, both plants and animals, are more spiny in the tropics to give them a greater surface resistance. It has been observed that species of *Ceratium* have the power of adjusting the length of their horns to the varying viscosity of the water; on being carried by currents into warmer water they grow longer horns, conversely if carried into a colder area they can shed parts of them. (Gran 1912, p. 321).

Several species of Dinoflagellates can produce a brilliant phosphorescence. Many plankton animals are luminous and produce the sparks of light we often see in the water at night. In addition to such displays, however, we may also see a more general ghostly light or sometimes when out in a rowing boat our oar as it cuts the water may leave a trail of blue-green flame behind it; and even from the shore we may see the waves breaking in a flash of light. Such displays are caused by countless millions of dinoflagellates each glowing by an oxidation process as it is agitated in the water. *Noctiluca* is the most celebrated for this, but although a dinoflagellate it is curiously modified to be entirely animal in its mode of life and so will be described in a later chapter

(p. 87); other members of the group, however, particularly *Ceratium*, give almost as good a display. Once on a fisheries research trawler, having stopped at night to make some observations in the Channel, I looked over the side to see a small shoal of fish, most likely mackerel, lit up by each individual being covered by a coat of fire; they were being chased this way and that by some much larger fish similarly aflame. On putting over a tow-net, which came up brilliantly illuminated, the sea was seen to be full of a very small *Peridinium*-like dino-flagellate of the genus *Goniaulax*.

Two other genera of dinoflagellates occuring in our waters, and shown in Fig. 15, p. 45, will just be mentioned. *Dinophysis* has the part of the body in front of the girdle reduced to a minimum so that the girdle itself, with very pronounced margins to its groove, appears like a band round its very front or top; the posterior part bears a marked keel at one side as if designed to prevent the rotation which is normal to the group. Instead of spinning round it is thought to set up a vortex current by which it draws into the groove still smaller organisms as food. *Polykrikos* is a remarkable genus having a number of girdles, usually four or eight, placed in regular succession down the body; it is often spoken of as a 'colonial form' as if made up of several individuals which have failed to separate on division, but this can hardly be the correct view since the number of nuclei is always smaller than the number of girdles present. It appears to be an individual with a repetition of organs similar to the segments of an animal like an annelid worm. They are also said to feed like animals as well as like plants; they possess remarkable little capsules containing coiled threads which can be shot out like those found in the stinging cells of sea anemones and jelly-fish, and may possibly be used for a similar purpose—the capture of prey. In addition to all these forms with their different characteristic patterns of armour plating and spines, there are a great many so called 'naked' dinoflagellates which lack all such coverings; many of these are, for part of their lives, internal parasites in a number of different marine animals.

Among the small shells of *Globigerina* first brought up from the ooze of the ocean bed were found numbers of still smaller calcareous bodies, little plates, some oval and perforated, others round and bearing stout blunt spines; they were called coccoliths and rabdoliths respectively, and presented naturalists with a puzzle as to what they were. It was Sir John Murray who discovered their real nature by showing them to be plates which had covered the bodies of other little plank-tonic flagellates which were given the name of Coccolithophores.[1] *Coccosphaera* and *Coccolithus* (Fig. 15) occur in our seas. They are commoner in the tropics, although in the Atlantic water coming into the northern North Sea they may occasionally be so numerous as to give a milky appearance to the water and cause a chalky deposit to be left on the fishing nets as they dry. This is what the herring fishermen

[1] They were subsequently well described by G. Murray and V. H. Blackman (1898).

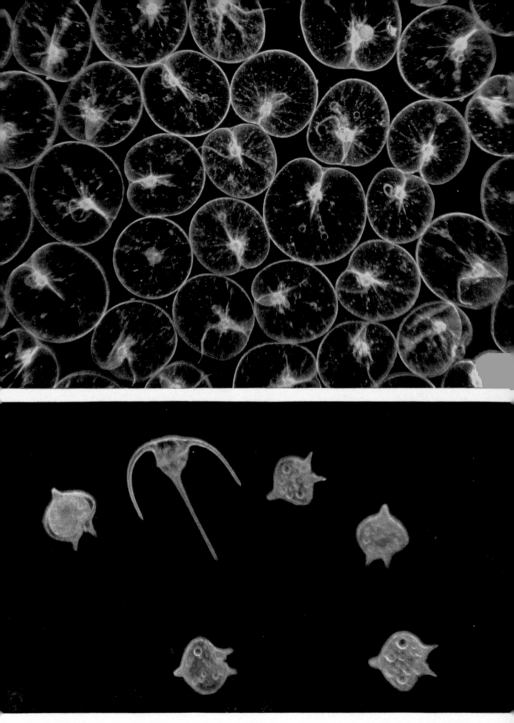

Plate III (above) a. Living dinoflagellates: *Noctiluca scintillans,* x 50. *(below) b.* More dinoflagellates: *Peridinium depressum* (5 specimens) and *Ceratium tripos,* x 140. (Both by electronic flash). *(Douglas Wilson)*

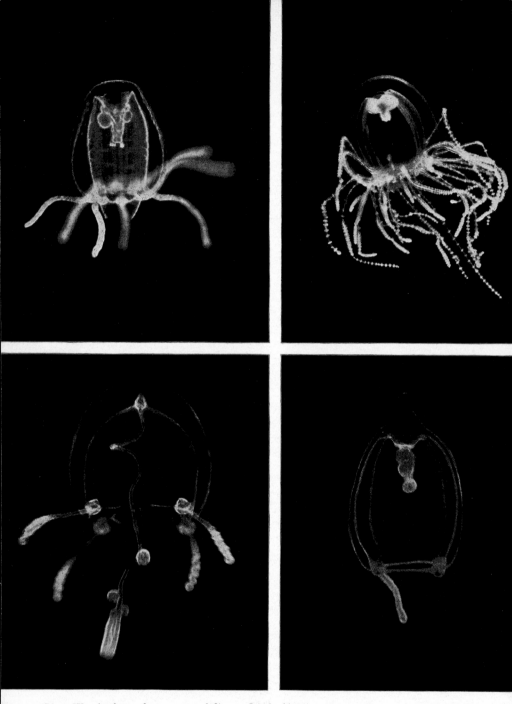

Plate IV. Anthomedusae: (*top left*) *a. Lizzia blondina*, contracting and showing action of velum in forming jet for propulsion, x 50. (*top right*) *b. Gonionemus vertens* actively swimming, x 7. (*bottom left*) *c. Sarsia gemmifera*, showing secondary medusae budding from long slender manubrium, x 28. (*bottom right*) *d. Steenstrupia nutans* (medusa of the hydroid *Corymorpha nutans*), x 17. (All by electronic flash). (*Douglas Wilson*)

call 'white water' and generally believe to be a good sign for the presence of herring. A well-known herring skipper, Mr. Ronald Balls, who is also a keen naturalist, has recently written, under the pen-name of "Peko", an excellent article on this white water in *World Fishing* (July 1954). He describes how this water gives 'the queer impression of whiteness coming upwards: as if the light was below the sea instead of above it'. He then refers to recent views that the coccoliths are shields reflecting light from their owners which normally live in tropical seas where the illumination is too strong; 'and here', he writes, 'was the perfect explanation of the fairy glow or white reflection that I had experienced long ago, and wrote about before I knew even that this organism existed'. As with the cell walls of diatoms, the electron microscope is showing that each little plate or coccolith has a much more complicated structure than was originally supposed; its base consists of radiating ribs like the spokes of a wheel and its rim is decorated with a frill like that with which a chef may decorate a ham. There are other similar little creatures, the Silicoflagellates, which form a delicate siliceous skeleton with radiating spines (Fig. 15, *i* and *j*).

Herring nets, although they hang in the water near the surface, may often come out of it in a very slimy condition; this may be due to an excessive number of diatoms; more usually, however, it is due to globules of jelly large enough to be seen quite easily by the unaided eye. These slimy blobs are produced by aggregates of microscopic flagellates called *Phaeocystis* which colour the surface of the jelly in green patches (Fig. 15k, p. 45). All the meshes of a tow-net may be blocked with them. Dense concentrations of *Phaeocystis*, like those of diatoms, which cover wide areas of sea, have also been thought to have a deleterious effect on the shoaling of herring and at times to have led to a poor or delayed fishery. We shall refer to this again in Chapter 15 (p. 293).

This completes the review of those planktonic plants we shall mention by name; but we have so far left out of account a vast number of still much smaller flagellates which have escaped capture by passing through the meshes of the finest net we can use. It is only comparatively recently that their influence in the economy of the sea has been realised. Their prominence was first demonstrated by the German naturalist Lohmann who examined the remarkably fine filtering mechanism, far finer than any gauze that man can make, used for their capture by some little plankton animals, the Larvacea, to be described on page 153. They may, however, be extracted from a

TOS—E

sample of sea water by centrifuging[1] small quantities of it in tapering tubes. If after such treatment the greater part of the water is carefully decanted, a drop of the remaining fluid may be taken up in a pipette and examined on a slide under the high power of a microscope; then they will be seen as tiny yellow specks jigging in the water Today a great many of these minute flagellates are being successfully cultured in the laboratory, notably by Dr. Mary Parke at Plymouth (1949). Five examples, sketched in line from her beautiful coloured drawings, are shown in Fig. 15 *n-r* (p. 45).

Smaller still, of course, are the bacteria which really lie outside the scope of this book; at present, very little is known about their occurrence in the plankton. Dr. H. W. Harvey (1945) states that their population density decreases on passing from inshore waters to the open sea and that in the ocean the greatest numbers are found where phytoplankton is abundant and in the water immediately above the sea-floor. They are found particularly in dense phytoplankton regions because of the undigested organic matter passed out by the animals which are eating more of the plants than they really require.

It is exceedingly difficult to get an accurate measure of the amount of plant life in a given quantity of sea water, even of the larger forms which are captured by a net. Although we can calculate the filtering efficiency of the net and know the quantity of water it *should* filter, we cannot be sure that it actually does filter this amount; in fact it rarely does, for to a varying extent under different conditions the meshes of the net become clogged by the organisms themselves and the filtering is much reduced. However we can get an *approximate* idea of the number of larger forms—the diatoms and dinoflagellates— in a given volume of water by using a net. Let us take an example. In 1907 Sir William Herdman and his co-workers began an intensive study of the plankton of Port Erin Bay in the Isle of Man which they continued until the end of 1920. Usually six times a week, every week for fourteen years, two standard nets of coarse and fine mesh were towed in exactly the same way over the same distance—half a mile—across the bay. Johnstone, Scott and Chadwick, who describe the results in their book *The Marine Plankton* (1924), estimate that for each such double haul "taking the two nets we shall not be very much in error (when all the conditions are considered), in assuming that 8 cubic metres of water were filtered through both nets." The following figures, taken from their book, give the *average* number of the principal

[1] Subjecting to a force greater than gravity by spinning in a rotary apparatus: the centrifuge.

plant forms taken in such a catch during the month of April, i.e. the average of all the April hauls made over fourteen years:

Chaetoceros spp.	4,969,809	*Landeria borealis*	324,628
Rhizosolenia spp.	20,585	*Thalassiosira spp.*	157,666
Coscinodiscus spp.	206,689	*Ceratium tripos*	2,968
Biddulphia spp.	122,543	*Peridinium spp.*	1,307
Guinardia flaccida	18,998		
		TOTAL :	5,815,193

That in round figures is 727,000 per cubic metre or about 20,000 per cubic foot. The number of plankton animals taken at the same time is given in Chapter 5 where the zooplankton is considered and may be compared by turning to page 71. The actual numbers present are estimated by using a specially calibrated pipette which takes up a known fraction of the sample; the fraction is spread out on a glass slide ruled in squares so that the number of plant cells can be counted below the microscope just as the corpuscles are counted in a sample of blood. We must remember two important things about the figures just given. Firstly they are for the *larger* microscopic plants; the very small ones are present in far greater numbers as we shall see in a moment. Secondly they are average figures for April over fourteen years; those for one year may be very different from those of another and the average figures for other months of the year will show still greater differences. There are marked seasonal changes in the plankton; but that is the subject of our next chapter.

The difficulty of knowing exactly how much water is filtered by a net when its meshes are becoming clogged by the organisms sampled, has been got over by an ingenious device invented by Dr. Harvey of Plymouth (1934). At the mouth of the tow-net he has fixed a little propeller which is turned by the water flowing into it; the number of revolutions it makes are recorded on little dials which measure the amount of water actually passed through the net. There is still, however, the difficulty of forming a true quantitative estimate of the plant life present. We can calculate, as we have just seen, the number of plant-cells in the sample; but these vary so enormously in size it is difficult to convert such an estimate into a measure of the total bulk of planktonic vegetation. Measurements of the volume of the sample can be made after all the plants have been killed by the addition of formalin and allowed to settle for several days in the bottom of a measuring jar; but this too is a very misleading estimate, because the various kinds, having different shapes, may pack together very differently: for example spiny forms take up more space than round or flat ones. However, these various methods do enable us to say broadly that one area is relatively so much richer in phytoplankton than another—always excluding the small flagellates which escape the net and must be estimated with the centrifuge. A more recent method of estimating the quantity of plant life caught in a plankton sample is to extract the plant pigment by acetone and measure the quantity present by matching up the samples obtained with a standard colour scale and expressing it in so many pigment units.

The late Dr. E. J. Allen (1919), when Director of the Plymouth Laboratory, made a simple but important experiment that gives us some idea of the vast numbers of little plants there are in the sea which are not caught by our ordinary methods. He had first perfected a method of growing them in bottles in a special culture solution, i.e. in sea water enriched with the addition of certain beneficial chemicals. He then took a sterilised quart-sized bottle and filled it with sea water from just below the surface about half a mile outside the Plymouth breakwater. This water he treated in two ways. The procedure may seem a little involved but it is worth following. Firstly he took four 10 cc samples of it and centrifuged them each twice with the result that he obtained an average of 14.45 organisms per 1 cc of water which gives us an estimate of 14,450 per litre. Secondly he took just ½ cc of the water he had collected and added it to 1,500 cc of his culture solution which he had previously sterilized; then after it had been thoroughly shaken up he divided this between 70 small flasks—a little over 20 cc in each—and placed them against a north window. After 10 days signs of growth were apparent. When they were finally examined there was not a flask that had not had some growth in it. He now recorded the different *kinds* of organisms in each. In two flasks there was only one species; in all the others there were from two to seven different species present, giving an average of 3.3 different kinds per flask. Thus at least 70 x 3.3 or 231 *separate* plants must have been taken up in the ½ cc originally added to the culture solution; that makes 464,000 per litre as compared with the 14,450 estimated by using the centrifuge! For comparison with the larger plant forms caught by the net in the former example we must express the number as per cubic metre: i.e. 464 million, or about 12½ million per cubic foot. Now this must be regarded as an absolute minimal estimate, for it is made by assuming that only *one* individual of each kind of plant recorded in a flask went into that flask at the beginning; this is most unlikely.

We begin to have some idea of the great wealth of plant life there is in the sea. Can we make it still richer by adding fertilizers in the same way as we increase our crops on land ? Experiments have been made in that direction, but a discussion of them will come better in the next chapter, where we will deal with the various factors which govern phytoplankton production. For a more detailed and fuller account of the pelagic plants in general I would recommend for further reading the splendid chapter by Professor H. H. Gran in Murray and Hjort's *Depths of the Ocean* (1912).

CHAPTER 4

SEASONS IN THE SEA

THE NATURALIST with a tow-net, if he can sample the plankton at different times of the year, will find contrasts between spring, summer, autumn and winter in our seas almost as striking as those in the vegetation on the land. These seasonal changes in the plankton have a profound effect on the lives of many fish. Just as we can tell the age of a felled tree by the number of concentric rings in its trunk representing summer and winter growth-zones, so we can tell the age of a herring by similar rings on its scales; these mark summer growth-periods, when its planktonic food was abundant, separated by lines showing where the scale, and the fish, had ceased to grow during winter when the plankton was scarce.

There is not, however, a simple and gradual increase in the plankton as spring advances into summer followed by a gradual decline in the autumn. Our naturalist with a tow-net will find some of the changes very puzzling at first sight. In British waters in the winter there is a general paucity of both animals and plants in the plankton; then as the sunlight grows stronger (the date varying in different years, but usually in March) there is a sudden outburst of plant activity. The little diatoms start dividing at a prodigious rate: in a week they may have multiplied a hundred-fold and by a fortnight perhaps ten-thousand-fold. The meshes of the tow-net are clogged by them and the little jar at its end is filled with a brown-green slime, a slime which under the microscope resolves itself into a myriad forms of beautiful design. Then as spring advances into summer the number of little

floating plants steadily declines until by late summer there are surprisingly few. Some reduction in their numbers might indeed be
expected, for the little animals in the plankton which feed on them are
also multiplying as the season advances and the waters are warming
up; with the increasing sunlight, however, we might have thought
that the plants' remarkable power of increase could largely keep pace
with the grazing of the animals. Something seems to be preventing
the diatoms from keeping up that rapid multiplication. In the autumn
comes another surprise. As the days begin to shorten and the sunlight
is getting less intense, when in fact we might least expect a renewal
of plant activity, there comes a second phytoplankton outburst; it is
not as spectacular as the spring maximum and not in every year is it
of an equal intensity, but there it is—a definite surging up again of
reproductive power. From this second peak of production, as winter
approaches, the numbers fall again to the lowest level of the year.

This sequence of events was known for a long time before it was
properly understood; it was only after much more had been discovered
about the physics and chemistry of the sea that it was possible to see
at all clearly the chain of cause and effect throughout the year. So
important are these events that we must devote a little space to considering some of the more important elements in the physical and
chemical background that will help us to explain them.

Let us first consider some of the physical properties of the water.
At the very beginning, in the introductory chapter, we referred to the
limited transparency of the sea and some figures were given to show
how quickly light is actually absorbed on its passage below the surface.
As might be expected absorption of light will be found to vary considerably according to the amount of suspended matter, either sediment
or plankton, in the water. Far out from the land the water is usually
much clearer than in the shallower regions against the coast where
detritus and mud may continually be stirred up by the tides or brought
in by drainage from the land. If we compare measurements of light
made at different depths below the surface in the waters near Plymouth
we find the penetration in inshore waters at Cawsand Bay to be only
half what it is at a point some 10 miles S.W. of the Eddystone Light-
house (Poole and Atkins, 1926). Taking the light entering the sea, i.e.
just below the surface, as our standard, we find in Cawsand Bay that
half of it has been absorbed at 2 metres depth (i.e. 1 fathom), some
75% at 4½ metres and 90% at about 8 metres depth; whereas at 10
miles out the same percentage reductions in light intensity are found
at depths of about 4¼, 9½ and 17 metres respectively. At points in the

open sub-tropical or tropical Atlantic, where the phytoplankton is very sparse, as in the Sargasso Sea, the corresponding depths might be increased four or five times. A very simple bit of apparatus known as the Secchi disc, which can easily be home-made, will enable you to compare the transparency of the sea at different points; you may well be surprised at some of the results you will get with it. Take a white painted metal disc, say two feet in diameter, and drill three equidistant holes near its margin; now take three cords each about 6 feet long, tie them to a weight, say a 7-lb lead, and then tie a knot in each at 3 feet from the lead; next pass the cords through the holes in the disc and bring them together as supporting bridles to be tied to a loop or eye at the end of the line which will suspend the whole device in the water as shown in Fig. 16 (p. 56). If a metal disc cannot easily be obtained, a white dinner plate, with wire clips behind it to take the cords will serve the purpose quite well. To compare the transparency of the sea all you have to do is to stop your boat and lower the disc over the side at different places and find how deep it must go before you can no longer see it. The weight not only carries the disc down but, if the cords are properly adjusted, ensures that it is always kept horizontal. The line can be knotted at metre intervals to facilitate measuring the depth. It is well to raise and lower it about the disappearance point several times in order to make quite sure just at what depth it goes out of sight; at some places it may vanish in only 5 metres, at others it may be seen for as much as 12 metres or more.

In passing I may say that the Secchi disc is also valuable in helping us to compare the varying colour of the water. Here, of course, I am not thinking of the often striking and delightful changes of hue which we may see as light of different quality is reflected from a changing sky: when for example dark grey clouds give place to an open space of blue or when cumulus clouds dapple the sea with purple shadows. Reference was made in Chapter 2 to the contrast between the green water of the Channel and the deep blue of the Atlantic; such differences are due to the nature of the contents of the sea itself and are examples of what I mean by the varying colour of the water. A great variety of shades and hues may be found at different times; these are mainly due to the presence of different kinds of very small plankton organisms in exceptional numbers. The light reflected back through the water from the white background of the disc enables us to judge and compare these differences more easily than by just looking into the depths. Dense concentrations of diatoms such as *Rhizosolenia* and *Biddulphia*, or of the colonial flagellate *Phaeocystis*, may give a brown appearance to the water over large areas; the North Sea fishermen often call such patches of water "Dutchman's baccy juice". Some dinoflagellates may make the water almost red, coccolithophores may give it the white milky appearance referred to in the last chapter, and other small flagellates may occasionally make it a vivid green.

But let us return to the sunlight and these little plants of the sea; in order to flourish and grow they must produce more oxygen in the

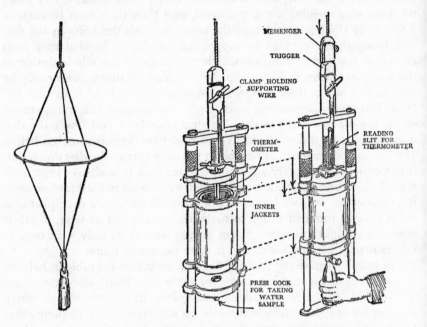

FIG. 16 (*left*). The Secchi disc for measuring the transparency of the sea.
FIG. 17 (*right*). The Nansen-Pettersen water sampling bottle: shown open and closed.

process of photosynthesis (see p. 3) than they use up in respiration. Plants, of course, breathe as well as animals. Some very significant experiments were performed in the Clyde sea-area by Drs. Marshall and Orr (1928) of the marine biological station at Millport on the Island of Cumbrae. They grew cultures of diatoms in glass bottles in the sea at different depths; these they suspended on strings from a long thin rod between two buoys at the surface and so kept them free of shadow. All their bottles were in pairs; one of each pair was exposed to the light and the other covered with a black cloth. In each bottle the oxygen-content of the water was measured at the beginning of the experiment and again at the end of twenty-four hours. An increase in the oxygen in the uncovered bottles showed the amount produced by photosynthesis *less* that used up in respiration; a fall in oxygen-content in the 'blacked-out' bottles measured respiration alone. By adding this oxygen-loss to the oxygen measured in the uncovered bottles the total oxygen-production as a record of photosynthesis could be estimated. The experiments were repeated as the spring passed into summer, and were also made on days which were overcast and on others which were sunny. As the sun went higher in the sky and the light became more intense the depth at which diatoms could produce more oxygen than they used in respiration increased from a depth of less than 10 metres on an overcast day in March to nearly 30 metres on a sunny day at midsummer. By far the greatest photosynthetic activity—on which their growth depends—took place, however, in the top 5 metres. In the waters round Great Britain we may now say that practically all the plant-production that matters takes place in the top 10 or 15 metres. This is one important clue in the puzzle of the seasons; we must now turn to temperature.

The water round our coasts varies in temperature from about 8°C in winter to sometimes as much as 17°C in the Channel in a warm summer. It is, of course, because the sea loses and gains heat so much more slowly than the land that we in Britain have so equitable a climate compared to that of an area in the middle of a continent. Two methods are used in taking the temperature of the sea. Down to moderate depths, say to 50 metres, the insulated Nansen-Petterson water-bottle, which is shown in Fig. 17, opposite, is used; it is of metal and is sent down suspended on a wire to obtain samples of water both for chemical analysis and for temperature determination. It goes down with the bottom and top open so that water can circulate through it; then at the required depth a small 'messenger' weight is sent sliding down the wire to hit a trigger which releases springs to close it. Projecting through the top, in a protective casing, is the stem of a thermometer whose bulb is in the centre of the sampling bottle; its scale and mercury thread are visible through a slit in the upper casing so that it can be read as soon as the bottle is brought back to the surface. There are actually three walls to the cylindrical bottle, one inside the other, with a little space between; when the top and bottom are firmly closed there are thus two water jackets outside the bottle proper and these act as insulating chambers preventing loss or gain of heat in the water sample while it is coming up and the thermometer is being read. As soon as the temperature has been noted the water is run out from a cock at the bottom to be stored for later analysis and the bottle is opened ready to be sent down to another level. From much greater depths the bottle would take so long being drawn up that the insulation just described would not be adequate to prevent a change of temperature in the process. To get over this, special so-called reversing thermometers and bottles have been devised. The mercury tube of the thermometer, just above the bulb, has a loop and a kink in it, so that when it is swung rapidly upside down the thread of mercury breaks; as soon as this happens all the mercury that before was *above* the kink now runs to the opposite, and now lower, end of the tube. When it is brought up the height of this *inverted* column of mercury is seen against a scale which can only be read when the thermometer is upside down; it tells us the temperature that the thermometer was recording at the moment it was turned over. The bottle and thermometers (there are usually two to give check readings) are mounted in a frame which rotates when a trigger is hit by a messenger weight; the bottle, which before was open, is closed as it swings over.[1]

After this digression on thermometers, let us return to consider the temperatures of our seas with the passing of the seasons. Water, above 4°C, expands when warmed and contracts when cooled; so its density is altered: a given volume of cold water weighing more than the same volume of warm water. In winter the atmosphere is colder than the sea so that the surface waters are cooled and therefore sink beneath the warmer and less dense layers which were below; this is repeated again and again until after a time there is an almost uniform low temperature from top to bottom. The winter gales help in the process of mixing up the layers too. The sea, of course, is rarely so cold in winter or so warm in summer as is the atmosphere; as we have already noted, it gains and loses heat much more slowly. As spring passes into summer the air warms up and the radiant heat of the sun gets stronger, so we find the upper layers of the sea becoming warmer too; as they heat up they become increasingly lighter than the layers below and thus tend more and more to remain separated on the top because less and less are they likely to be mixed with the heavier waters beneath. This division between the upper and lower waters is called a

[1] When taking samples from a series of levels in very deep water several of these reversing bottles are generally used together, one above the other, on one wire at intervals of perhaps a hundred metres or more; as the messenger weight hits the trigger to reverse the first bottle, it also releases from below it another messenger which now slides down the wire to operate the second bottle and this again liberates a third messenger and so on to the bottom.

discontinuity layer (or *thermocline* in still more technical language) and is usually set up at a depth of round about 15 metres. Let us take an actual example from the summer temperatures in the English Channel in July as found by the hydrologists of the Plymouth Laboratory. At depths from just below the surface down to 15 metres the temperature only varied from 16.5° to 15.82°C; but at 17½ metres it had dropped to 12.09°C and then, as it was sampled deeper and deeper, it remained practically constant to read 12.03°C at 60 metres. The upper layer was effectively cut off from the lower by this sudden drop in temperature of nearly 4°. A strong summer gale may destroy this discontinuity layer, but if it is not too late in the season it will soon form again. It is in the autumn that the air cools again and so the surface water loses heat; also the equinoctial gales stir up the sea and the more uniform temperatures of winter again become established from top to bottom. It will be noted that this warm summer upper layer corresponds very closely to the region (sometimes called the photic zone) in which the little plants get sufficient light to carry out effective photosynthesis. Two bits of the puzzle seem as if they would fit together; we require, however, yet another piece to go with them before we can see the explanation of the seasonal changes in the plankton. This last link concerns certain salts in the sea, and to them we must now turn.

First we must consider the general saltness of the sea; this, of course, is mainly due to the abundant sodium chloride which accounts for almost 77.8% of the total salt content. However there are many other salt constituents, of which the next more important, in order of descending quantity, are magnesium chloride (10.9%), magnesium sulphate (4.7%), calcium sulphate (3.6%), potassium sulphate (2.5%), calcium carbonate (0.3%) and magnesium bromide (0.2%). These proportions are actually those in which these different salts would be recovered from the sea on evaporation; their molecules as dissolved in the sea, however, would largely—some nine out of ten—be split up into their respective parts or ions: sodium and chlorine or magnesium and sulphate ions as the case may be. It is better to think of the salt constituents of sea water, as they mostly are in the sea itself, in terms of separate ions. We can tabulate the percentage proportions as follows, based upon a mean of 77 samples collected from different localities by the *Challenger* Expedition:

Sodium	30.59	Chlorine	55.29
Potassium	1.11	Bromine	0.19
Calcium	1.20	Sulphate	7.69
Magnesium	3.72	Carbonate	0.21

In addition there are minor constituents, for example iron, strontium, silicates, phosphates and nitrates, which constitute together only 0.06%. The degree of saltness of the sea, or its *salinity*, is usually expressed in terms of the total weight of salts in grams per thousand (°/oo) grams of sea-water; it varies in the open ocean from 34°/oo in polar waters, where it is low on account of additions of fresh-water from melting ice, to 37°/oo near the equator where it is high because of excessive evaporation of water. The North Atlantic surface water as it flows round our islands has a salinity of about 35°/oo, but in the southern North Sea it is diluted to some 34.5°/oo by water-drainage from the land.

Now the important salts for our little plants are those which have only been mentioned among the minor constituents: they are the phosphates and the nitrates. Because they are present in such small quantities, it was a long time before accurate methods for their estimation could be devised; these were developed largely through the work of Drs. Atkins and Harvey at the Plymouth Laboratory just after the first world war. It had been realised that the plants of the sea must be limited, as are the plants of the land, according to Liebeg's Minimum Law; i.e. so long as *any* really essential nutritive substance occurs in minimum quantities, plant production will be proportionate to the available quantities of it, even though there is a super-abundance of all other essentials. This seemed obvious enough but could not be proved until we had these more refined methods. It now became possible to measure the amounts of phosphates and nitrates taken up from the water by the little plants; it was shown that in our waters these salts could and did in fact limit their growth. The reproductive rate of these little plants grown in culture solutions was seen to fall off as the phosphates and nitrates were depleted and finally growth would stop altogether when they were entirely used up.

It is now possible to explain the seasonal cycle of events. In the winter, as we have seen, the waters from top to bottom are well mixed and their temperature is almost uniform. As the length of the days and the intensity of the light increases there comes a point at which the little plants can begin to multiply and they find a comparatively rich supply of phosphates available—about 40 milligrams per cubic metre of water. We have seen how rapidly they undergo fission when once they start. They are multiplying only in the upper, well illuminated zone; in the early spring these upper waters are being well mixed up with the lower layers by the equinoctial gales and there is a general reduction of the free phosphates as they pass into the plants. Presently, however, there develops an upper warm layer which becomes more pronounced as spring advances into summer; this is also the photic zone, in which alone the plants can multiply. The phosphates and nitrates are now being used up by the plants in this upper zone and are not being replaced by any mixing with the lower waters because of the difference in density between them. In fact the phosphates and nitrates are continually passing from the upper to the lower layers. The plants, which have taken up the salts, may either eventually form resting spores and sink, or may just die and sink, or more likely be eaten by the animal members of the plankton; these may themselves just die and sink or in turn be eaten by other larger animals and so on. The nutritive salts which were once present in the upper zones are now by late summer reduced to a minimum; they are carried in the falling bodies to the bottom or still locked up in animal life. Actually a good deal may be excreted back into the water by the animals,[1] but as most of the animals only make comparatively short visits from the lower into the upper layers to feed on the plants at night, most of the excreted phosphates and nitrates will pass into the lower waters. Thus we see that so long as the discontinuity layer lasts, the plants, such as have not been eaten by the animals, are cut off from the richer phosphates and nitrates below. That is why their numbers decline so markedly as the summer advances and why they cannot reproduce at a rate sufficient to counterbalance the inroads made upon their population by the grazing animals. Down below, the supplies of phosphates and nitrates are being to some extent built up again by their return from dead animals broken down by bacterial action. Now, as the summer wanes, the upper layers are cooled again and the autumn equinoctial gales assist in a general mixing; the water richer in phosphates and nitrates is brought up from below towards the surface where once again we have a fertile layer while the sunlight is still strong enough to encourage photosynthesis.

Here at last we have the explanation of that autumnal outburst of phytoplankton which had for so long been such a puzzle. The time of its appearance and the quantity produced vary markedly in different years; it is usually not very long-lived and eventually the population of plants dwindles to a winter minimum as the light gets too weak to allow much active reproduction. The winter gales stir up the water and the nutrient salts are once again more or less evenly spread through the different layers of water. The temperature, too, is more or less uniform; the cycle is complete.

[1] See Gardiner (1937).

This brief account of the events throughout the year has dealt with the phytoplankton as a whole. If it suggests, as well it might, that all the different kinds of little plankton plants are increasing and declining together, as the seasons come and go, it would be giving a very false picture. There is in truth a succession of different forms which wax and wane in turn within this larger framework. As the summer advances and the quantity of the phytoplankton is declining, the dinoflagellates come to occupy a much more prominent part in the community; in late August species of *Ceratium* and *Peridinium* may be much more evident in the fine net samples than the diatoms. At the second autumn outburst the diatoms will swing back into prominence again. Then within these spring and autumn periods of production there is usually a fairly definite order of appearance of different species of diatoms as the weeks go by; not that one kind disappears entirely of course, but after a period of abundance the reproduction falls to a low ebb and the stock is maintained by only a few individuals or by the resting spores already referred to. The intensive work, already referred to (p. 50), carried on week by week for fourteen years at Port Erin in the Isle of Man, has furnished us with a mine of information about these detailed seasonal changes at one place; and now the monthly plankton recorder surveys which will be described in the last chapter (p. 309) are giving us similar information for a very wide area.

What makes one species give place to another? Why for example should *Chaetoceros decipiens* give way to *Ch. debilis* and *socialis* as the season advances or *Rhizosolenia semispina* be replaced by *Rh. shrubsolei* which in turn may leave the stage to *Rh. stolterfothii*? Whilst the grazing of the little plankton animals coupled with the reduction of phosphates and nitrates in the upper layers is bringing about the general decline in the planktonic vegetation it can hardly be controlling the rise and fall of the different species. Johnstone, Scott and Chadwick (1924) in discussing this seasonal sequence of species which they found in their long series of tow-nettings at Port Erin, made an important suggestion as to its cause.

"It is known that some bacteria are incapable of producing their typical effects (say in fixing elementary nitrogen from its solution in sea water) if they are present in pure culture. In order to function effectively they must be associated with some other organism which, by itself, cannot produce the effect in question. Probably such symbiotic relationships may exist on the great scale in the sea. The work of Allen and Nelson (1910) on the artificial culture of diatoms suggests this. In mixed cultures there is always a certain succession of species, one attaining its maximum when another has ceased actively to reproduce. The succession of diatom species during the period of the spring growth suggests that something of the same kind occurs in the sea."

A similar effect has, of course, now been demonstrated by Sir Alexander Fleming's great discovery that moulds such as *Penicillium* produce substances which inhibit the growth of bacteria. In my hypothesis of animal exclusion (in Hardy and Gunther, 1935), which will be referred to again in a later chapter (p. 216), I have suggested that dense concentrations of planktonic plants may produce an effect in the water which is uncongenial to animal life and so account for the fact that animals are usually scarce in regions of great phytoplankton abundance. Dr. C. E. Lucas, my former pupil and colleague, now Director of the Scottish Fishery Laboratory at Aberdeen, has developed much further the idea of chemical interaction between organisms and stressed the possible importance of various substances given out into the water by different plants and animals as a result of their internal activities. Just as cells inside the body of an animal produce those various substances called hormones (or endocrines) which circulate in the blood stream to have profound effects on other parts of the body, so also may substances (ectocrines) be liberated from the body to have their effects on other organisms in an aquatic environment. The changed conditions set up in the water by one species may perhaps become both injurious to itself and at the same time more suitable to another kind which will follow it. Thus, among other interesting ideas, he gives strong support to this idea of seasonal succession: a chain of action, a conditioning and reconditioning of the water, as the year advances (Lucas, 1938, 1947 and 1956*a*).

Among the animals of the plankton there are also successive changes; particularly noticeable are the various broods of different species which follow one another, giving us in one month mainly adults and in another the young developing stages. The seasons are marked too by the throwing up into the plankton of the young larval stages of various bottom-living invertebrates, and also by the eggs and fry of different species of fish, all of which have their own distinct breeding times.

To return to more general considerations, it is interesting to compare the conditions as found in our waters at mid-summer with those in the surface waters of the tropics; we have the same heating up of the surface to form a discontinuity layer, but there it tends to be a permanent feature. In the open ocean in the tropics the phosphates and nitrates in the upper layers are thus reduced to a minimum all the year round and as a consequence the plankton of those regions is extraordinarily sparse compared with the more temperate or polar seas. This relative poverty of the tropical seas compared to our own was one of the surprising discoveries made by Victor Hensen's German Plankton Expedition in 1889 and was at first disbelieved by many who thought, on false *a priori* grounds, that the warmer tropical seas must be richer in life than our own or the cold polar regions. A tow-netting in the tropical ocean may yield many more species than one in our own waters, but the total quantity of life is very much less; when we glance at a tropical sample at first sight nearly every specimen seems a different kind, whereas in one from our own waters there will be thousands of representatives of the same species. Plankton at certain places in the tropics, however, can be remarkably rich; this happens when deeper water with a supply of nutritive salts comes welling up into the sun-lit zones as against the coast or where a submarine bank comes near the surface and gives rise to disturbed water conditions.

The upwelling of water rich in phosphates and nitrates may well be very important in producing a more prolific phytoplankton in our own waters. There can be no doubt that the abundance of fish-food on the sea-bed which makes the Dogger Bank so renowned as a fishing ground, is due to the heavy rain of plankton showered upon it from above; there can also be little doubt that this rich phytoplankton in the surface layers is in turn produced by the upwelling of the richer phosphates and nitrates from below as the Atlantic inflow into the North Sea meets this large submarine bank set across its path (Graham, 1938).

A prolonged off-shore wind may have the effect of producing a heavy crop of plankton near the coast: it pushes the surface water away from the land so that its place has to be taken by water from below which wells up near the coast and thus again brings the desired nutritive salts up into the sunlit upper layers.

There are indications that at times other minor constituents of sea-water may have an effect upon phytoplankton. There is some experimental evidence that organic salts of iron and manganese will stimulate phytoplankton production; and it is suggested that such salts carried out by the drainage from the land may lead to an earlier outburst of reproductive activity among planktonic diatoms in coastal waters than among those further out to sea. Much information about these minor constituents will be found summarised by Dr. H. W. Harvey (1942, 1955) who has himself done so much of the experimental work.

For a long time there has been evidence that suggests that there is in the sea some trace substance at present unknown which is necessary before life can exist—some substance rather like the vitamins in our diet. Artificial sea-water has been made up to contain precisely the same proportions of chemicals that are known to be present in natural sea-water by the most exact analysis; yet planktonic plants will not grow in it unless about 1% of natural sea-water has been added to it (Allen, 1914). Quite recently the vitamin B^{12} (Cobalamin) has been shown to be necessary for the growth of several marine flagellates and a diatom (Droop, 1954 and '55) and its presence in natural sea water has now been demonstrated.

In addition to 'trace substances' which affect plant production, it now appears that there are some which are essential for the healthy development of delicate young animals. D. P. Wilson (1951) has recently shown a remarkable difference in this respect between the two types of water which we discussed in Chap. 2 (p. 29): the more oceanic water characterised by one species of arrow-worm *Sagitta elegans* and the coastal water by another species *S. setosa*. It will be remembered that in the 1920's there was usually *elegans* water off Plymouth; at that time Dr. Wilson had no difficulty in rearing the planktonic young of some of the bottom-living worms he was interested in. In the 1930's, however, when the *setosa* water was over the area, he experienced much frustration in doing so. Thinking that there might be some subtle difference between the two waters, he made experiments to test his suspicions. He took the fertilised eggs of two different kinds of worm and of a sea urchin, and then divided each lot into a number of smaller batches; some he put into *elegans* water collected out in the Celtic Sea to the west of the Channel and the others into *setosa* water taken off Plymouth near the Eddystone. In the former water most of the larvae developed well, but in the latter they were abnormal or in poor health. "The experiments indicated," writes Wilson, "that the Channel water lacked some unknown constituent, essential for the healthy development of these species, present in the Celtic Sea."

Not infrequently a particularly rich outburst of phytoplankton is reported at a place where the waters of two current systems meet and mix. I have seen it particularly in the region of South Georgia in the sub-Antarctic where waters from the Weddell Sea and the Belling-hausen Sea meet in eddies on each side of the island. It is said to be a feature too of the boundary separating the North Atlantic current from the arctic water and it is not uncommon generally where oceanic and coastal waters meet and mix. Perhaps, on account of different plankton communities, each water has become deficient in some differ-ent but vitally important minor constituent; then on their coming together each will fertilize the other with the missing ingredients and so release an outburst of reproductive activity.

A good deal of interest was aroused in experiments performed during the war by the late Dr. Fabius Gross and his co-workers (1944) to see to what extent the growth of fish could be accelerated by increasing the quantity of plankton in an enclosed sea loch by the addition of nutritive salts. The plankton was certainly enriched and an increased growth of the fish was recorded. It was in fact doing in a confined part of the sea what had already been successfully done in fresh-water fish-ponds; it is indeed a practice dating from ancient Chinese days. It has been suggested that it might be possible to add fertilizer to parts of the more open sea to increase the plankton in a limited area to provide a better chance of survival for the hosts of young fish that are expected to be developing there. But the open sea is a very big place and to do anything at all effective would need the provision of fertilizers on a scale perhaps too vast to be contemplated as a feasible proposition. Yet it seems that man does unwittingly influence the production of phytoplankton in the sea and consequently the yield of fish. In the southern North Sea opposite the opening of the Thames estuary there is frequently developed an area of a particularly rich growth of phytoplankton and here Mr. Michael Graham (1938) has shown an abundant source of phosphates and nitrates derived from the sewage of London. Dr. K. Kalle of the Oceanographic Institute at Hamburg has recently written a paper on the influence of this drainage from the Thames upon the fish population of the southern North Sea and this has been conveniently summarised in English by Dr. J. N. Carruthers (1954). He points out that the water from the continental rivers is carried quickly to the north-east by the current from up the channel; whereas that from the Thames is held up, wedged between two streams of oceanic water of higher salt content: i.e. that just mentioned and the Atlantic influx from the north. He estimates that 2,900 tons of phosphorus a year are carried from the rivers and when spread through the 171 cubic miles of English coastal waters south of the Humber amounts to an increase of 4 milligrams (0.004 grams) of phosphorus per cubic metre. Dr. Kalle then shows that the catch of fish in this region is per unit area 'about double the corresponding catch made in the rest of the North Sea, in the English Channel and in the Kattegat/Skagerak region . . . and is about 25 times the catch reckoned for the Baltic Sea as a whole.' He holds that two-thirds of this higher average catch may be attributed to the rich supply of nutrients from the population of our metropolis.

For a comprehensive treatment of the physics and chemistry of the sea in relation to plankton production the important books by Dr. H. W. Harvey (1945 and 1955) should be studied.

CHAPTER 5
INTRODUCING THE ZOOPLANKTON

THE ANIMALS of the plankton are by definition those which are passively carried along drifting with the moving waters; those other inhabitants of the open sea which are powerful enough to swim in any direction—the fish, whales, porpoises and the squids or cuttle-fish—are in contrast referred to as the nekton (p. 5). The vertebrates therefore, except for certain primitive relations, will only be represented in the plankton by the floating eggs of fish and the young fish themselves up to the time when they become strong enough to migrate at will instead of being just helplessly transported.

In spite of this limitation, and the absence of insects, I believe it is no exaggeration to say that in the plankton we may find an assemblage of animals more diverse and more comprehensive than is to be seen in any other realm of life. Every major phylum of the animal kingdom is represented, if not as adults, then as larval stages with the partial exception of the sponges; the sponges do indeed send up free-swimming larvae but they are in the plankton for so short a time that they can only be claimed as very temporary components of it. In no other field can a naturalist get so wide a zoological education and in few others will he find a more fascinating array of adaptational devices.

It is this great variety of forms, and the unexpected finds which are always turning up, which make hunting in the plankton such an exciting occupation. Except for the jelly-fish and some of the larger crustacea, it is of course hunting with a lens. Nearly every member of the zoo-plankton can be seen with a $\times 6$ hand-lens or a simple dissecting microscope, and the most effective searching can be done with these. Before transferring any specimen to a slide for examination under the more powerful compound microscope, it is well to watch it for a time swimming in its own characteristic way in a small glass dish under the simple magnifier. To anyone who has never seen this life before it is difficult to convey in words a picture of the delights in store for him. I am indeed lucky to have the privilege of having my account illustrated and enriched by the beautiful photographs of living plankton animals taken through the microscope by my friend Dr. Douglas Wilson of the Plymouth Laboratory; they are quite unique and many of them have been taken by that remarkable new device, the electronic flash, which has for the first time made the photomicrography of such small and rapidly moving creatures possible. The naturalist will soon forget the absence of the insects in the wealth of variously shaped and often beautifully coloured crustaceans which are to be seen swimming rapidly in all directions. Tiny pulsating medusae—miniature jelly-fish—swim into view; and here and there can be seen the transparent arrow-worms *Sagitta* which remain poised motionless for a time and then dart forward at lightning speed to capture some small crustacean. Then there may be delicate comb-jellies propelling themselves by rows of beating iridescent comb-like plates and trailing long tentacles behind them. These comb-jellies and the arrow-worms belong to two phyla—i.e. major groups of the animal kingdom—which are found nowhere else but in the marine plankton. There are many different kinds of Protozoa, among which one order (the Radiolaria) is also entirely planktonic. The segmented worms may be represented by beautiful pelagic polychaetes and the molluscs by the so-called sea-butterflies (pteropods) which are really small snail-like animals with the foot drawn out into wing-like extensions to assist in their swimming and support. Most of these animals are permanent members of the plankton, spending all the stages of their life-histories drifting in the open sea; in addition there are, however, a vast number of the young, or larvae, of the bottom-living invertebrates which ascend to live for a time in the plankton and so distribute the species far and wide. These temporary members present us with some of the most striking adaptations to this floating life. Some of them are nearly always to be found in a tow-net sample from our surrounding seas which have such a rich fauna on their floor. Group by group—flatworms, seg-mented worms, different kinds of polyzoa, starfish, sea-urchins and, of course, the bottom-living crustacea—each has its own characteristic way of solving the problems of pelagic life. The plankton indeed pre-sents a paradise for the student of invertebrate development; we shall devote a special chapter (Ch. 10) to a consideration of these larval forms.

Hitherto only a small minority of amateur naturalists has shared the delights of exploring the *living* plankton. Preserved samples, such as are often obtainable from marine laboratories for examination, are certainly full of interest; they can, however, never give the observer the same satisfaction as seeing this teeming world all alive. The professional marine biologist, engaged in investigating the relationship between plankton distribution and the fisheries, finds it very tantalizing to be able only very rarely to find time to stop and look at his captures before he must kill them; he travels to and fro across the sea taking as many samples at intervals as he can, in order to get the most comprehensive picture of conditions in the time available. Usually he only just has time to deal with the concentration, labelling and preservation of one set of collections before the ship arrives at the next position where another set must be taken; for the sake of understanding the fisheries he must always hurry on. In the past the amateur has often had an even more disappointing experience: having obtained a tow-net and hired a boat to take him out in the bay, he has returned home only to find that the wonderful sample of plankton he collected is now just a mass of dead or dying creatures crowded together at the bottom of the jar. Two modern inventions have altered all this: the Thermos flask and the refrigerator. If you have a Thermos flask, or preferably two, you can go to the sea, travel back by train for several hours and still have your plankton alive; if you have a refrigerator, or know a kind neighbour who will allow you to keep one or two 4 lb or 7 lb preserving jars in his, then you can keep your animals healthy for several days to be studied at your leisure.

I believe there are a great many people—and not only those who would call themselves naturalists—who would like to see something of this strange planktonic world, or show it to their children, if only they knew how. Anyone who goes to the sea can catch plankton quite simply. Those who can take a yachting cruise are particularly fortunate; they can study the changes in the plankton as they move from one area to another, can see the difference between the animals at the surface at night and in the daytime, and can try and find out just what organisms are making the flashing lights around their vessel in the darkness. Those, however, who can only take out a rowing boat may make very good collections, especially if there is water from the open ocean bathing their coast. If there is a pier sticking out into the sea and sufficient tidal current, as there usually is at some time of the day, quite good samples may be obtained by streaming out a net on a line and allowing it to fish for a quarter of an hour or so. Some may even think this preferable to a boat if the sea is a bit choppy! If you can only collect from a pier, or from a confined area in a rowing boat, you need not be too envious of your friends in the yacht, for fortunately the water is always on the move; a sample taken at the pier today may be very different from one taken only a few days ago and quite different again from one you may get next week. I have taken very good samples from some of the many piers built out to receive the steamers plying in the Firth of Clyde area.

To help those who do not know how to proceed I will give a few instructions. It is a good thing to have at least two Thermos flasks, so that you can keep at least two different plankton samples separate from one another. If you can manage it, it will be an advantage to start out with one of your flasks filled with sea-water that has stood in a jar in the refrigerator over night. Half of this you can pour into the other thermos just before you add the plankton sample collected. Thus in each flask the animals will be added to sea-water that has been chilled; it will keep them cool, inactive and in good condition whilst they are brought home. Details of how to make and use a tow-net have already been given in Chapter 3 (p. 38). The net of very fine gauze suitable for collecting the small plants will also at the same time catch the very small animals, particularly the protozoa and small larval forms. For the capture of most of the zooplankton a coarser net having some 60 meshes to the inch is the most useful. If more of some of the larger animals are required, for example the larger crustacea and medusae, a still wider mesh net, say 25 meshes to the inch, should be used; this will filter much more water but let nearly all the smaller animals escape. The three nets of 200, 60 and 25 meshes to the inch will provide a very good equipment. Remember, as stressed in Chapter 3, to tow *slowly*, at a speed of not more than 1½ knots. It is best to tow only for short periods—not more than five minutes at a time—which can be repeated if too small a sample has been collected. If the plankton is very abundant a longer haul will give you much too much so that all the little animals will be far too crowded together to live healthily for more than a very short time. If you have too thick a sample, pour a lot of it away and only take home in your flask a small part of it, diluted as much as possible with more sea-water. It seems hard to pour most of it back, but you will be sure to have sufficient of the commoner kinds and a few kept in good shape will be better than a great many in poor condition.

I must now give some idea of the actual numbers of animals you may expect to get. On page 51 I gave the figures for the diatoms and dinoflagellates taken in two 14-inch diameter tow-nets hauled for half a mile across the bay at Port Erin in the Isle of Man; they were averages for several hauls a week during the month of April over a period of fourteen years. For comparison I now give in the accompanying table the corresponding figures from the same source (Johnston, Scott and Chadwick, 1924) for the more important elements of the zooplankton in the same series of hauls.

Noctiluca scintillaus	29	Copepoda (Crustacea):	
Medusae	54	*Calanus finmarchicus*	131
Sagitta	15	*Paracalanus parvus*	14
Various larvae:		*Pseudocalanus elongatus*	3,261
Polychaete	5,369	*Centropages hamatus*	25
Gastropod	251	*Acartia clausi*	458
Lamellibranch	2,440	*Temora longicornis*	1,012
Cirripede	11,061	*Anomalocera patersoni*	97
Decapod Crustacea	14	*Oithona similis*	1,370
Echinoderm plutei	590		
		TOTAL:	26,191

The corresponding average totals for the months of June, August and October were 39,105, 38,812 and 35,631 respectively. Since it was calculated that approximately 8 cubic metres of water were filtered by the nets during towing, this gives an average of about 4,500 animals per cubic metre or some 120 per cubic foot of sea-water during the summer months. It must be remembered that these figures are averages and that individual samples may vary enormously from week to week. For comparison it may be interesting to give the average figures for the total plants of the plankton—the diatoms and dinoflagellates—recorded from the same series of net hauls for the four months April, June, August and October; they are in round figures 5,815,000, 6,674,000, 107,000 and 485,000 respectively. It must be remembered, however, that there will have been much larger numbers of the still smaller plants, the tiny flagellates referred to on p. 49, which will have passed through the meshes of the net and so not been recorded. To give the number per cubic metre we must again divide by 8.

If you have time, and the sea is calm enough, you should pour your plankton haul into a dish and examine it with a pocket lens as soon as it comes up; then with a wide-mouthed pipette you can pick out from it into another jar some of the rarer animals that you particularly want to study. After that you can more light-heartedly pour away most of the sample before putting the remainder, together with the rarities you have picked out, into your Thermos for transport home. The most useful dish from which to pick out specimens is one of the large oblong photographic dishes made of white porcelain and used for washing whole-plate negatives; half of the bottom of this can be covered with black paper so that you have a contrast of backgrounds to enable you to see both the darker and lighter forms more easily. All the jars, dishes and pipettes you use for living plankton must be kept thoroughly clean and *never* be used for samples that have been preserved with formalin or other chemical fixatives. These small animals are delicate in constitution as well as in form.

The majority of plankton animals tend to come up towards the surface at night and sink down into the deeper layers during the day (Chap. 11, p. 199). Very rich samples of plankton may be collected by simply towing the net just below the surface at night; in the daytime, however, if you are over deep water you may have to send your net down to 15 to 20 fathoms to get a good haul. To reach this depth you will require a good length of line—50 to 60 fathoms—and you will also require a much heavier weight, say a 20 lb lead, to take your net and all this line down. Care must be taken, of course, to know just how deep the sea is at the point where you are working so as not to run the risk of trawling the bottom with your net and either bursting it by filling it with mud or tearing it to ribbons by dragging it over a rough bottom. If you have not a chart you can consult, you should take a sounding with your lead and line before starting.

It is often very interesting to take a series of samples from different depths at the same place as near together in time as possible to enable you to study the depth distribution of the various animals; if you repeat the series again at night you may be very surprised at the different results you will get. As you let the net run out on its line to a deeper level it will fish very little on its way down, for it is moving backwards with the water as it runs out and sinks; when you haul it up, however, at the end of a tow, it will of course fish all the way up. This difficulty is got over in modern oceanographic practice by having in front of the net a special closing mechanism which is operated by a brass messenger weight sent sliding down the cable; this releases the bridles when a trigger is struck and the net falls back to be closed and held by a throttling noose which passes round it behind the mouth. The net is thus hauled up to the surface closed like a sponge-bag with the strings drawn tight and you know that all the animals in it must have been caught at the actual level at which it was fishing. A simple example of this arrangement is shown opposite in Fig. 18. These divices, however, are perhaps rather elaborate to be practised by the amateur, especially as a smooth steel cable is required for the messenger and this means the use of a winch; they will not be further described but full information about them will be found in the descriptions of the equipment used on the *Discovery* Expeditions (Kemp and Hardy, 1929). To minimise the effect of catching whilst hauling up an open net, it is well to make rather longer hauls with it down; the time taken in coming up will then only be a small fraction of that during which the net was fishing at its proper level.

If you are going to take a number of such hauls for study you will soon accumulate far more material than you can hope to keep alive successfully; in this case it will be best to keep only a small part of one or two samples fresh and preserve the rest for study dead. The living plankton will give you the greatest pleasure in studying the swimming

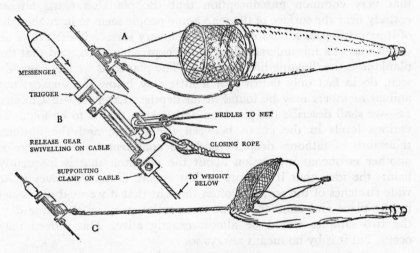

FIG. 18

Sketches of a simple release mechanism for closing the mouth of a tow-net before hauling it to the surface. A, the rig of the net when towed; B, enlarged view of release gear about to be struck by messenger weight; C, the towing bridles released and the net closed by throttling rope.

movements and behaviour of the animals; the dead samples may nevertheless give you interesting information about the depth distribution of the same animals, which you could not otherwise obtain. The best general preservative for plankton, and the easiest to use, is formalin, i.e. a 40% solution of formaldehyde in water; this, which can easily be obtained at any chemist, can be added to the sample in quite small quantities to give a mixture (about 5% formaldehyde) strong enough to keep it indefinitely in good condition. Remember always to reserve *separate* jars for preserved samples—never mix them with those used for fresh; a good plan is to stick a red label on them for danger! The dead formalined samples can of course be concentrated into a smaller space; 1lb honey jars with screw-on tops are convenient for their storage. If you are going to keep the samples for any length of time it is well to use what is called *neutral formalin*, i.e. that to which just sufficient borax—from 5 to 10 grams per litre—has been added to neutralise its acidity; ordinary formalin nearly always contains formic acid which if not so neutralised will very soon dissolve away the calcareous shells and skeletons of many of the animals.

What has just been said will have been sufficient to have corrected that very common misconception that the plankton exists almost entirely near the surface of the sea. Some people seem to have thought of it as existing as a kind of scum on the very surface itself; this is no doubt due to a misunderstanding of the expression often used that the plankton is the 'floating life' of the sea. The plants, as we have already seen, do in fact only flourish for a little way below the surface; but animal members may be found at all depths. Later on—in Chapter 12—we shall describe the plankton and nekton that is to be found at various levels in the ocean between the surface and the bottom, thousands of fathoms deep beyond the continental shelf. There is another erroneous impression about the plankton that is frequently held: the idea that it is more or less evenly distributed over quite wide stretches of the sea; it is often thought that if we used a tow-net in one place and another two or three miles away on the same day, the two samples would be almost exactly alike. This indeed may occur, but it is by no means always so.

A great many surveys have been made in the past, often in relation to some fishery problem, attempting to give some idea of the varying quantities of the major plankton organisms over a particular area. I have already described how a research ship will proceed in such a survey to traverse the area, stopping or slowing down to take tow-net samples at regular intervals. If the area to be covered is a big one, the stations—as the different points of observation are termed—cannot be very close together or the survey would take much too long; they are frequently spaced twenty miles apart. It has usually been assumed that a sample at one point, will, within a reasonable range of error, give a fair representation of the plankton in the area for ten miles around it; thus it has been felt that a series of such stations twenty miles apart will give an adequate quantitative survey. Some plankton organisms are much more patchy in their distribution than others; for some kinds such a method may give quite an adequate picture, but for others it may be hopelessly misleading. Very early in my career as a marine naturalist I had an experience which I will recall because it so well illustrates this very point; it was an episode which had a marked effect on much of my later work. In 1921, soon after leaving the University, I was appointed as Assistant Naturalist on the staff of the Fisheries Laboratory of the Ministry of Agriculture and Fisheries at Lowestoft and was delighted to be allowed to study the plankton in relation to herring. In March of the following year, through the illness of a senior, I found myself, at the last moment of sailing, as naturalist in charge of a cruise on that grand old research trawler the *George Bligh*. Our task was to make a survey of the distribution of the very young herring, which would at that time be spreading up into the southern North Sea from the winter spawning grounds in the Channel, and at the same time to chart the varying quantities of planktonic food available for them. The plan was to steam backwards and forwards on a series of parallel tracks across the area, stopping at intervals of twenty miles to take observations and samples. At each of these stations we took the temperature of the water and samples of it for salinity determination at different depths and then steamed slowly forward towing three plankton nets at different levels: one at the surface, one in midwater and the third near the bottom; finally we made hauls at these same three levels with a larger net—the Petersen young fish trawl—specially designed to capture large quantities of fry. Exactly the same procedure was followed at each station. As was always our practice, we steamed day and night so that we came upon our stations, which were separated in time by about 2½ hours, at any hour of the twenty-four. The weather was unusually kind so that we had no delays and I found that I could finish the cruise in six days instead of the seven allowed. I was at this time making a special study of the food and feeding habits of the young herring and, as they were so plentiful at one station, I decided to spend a day in one place devising methods of capturing them unhurt, so as to watch them feeding upon plankton in tanks on deck. I also thought that at the same time I would repeat the whole routine of the station at four different times during the twenty-four hours: partly as a check on the validity of the methods and partly to see if there was much difference between the vertical distribution of the young fish and plankton at different times of the day. We had arrived at this station and begun our first observations at 7 o'clock in the evening. I repeated the whole set of sampling again at 7 o'clock the next morning, at 2.30 in the afternoon and again at 7 o'clock the next evening.

During this time the ship at the surface had no doubt drifted slightly in relation to the main body of the water below and so we were sampling slightly different parts of it. The numbers of young (post-larval) herring taken respectively at the four times of sampling were: 2,448; 24; 7 and 341. The range of difference between these numbers taken at the one station was greater than that between the largest and the smallest numbers taken at any of the other stations on the cruise. The variation in the total quantity of plankton other than young fish was considerable but not as great, the highest value being just about three times that of the lowest value. It showed how very patchy can be the distribution of both young fish and plankton at quite small distances apart. It showed also that the figures representing the numbers of young herring present at all the other stations of the cruise—and upon a number of similar previous cruises—were quite valueless. It was indeed a lesson. It showed how important it is, before designing any such survey, to make a number of repetitive hauls at short intervals to see if the nets do repeatedly give a sufficiently reliable result to be used only once at any one place.

An interesting experiment may easily be made to test how even or how patchy the distribution of the plankton is in any particular area; to do this you require two identical tow-nets. You steam slowly forward, or row, keeping as constant a speed as you can; as you go, you tow first one net and then the other, each in exactly the same way and for the same period of time: one coming in over the starboard quarter as the other is going out over the port quarter, and so on *vice versa*. Thus you have a continuous line of sampling through the sea, but one broken up into a series of equal parts which may be compared with one another. If you do this you will probably get some very striking fluctuations in the quantities of some of the animals over quite short distances; often they appear to occur in swarms.

It was this experience on the *George Bligh* in 1922 which led me to devise what I have called the continuous plankton recorder: a torpedo-shaped machine which can be towed like a tow-net but at full speed behind any ordinary ship. It automatically samples the plankton mile by mile as it goes along. Since I want to show a record made by its use to illustrate the typical fluctuations in the plankton I must very briefly describe how it works. Fig. 19, opposite, shows a simplified sketch of it and a photograph is given in Plate XXIV (p. 289). It is fitted with planes which, when it is towed, make it dive below the surface and ride at a depth which may be determined by the amount of towing cable veered out. As it is towed along, the sea enters the machine by a small hole in front, passes through it in a tunnel and out at the back; the cross-section of the tunnel increases in size so that

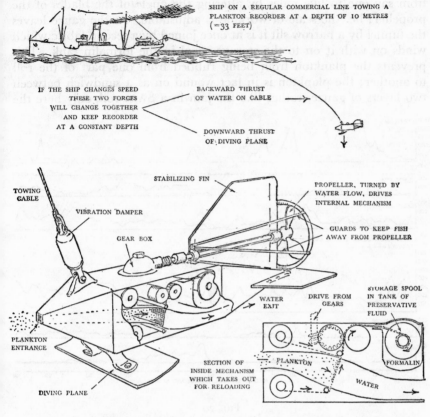

SHIP ON A REGULAR COMMERCIAL LINE TOWING A PLANKTON RECORDER AT A DEPTH OF 10 METRES (=33 FEET)

IF THE SHIP CHANGES SPEED THESE TWO FORCES WILL CHANGE TOGETHER AND KEEP RECORDER AT A CONSTANT DEPTH

BACKWARD THRUST OF WATER ON CABLE

DOWNWARD THRUST OF DIVING PLANE

STABILIZING FIN

TOWING CABLE

VIBRATION DAMPER

GEAR BOX

PROPELLER, TURNED BY WATER FLOW, DRIVES INTERNAL MECHANISM

GUARDS TO KEEP FISH AWAY FROM PROPELLER

WATER EXIT

DRIVE FROM GEARS

STORAGE SPOOL IN TANK OF PRESERVATIVE FLUID

PLANKTON ENTRANCE

DIVING PLANE

SECTION OF INSIDE MECHANISM WHICH TAKES OUT FOR RELOADING

PLANKTON

FORMALIN

WATER

FIG. 19

Diagrams explaining the working of the continuous plankton recorder; see also the photograph on Plate XXIV (p. 289).

water entering it at some 12 knots is slowed down, as it passes along, to about a tenth of its original velocity. The plankton is sieved out from the slowed-down water stream by a continuously moving banding of silk gauze which is slowly wound across the tunnel and into a storage tank of preservative fluid by a system of rollers geared to a propeller on the outside of the machine. The propeller is turned by the water flowing past it as it is towed along and so the gauze banding is wound on in direct relation to the distance travelled through the water; the collecting banding is ruled in transverse numbered divisions two inches apart and these can be made to represent various distances from say one to five miles by altering the angle of the blades of the propeller; i.e. the scale of working is adjustable. As the gauze leaves the tunnel by a narrow slit it is at once joined by a second fabric which winds on with it on to the storage spool in the formalin tank. This prevents the plankton from being rubbed from one part of the roll to another; the plankton is in fact wound on as a sandwich between two layers of gauze, rather like the jam in a Swiss roll, only here the

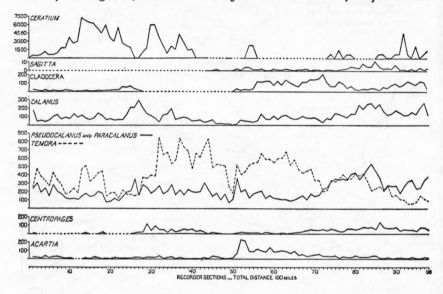

FIG. 20

A graph showing the mile to mile variation in the numbers of dinoflagellates (*Ceratium*) and several plankton animals: arrow-worms (*Sagitta*) and crustaceans (Cladocera and Copepods: *Calanus,* Temora, etc.) as shown by the analysis of a typical plankton recorder roll representing a distance of 100 miles in the North Sea. From Hardy (1939).

spongecake, i.e. the gauze, is double. At the end of a run of perhaps two or three hundred miles the spool is taken out from the machine, mounted on a special stage below a microscope, and unwound section by section so that the varying plankton may be examined, estimated and recorded mile by mile. It presents a continuous line of observation right across the sea to show us just how patchy different kinds of plankton may be. Fig. 20 above shows, graphically a typical hundred-mile record in the North Sea obtained by its use; the top curve gives the fluctuations in the numbers of the dinoflagellates *Ceratium,* the next of the arrow-worm *Sagitta* and the remainder of various kinds of small crustaceans which are important as herring food. It shows clearly how the composition of the plankton may often be very different at points only a few miles distant from one another. Actually in the sea itself the fluctuations are likely to be much more abrupt than shown in our record, for the gradual winding on of the gauze banding has a marked smoothing effect on the results.[1] The regular use of these machines on commercial steamships to provide information on the changes in the plankton over wide areas in relation to the fisheries will be described in the final chapter. One further particular about their design should be mentioned. The diving planes, which are based on the principle of the paravane as used in the first world war, give the machine a very valuable quality: provided the length of the towing cable is kept constant, the depth at which the recorder rides does not vary at all with changes in the speed of the ships from say 5 to 20 knots. Below 5 knots the weight of the recorder will make it sink deeper. The secret of its swimming at a constant depth is really very simple. Its position in the water is determined as the result of two forces: a backward one due to the pressure on the towing cable and machine body as they are pulled through the water, and a downward one due to the pressure on the inclined diving plane; if the speed of the ship varies, so do these two forces in exactly the same proportions. A full description of the machine and its use will be found in the *Discovery Reports* Vol. XI and the *Hull Bulletins of Marine Ecology* Vol. I (Hardy, 1936 and 1939).

But I am digressing too far: all I want to do here is to present a general picture of the zooplankton and the methods of collecting it, as an introduction to the more systematic account of its different members which is to follow. While a few plankton animals can tolerate considerable ranges of temperature, most species can be classed as either oceanic or neritic according to whether they are confined in their distribution to the more oceanic waters of higher salinity or to those coastal waters in which the salinity has been slightly lowered by fresh-water drainage from the land. We have already given a good example of an oceanic and neritic species of the same genus in discussing the use of *Sagitta elegans* and *S. setosa* as indicators of Atlantic and Channel waters respectively (p. 29). In addition to there being neritic members of the permanent plankton characteristic of the shallow

---

[1] This requires rather a technical explanation which is given in the full account of the apparatus (Hardy, 1936); the gauze winds on continuously and slowly so that a section representing one mile of sea passing through the machine also samples part of the mile in front and part of that behind.

coastal waters, there are usually also a number of larval stages of bottom-living invertebrates or the medusoid stages of hydroids which help to distinguish such plankton from that of the more open ocean. At the mouths of large rivers where there is an even greater dilution of the salt-water we find an estuarine plankton with its characteristic species, as in the Thames Estuary, the Humber and the Wash. I spoke of taking samples from piers jutting out into the sea; those who cannot visit the sea, but have an estuary close at hand, may obtain some quite interesting plankton animals to study, as I have often done from that ferry pier at Hull some fifteen miles up the Humber from the sea. While this can hardly be called the open sea it may form an introduction to the wider planktonic world beyond.

 ' Let us come back to the samples which you should now have safely in your Thermos flasks to take home—just be quite sure once again that you haven't put too much into them! It is well to take a quantity of extra sea water home with you so that you can dilute the samples still further before you put them into glass jars in the refrigerator. Two 7 lb sweet-jars are very good for this purpose; they can be conveniently carried in an oblong basket with a space just big enough to take one jar on either side of the handle and with a partition to keep them apart. When you reach home divide your samples out into as many large sweet or preserving jars as your wife—or your neighbour!—will allow you to keep in the refrigerator. Before finally putting them into it, examine them to see if there are any dead or obviously moribund creatures which will poison the sample if they are allowed to remain; they should be removed with a wide-mouthed pipette. The fine net sample need not be stored in such a big jar as that required for the larger zoöplankton; a small quantity kept in a 2 lb jam-jar will be quite sufficient. Now keep the plankton in cold storage until you are ready to study it and then take out only small quantities at a time.

Plate 1. PLANTS OF THE PLANKTON ($\times$ 400)

DIATOMS:

1. *Asterionella japonica*
2. *Rhizosolenia stolterfothii*
3. *Rhizosolenia alata*
4. *Grammatophora serpentina*
5. *Coscinodiscus excentricus*
6. *Biddulphia regia*
7. *Biddulphia sinensis*
8. *Lauderia borealis*
9. *Skeletonema costatum*

10. *Chaetoceros decipiens*
11. *Ditylium brightwelli*
12. *Guinardia flaccida*
13. *Eucampia zoodiacus*
14. *Thalassiothrix longissima*

DINOFLAGELLATES:

15. *Peridinium depressum*
16. *Ceratium tripos*
17. *Ceratium furca*

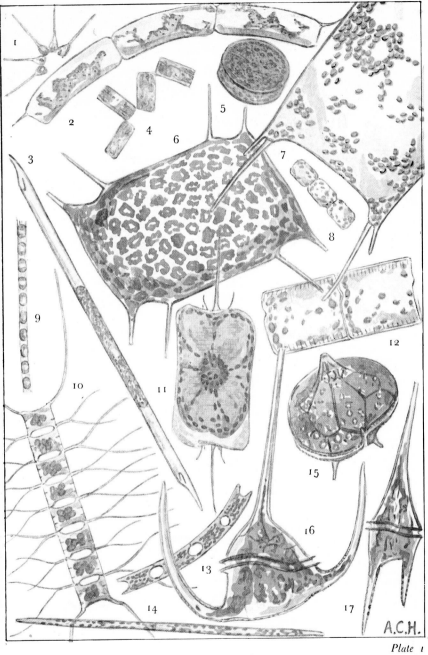

Plate 1

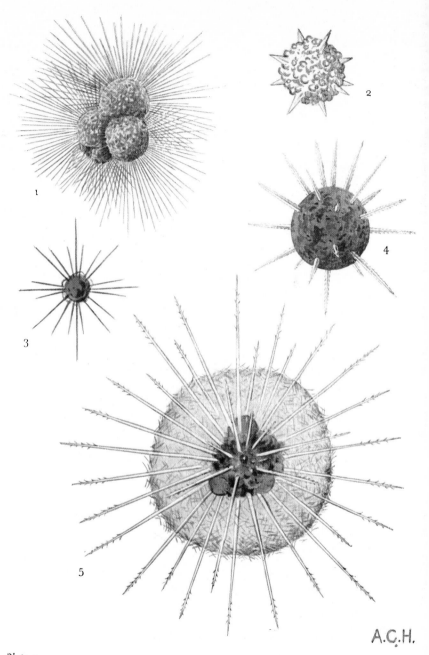

Plate 2

If the samples are not too crowded they should keep healthy for three or four days or even a week; most of the animals seem to sink into a torpor and remain motionless so that their internal activity and respiration are reduced to a minimum. On bringing them out they may appear dead, but after a short time at a normal temperature they will gradually become active again and move about as if freshly caught. A few plain glass finger-bowls—almost forgotten objects now—are very useful for holding small quantities of the plankton, which may be decanted into them from the larger jars. From these you may pick out still smaller quantities into little glass dishes—such as watch-glasses —for examination under the simple magnifier. From here you may pick out individual specimens with a finer pipette for more detailed study on a slide under the compound microscope.

You must not of course expect to find all the different kinds of animals I have indicated in just one sample—the plankton will vary from place to place and from season to season; but you will usually find more than enough in each to keep you busy for quite a long time. If you want to build up a little reference collection of the different plankton animals you have examined and identified, a good way of keeping them is in small glass specimen tubes, say 2 in. x ½ in., which can be obtained, complete with corks, from most biological supply dealers. The specimens can be kept in 5% formalin with the label, written in Indian ink or black pencil, curled round inside the tube so as not to hide them (indelible pencil should not be used as it may stain the specimen purple). The corks may then be sealed to prevent evaporation by being dipped in molten wax and the tubes stored flat in shallow drawers.

If you wish to keep living plankton animals under observation in as natural conditions as possible over a considerable period of time, you must construct what is generally known as a plunger-jar: a most ingenious form of aquarium invented by that pioneer student of the smaller jelly-fish, E. T. Browne (1898). In this a plunger, consisting of

---

Plate 2. PLANKTONIC PROTOZOA

FORAMINIFERA:
  1. *Globigerina bulloides* (× 50)

RADIOLARIA:
  2. *Acanthosphaera sp.* (× 100)
  3. *Acanthometron bifidum* (× 50)
  4. *Acanthonia muelleri* (× 50)
  5. *Aulacantha scolymantha* (× 25)

From living specimens taken on the R.R.S. *Discovery II* to the southwest of Ireland, September 1954

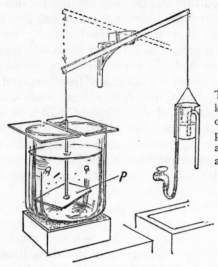

FIG. 21
The Browne plunger-jar apparatus for
keeping plankton animals healthily in
captivity. Note that the glass plunger
plate (p) is made to hang at an angle so
as not to trap the small plankton animals
at the surface as it rises. For full
explanation see text.

a glass plate, is made to rise and fall at intervals so that the water is
stirred up and, by constantly presenting a fresh surface to the air, so
prevented from becoming foul and stagnant; it is entirely automatic.
The plunger is suspended from one end of a long lever which see-saws
up and down as a tin can, hanging from the other end, is alternately
filled with water and emptied by a very simple yet cunning device;
it is filled by a narrow rubber tube from a tap and when the water has
risen nearly to the top it is suddenly emptied by a siphon tube having
a much wider bore. Thus the plunger rises and falls every few minutes.
With such equipment small medusae and other plankton animals may
be kept alive for several weeks and their methods of feeding studied.
Fig. 21, above, will show how to make the apparatus quite simply.
For feeding the medusae and other carnivorous plankton animals,
brine shrimps (*Artemia*) may be used if you cannot readily get a supply
of fresh copepods etc.; the eggs of these shrimps may be obtained
dry from some aquarium dealers and easily cultured in sea water. Like
many other members of their group, the brine shrimp normally lives in
pools which from time to time dry up; this explains why the eggs, which
have become adapted to desiccation, can so conveniently be kept dry
for many months. If put in a glass jar near a window, the sea water—
provided it has been taken from the sea only a day or two ago—should
develop a flora of microscopic algae more than sufficient to support
the shrimps.

In the next four chapters I shall aim at giving some account, group by group, of the planktonic animals of our seas; it is intended as a guide to the naturalist who is coming to this field of study for the first time. I cannot of course describe every species; to do that would require a set of large volumes like those of the *Nordisches Plankton* (Brandt and Apstein, 1901–28), an important work, edited in Germany, which attempts this very task. I shall hope to include nearly all the common forms usually met with, as well as the more interesting rarities. Where there are a number of species all very much alike I shall only refer to one or two. I will not consciously leave out any form that cannot easily be recognised as belonging to one of the different groups I shall describe; I shall hope from my account that it will be possible for the novice to decide at least to what class in the animal kingdom any creature he may find belongs. A general reader, who has not for the time being the opportunity of examining the plankton for himself, may perhaps find these next few chapters too concentrated a dose of descriptive natural history to be taken all at once. If this should be so, may I suggest that, after looking at the pictures to get a general idea of the range of the zooplankton, he should skip the sections which are printed in smaller type and return to them later when he can study the animals alive. I hope, however, that the accompanying photographs and drawings of living specimens may act as a leaven to make the text more palatable and in themselves lure him on to take a net and see for himself these different creatures actively alive and performing their various kinds of locomotion.

CHAPTER 6

LITTLE JELLY-FISH AND LESSER
FORMS OF LIFE

VERY OFTEN the waters round our coasts are teeming with tiny
medusae, no bigger than little buttons or even smaller; they are
the miniature relations, distant cousins, of the larger and more familiar
jelly-fish. They are among the very simplest forms of animal life to
have their bodies made up of definite tissues, or, in other words,
built up of many cells; lower still in the scale of life, in the realm of the
microscopic, are the Protozoa (Gk. *protos*, first; *zoön*, an animal). We
shall begin our survey of the zoöplankton with these so-called 'lowly'
forms and work upwards to the more complex.

The Protozoa are so different from the many-celled animals that
they should at least be regarded as being in a distinct sub-kingdom.
It must be admitted, however, that they form a somewhat arbitrary
group, for it is in fact impossible to draw any hard and fast line to
separate them from the Protophyta or simplest plants. We have already
seen in Chapter 3 how some of these small organisms, the flagellates,
may be at one and the same time both plant and animal in nature,
and again how others of the same group feed entirely as plants while
yet others, closely related, feed just as distinctly in an animal fashion.
Here we are at a point where the plant and animal kingdoms have not
yet become completely differentiated. Some biologists have preferred
to class all these primitive plants and animals together as the Protista
and to regard them as a kingdom of their own—linked on the one hand
with the metaphyta, or many-celled plants, and on the other to the
metazoa; this however is not usually followed and there is as much
to be said against it as there is for it. It is convenient to separate, as we
are doing here, the consumers from the producers—the animals from
the plants; we must just keep in mind that some of these simple plants
and animals are cousins—and not very far removed at that.

Here is not the place to enter at any length into that other academic
discussion as to whether we should regard these relatively simple
animals as uni-cellular or non-cellular; I believe, following my friend

Dr. J. R. Baker (1948) who has given the matter much thought, that both views are correct when applied to different kinds of Protozoa. A great many have a single nucleus which to all intents and purposes appears to have the same kind of relation to the surrounding protoplasm as has the nucleus of one single cell in a many-celled animal; these are surely single-celled animals. On the other hand there are many other Protozoa which have a different kind of organization—they may have a number of nuclei in one mass of protoplasm, or large nuclei which appear to be compound structures, the equivalent of many ordinary ones, or yet again they may have one such compound nucleus (the meganucleus) and one simple one (the micronucleus); these various modifications are clearly not comparable to just one typical cell of a many-celled animal and may conveniently be termed non-cellular. Whether they are to be regarded as the equivalent of a small many-celled animal in which the actual cell boundaries are absent is a matter we must not pursue here.

The sub-kingdom of the PROTOZOA is divided into four great classes:

1. MASTIGOPHORA (Gk. *mastix*, a whip; *phoreus*, a bearer) also called FLAGELLATA (L. *flagellum*, a whip)
2. SARCODINA (Gk. *sarkodes*, fleshy) also called RHIZOPODA (Gk. *rhiza*, a root; *pous, podos, a foot*)
3. SPOROZOA (Gk. *spora*, a seed or spore; *zoon*, an animal)
4. CILIOPHORA (L. *cilium*, a hair or hair-like process)

In works of popular science the Amoeba is often spoken of as if it were the very simplest form of life: symbolical of the very beginning of the whole process of organic evolution. With a body of naked and semifluid protoplasm, of no definite shape, but flowing this way and that, Amoeba *is* one of the very simplest of animals; it feeds by just flowing round its food, engulfing it with that part of its body which is nearest to it. It cannot, however, represent the most primitive form of life, for the very fact that it is an animal renders this impossible. Animals of course can only live by feeding upon other living things, or products of their life and we have already reminded ourselves that plants alone can build themselves up from the simple inorganic chemicals. Before animals could feed there must already have been plants for them to feed on. Animals too must have oxygen and it is considered very doubtful if there was free oxygen in our world until it was produced by the green plants; in support of this we know that the older Pre-Cambrian sedimentary rocks are in a relatively unoxidised condition as compared with the later ones. How life began we do not know. Dimly we can imagine some process whereby organic compounds such as sugars and proteins were built up under the energy of sunlight and then we must suppose that in time some of the protein molecules became so modified that they could add to themselves at the expense of other organic substances, as viruses appear to be doing at the present day. Gradually, we must imagine, more and more complex virus-like forms became simple plants which could build themselves up by trapping the energy from the sun; in doing this they would liberate oxygen from the carbon dioxide and so prepare the way for a world in which animal scould be evolved. It is indeed likely that it was in the plankton of those seas which are now so far off in time that animals came into existence. We know that some of the higher land plants have taken to capturing insects by lures and sticky secretions in order to feed upon their supply of phosphates and nitrates. In the plankton we have seen that a vast host of little single-celled plants have been evolved which are provided with an undulating whip-like process which enables them to keep up in the sunlit surface layers. Some of these actually live, as do the insectivorous plants we have just referred to, by capturing some still smaller organisms, digesting them and assimilating them into themselves; this is the beginning of feeding on particulate food—in other words, of the animal way of life. In this way they may stave off the inevitable decline due to that nitrate and phosphate shortage in the water which we discussed in Chapter 4 (p. 60). Some of these have gone further, losing their chlorophyll and power of photosynthesis, to live entirely as animals. Yes, it is among the flagellated organisms that we must look for the most primitive animals and not to *Amoeba* which, in spite of its apparent simplicity, must have travelled much further along the line of animal evolution than have the flagellates. There are actually some flagellates which become amoeba-like in form at one stage in their life-cycles.

By the beginning of the Cambrian period, which produced the first sedimentary rocks giving us really reliable fossils, we find a considerable variety of invertebrate animals already in existence; they include highly organised crustaceans (the trilobites) which are now extinct, and equally elaborate aquatic snail-like animals. Those early Cambrian rocks were, according to the latest estimates, laid down some 500 million years ago and by then most of the invertebrate groups were already well developed; their evolution from primitive protozoa must surely have taken at least as long again. It seems likely that the first animals must have come into existence in a marine plankton of something like a thousand million years ago.

Most of the marine flagellates are either true plants or forms which live both as plants and animals; these we have already dealt with in Chapter 3. One dinoflagellate, however, has been reserved for treatment here, for it is entirely animal in nature: the remarkable *Noctiluca* shown on Plate III (p. 48). There appears to be only one species in northern waters, which has unfortunately been called by two names: *Noctiluca miliaris* or *N. scintillans*. It is large for a protozoon, being the size of a pin's head or about 1/16 of an inch across; this bulk, however, is largely due to its being filled with a gelatinous substance through which strands of protoplasm radiate from a central mass which contains the nucleus. This gelatinous substance would seem to be slightly lighter than water so that the animal floats and drifts like a balloon and has lost all power of active movement. The flagella corresponding to those of the typical dinoflagellate are greatly reduced and lie in a deep groove at one side of the otherwise almost spherical body; at the bottom of this groove is the mouth and at one end of it is a curious thick prehensile 'tentacle' which appears to be an organ for seizing prey and pushing it into the mouth. Usually the body may contain a number of diatoms it has eaten but Dr. Marie Lebour (1925) has recorded also finding larval crustacea (copepods). I have never seen one actually catching these active prey and it is rather a puzzle to know how they succeed in doing so. As its name implies, it is brilliantly phosphorescent and the displays it gives when abundant are perhaps the most striking in our seas; when agitated, the very water itself appears to be aflame.

To the next class, the Sarcodina, belong the protozoa which capture their food as Amoeba does by sending temporary protoplasmic processes streaming out in different directions—the so-called pseudopodia ('false feet'). Indeed Amoeba moves by streaming more persistently along one path than another and by withdrawing those pseudopodia which have been pushed out in other directions. The class is represented in the plankton by members of two orders: the Foraminifera (*foramen*, a hole; *fero*, to bear) and the Radiolaria (*radiolus*, a ray or spoke of a wheel); both however are more characteristic of the warmer oceans than the seas around our coasts. Most of the first group secrete a calcareous shell, but some of the bottom-living forms may instead make a little house for themselves from fragments of sponge skeleton or cast off spines of sea-urchins; those making shells usually start with a small one and then as they grow bigger add to it larger and larger chambers which are often arranged in a spiral. They receive their name because their shells are usually perforated by a large number of small holes through which stream out slender strands of protoplasm (pseudopodia). Outside the shell these protoplasmic strands branch and run together again and again to form a living network; the animal in fact throws out a net, like a spider-web, to ensnare its prey. But what a devilish device it is!—for it is spider and web in one continuous living whole. Their captures are not drawn into the shell to be discreetly disintegrated in the dark interior; they are digested *outside* by the net itself. On being caught in this sticky web, the victim struggles but only adds to its entanglement as more protoplasmic streams flow round it; digestive juices now pour out from the very strands of the net and bit by bit it is converted into a solution or emulsion which on passing into the protoplasm is sent flowing towards the interior. Such are *Globigerina* and its allies, which occur as many species and in vast numbers in the plankton of the warmer oceans; when they die, their shells drop to the bottom to form the great deposits of Globigerina ooze. It was shells such as these which built our white cliffs of Dover.

The Foraminifera are examples of what we should call non-cellular protozoa, for they are multi-nuclear, at least at certain stages of their rather complex life-histories. *Globigerina* is only rarely encountered in the waters round our islands, but those making cruises to the west of the Channel or to the north and west of Scotland may expect to find *G. bulloides* when tow-netting in the North Atlantic current; it is shown in Plate 2 (p. 81). Skeletons from the ooze of the sea-bed are usually just a group of perforated calcareous spheres; but when alive near the surface, the spheres of *G. bulloides* are covered with long and exceedingly slender needle-like projections which are dissolved away as they fall to the depths. I have never been able to see them floating alive and undisturbed with their protoplasmic web spread out, but caught freshly in a tow-net the protoplasm within the spined shell is a beautiful rose-pink colour.

The other order just mentioned, the Radiolaria, is entirely planktonic. Its members differ from the Foraminifera in many respects. Their protoplasmic strands do not run together to form a net; they project in all directions as exceedingly thin, long and straight filaments which are said to possess a more rigid axis giving them support. These rays are sticky; when any little plant or animal is caught against them, two or more come together to hold it, and then, between them, they bear it in a stream of protoplasm towards the central mass to be digested. Apart from any other skeleton which they may possess, they all have a perforated central capsule of a horny substance, which surrounds either a single very large nucleus or a number of smaller ones. Immediately outside this capsule the protoplasm forms a dark layer where digestion takes place and outside this again it appears as a curious frothy mass from which the delicate radiating strands arise. In this outer frothy layer are to be found a large number of small yellow objects which have been shown to be small plant-cells living in that close partnership with their host which is known technically as *symbiosis* (living together). They obtain shelter and a supply of nitrogenous waste products from the animal; if they do not give some food in return—and this is not certain—they give oxygen and assist their host to get rid of its poisonous waste material. Symbiotic plant-cells are also found in the planktonic Foraminifera just referred to and in the tissues of a great many metazoan pelagic animals of the tropical seas. We have seen in Chapter 4 how difficult it is for the free-living planktonic plants to eke out an existence in the upper layers of the warmer seas which are almost cleared of nutrient salts; it would appear that there has been evolved a phytoplankton which has sacrificed its freedom for the chemical food found available in the prison cells of its hosts. Perhaps the animals do really feed upon them in moderation—as is indeed the case in many examples of symbiosis—so that this association is really a drawing closer together of the animal and plant inter-relationship in regions of difficult living for both.

Many Radiolaria produce delicate skeletons of silica of great beauty. Some have a series of perforated spheres one inside the other just like those fascinating balls of ivory that the Chinese are so fond of carving, some are helmet-shaped and others form complex lattice systems; but most of these are the warmer water forms whose skeletons, when they die, sink to form the Radiolarian ooze that covers the bottom in the deeper parts of the tropical oceans. *Hexacontium* is a form with perforated spheres which may sometimes be taken in the Atlantic along the western edge of the British Isles; when taken alive its beautiful skeleton is hidden, all but the projecting spines, beneath the frothy protoplasm which is yellow with minute symbiotic algae. Its skeleton is shown in Fig. 22 (p. 90) and the appearance alive of a closely allied form *Acanthosphaera* in Plate 2, together with the next three species to be mentioned. A more characteristic species of the waters to the west is the very large *Aulacantha scolymantha* with a skeleton of radial hollow rods and fine tangential needles. Also to the west, and sometimes coming into the northern North Sea in the Atlantic influx, are forms like *Acanthonia muelleri* and the little bright red *Acanthometron bifidum*; these belong to the suborder *Acantharia* (Gk. *acantha*, a thorn) which are characterised by having many nuclei and 10 to 20 (in a few species, more) long spicules of strontium sulphate radiating in a definite geometrical pattern from a central point. I once found swarms of an allied species, *Phyllostaurus quadrifolius* (Fig. 22), just to the north of the Dogger Bank on the first plankton collecting voyage I ever made, in the summer of 1921; I have not heard of its being recorded there in such numbers since. This kind of radiolarian has one very remarkable character: attached to each radiating skeletal spine are contractile strands, like miniature muscle-fibres, stretching to join the outer layers of protoplasm; when they

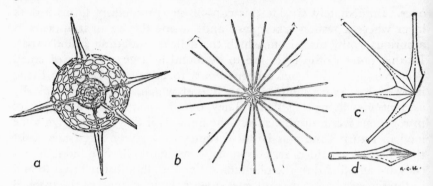

FIG. 22

The skeletons of Radiolaria from British seas: *a, Hexacontium,* with part of the outer shell omitted to show the inner spheres; *b, c* and *d, Phyllostaurus quadri-folius,* whole skeleton and enlarged details. Drawn from specimens in the British Museum (Natural History). Compare with Plate 2, (p. 81).

shorten the frothy protoplasm is drawn out to occupy a larger volume and when they relax it is allowed to contract, thus altering, so it is said, "the hydrostatic balance of the organism and enabling it to rise or sink in the water."[1] How this is actually achieved does not at once seem clear, for the increase or reduction of surface in relation to mass will only retard or accelerate its *sinking*; but it is probably linked with another hydrostatic device which is said to exist in the frothy external layer of all radiolarians. This is in the form of large vesicles filled with water saturated with carbon dioxide causing the animal to float and enabling it to regulate its position in regard to depth; in rough weather or other unfavourable conditiosn the vesicles burst or are expelled and the animal sinks to a lower level until in due course new vesicles are formed again and it then rises balloon-like once more. The little contractile fibres of the Acantharians may possibly assist in the emptying and filling of these little buoyancy chambers.

The members of the next great class of protozoa—the Sporozoa—are all internal parasites of other animals and whilst a number are found in the bodies of planktonic animals they will not be dealt with here; one of their effects upon copepod hosts will be referred to later (p. 161).

The last division of the Protozoa, the Ciliophora, are those forms, already referred to, which have both a meganucleus and a micronucleus; they have on their surface, as their name implies, numerous hair-like processes which beat rhythmically and act as locomotor organs. These cilia are usually arranged in rows and, as they beat one after the other, we see an effect very like that of wind blowing across a field of corn; just as the current of air makes the passive corn-stalks bend and rise alternately in waves, so, when the process is reversed, the active beating of the cilia creates a current of water which propels the animal forwards. In the sub-class Ciliata, cilia are present at all stages in the life-cycle; but in the smaller sub-class Suctoria, the adults are sedentary animals with sucking tubes as feeding organs, and only their young stages are ciliated. Feeding on very small organisms, most marine ciliates are to be found near

[1] Minchin, 1922, p. 253.

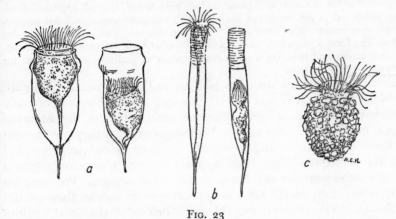

FIG. 23

Ciliate protozoa (Tintinnidae) from the plankton: *a, Cyttarocylis arcuata;*
b, Tintinnus subulatus; c, Tintinopsis sp., drawn from living specimens taken in
a fine surface tow-net near Millport, Firth of Clyde in late August, 1950. × 180.

the bottom or in pools along the coast where there is plenty of decaying organic
matter surrounded by clouds of bacteria. In the plankton of the coastal regions,
particularly towards autumn and winter when the community is dying down, there
may be some of the larger species such as *Spirostomum, Lionotus* or the long-necked
Lacrymaria spiralling their way this way and that; but these are not really planktonic.
One group is pelagic *par excellence*—the Tintinnidae (L. *tintinnus*, a bell). They are
very small ciliates not unlike in superficial appearance to the little stalked *Vorticella*
of our ponds; however they lack the spiral groove leading to the mouth and belong
to a different sub-order. They secrete round themselves a delicate and transparent
glass-like test, a casing or vessel, often shaped like an inverted bell or a wine-glass.
The body, which is rounded, is attached by the little stalk to the bottom of this test;
at the opposite or upper end there is a circle of cilia creating a vortex current to
swirl still smaller organisms—mostly the minute flagellates—down towards the
mouth which lies in a deep groove within them. The wine-glass-shaped cases
secreted by different species range through port and sherry styles to wider burgundy
or slender claret glasses, or to thistle shapes or again to very tall tubular vessels.
They are objects of great beauty. But what keeps them up in the water? Man,
in conquering the air, unwittingly copies the inventions of nature devised to overcome the difficulties of support in the pelagic world; after discussing forms like parachutes and balloons, we have here the counterpart of the helicopter. The beating
cilia create waves of motion that sweep round and round the upper rim; it is the
nearest nature can get to the rotating wheel or air-screw. It acts in fact just like the
blades of the rotor aircraft in creating a downward current which thrusts the body
upwards; or it may go into reverse and make the animal swim downwards. The long
slender forms illustrated in Fig. 23 above, whilst sometimes going "mouth" first,
usually propel themselves swiftly forwards, pointed-end first, like torpedoes with a
propeller at the stern; the drawing was made from living specimens all taken in
the same sample off Cumbrae in the Firth of Clyde in August. When taken from the

refrigerator where the sample of plankton had been stored, they all appeared dead, being contracted at the bottom of their 'glasses'; but gradually after a few minutes at room temperature they slowly awoke, stretched themselves and sprang into active motion. The form *Tintinnopsis*, which plasters its gelatinous test with small particles, fine sand grains and the like, is more characteristic of shallow coastal waters; the others are mainly oceanic.

From the Protozoa we must pass on to the many-celled animals: the Metazoa (Gk. *meta*, next to). The sponges, while composed of many cells, have such an entirely different make-up from the Metazoa proper, being much more loosely held together, that they are thought to have been quite independently evolved from the Protozoa; they are consequently regarded by most zoologists as belonging to a separate sub-kingdom: the Parazoa (Gk. *para*, beside). We have already noted that, except for a very short time as larvae, they are not planktonic. The simplest true Metazoa belong to the great phylum Coelenterata (Gk. *koilos*, hollow; *enteron*, bowel, or gut). In its most rudimentary form, one of these animals may be likened to an uncorked flexible bottle with a circlet of tentacles around its mouth for the capture of food; not only is this opening used for the passage of the food into the 'bottle' for digestion but also for the final expulsion of the undigested remains which are ejected by a contraction of the bottle-wall itself. Essentially each animal is made up of two layers of cells: an outer protective 'cover' and an inner lining to the 'bottle' for the secretion of digestive fluids and the absorption of food; in addition the cells of each layer are provided with muscle fibres, which bring about the contractions and movements of the animal, and then in between the two layers of cells—and secreted by them—is sandwiched a gelatinous substance, a packing material (called the mesogloea) which can be very thin or, as in the big jelly-fish, very thick.

Clearly in this book we cannot go into the detailed structure of all the animals we are considering, but some account must be given of the remarkable stinging cells which are possessed by all Coelenterates and are the most characteristic structures of the group; they are found in the greatest numbers in the tentacles and are used both for the capture of food and for defence against enemies. Each such cell contains a most remarkable 'explosive' weapon and projecting from the outside is a little trigger 'hair' which, on being touched by some unsuspecting prey, at once sets it off. It is almost incredible that the apparatus I am about to describe can be packed within a single cell and, more so, that it has been somehow manufactured by the protoplasm of that cell. There are many variations of its form, but I will describe one of the simplest and one which I have also illustrated in Fig. 24A, B and C.

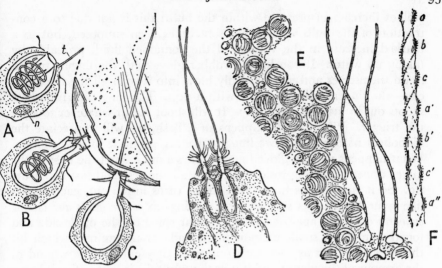

FIG. 24

Nematocysts, highly magnified. A, B and C are diagrams showing how a
nematocyst is shot out to penetrate some small animal prey which has touched
the trigger-hair t: in A the nematocyst is coiled up before release; in B it is
emerging and the spines, on coming out, are cutting the cuticle of the prey;
and in C the hollow thread has shot into the wound caused by the spines.
This is a nematocyst such as is found in the common freshwater *Hydra*. D,
two very large nematocysts (discharged) at the tip of one of the protective
bracts of the Siphonophore *Agalma elegans* (p. 117) × 170. E, a small part of the
surface of one of the stinging tentacles of the 'Portuguese Man-of-war,'
Physalia, showing many spherical nematocysts with their threads coiled up and
three of them discharged. (×210). F, a still more enlarged view of one of
the discharged hollow threads showing the three spiral and toothed ridges;
a-a'-a", b-b' and c-c' which give it a screw-like form. D, E and F are drawn
from specimens lent me by Mr. George Mackie.

Imagine a very long and slender rubber tube sealed at one end like
the finger of a glove and at the other opening into a large bulb, as the
finger of a glove opens into the palm. The long tube widens a bit just
before it enters the bulb and here it bears some sharp blade-like spines
on its outside. Now if you imagine the whole thing filled with a fluid
and further imagine that the long tube is turned completely 'outside
in' (as the finger of a glove may be turned in on itself) and coiled up
inside the bulb, then you will have an idea what the device is like when
it is at the ready. What is so remarkable is that this apparatus appears
to be made by the cell like this *to begin with* and not first made and then
turned 'outside in'! When the little trigger hair is touched there is a

violent increase in pressure within the bulb; this is not due to a contraction of the bulb wall as has sometimes been supposed, but to a
marked increase in the volume of the contained fluid, probably by
taking up water. If you have a rubber glove with the fingers turned
in on themselves and you suddenly blow into it with all the force you
can—the fingers will shoot out with a pop. The long thin tube now
shoots out in just the same way. It will shoot out at whatever touches
the trigger. Whilst it all happens in a flash, we will appreciate the
diabolical nature of it if we imagine we are seeing it in slow motion.
What happens when a glove finger is blown out? First the base of the
finger appears turning inside out, then all the rest and last of all the
tip. So it is with this tube. As its base turns inside out, each blade-
like spine fixed to its side will, on coming up *on the inside and turning
outwards* come up against another similar spine on the other side and
together make a small incision in the side of the animal which hit
the trigger. Then as they turn outwards they enlarge the wound in
the armour of the victim and into it the thin tube is shot with explosive
force as the violent pressure inside it keeps it straight and rigid. It
goes in like a rapier. When it is fully extended there is still a powerful
pressure on the fluid in the bulb; thus, as it snaps to its limits, the end
of the tube is burst and the fluid contents of the bulb—a paralysing
drug—*is injected into the victim.* It is in truth a hypodermic needle and
syringe! It is called a nematocyst (Gk. *nema, nematos,* a thread; *kystis,* a
bladder). It is not a living thing; it is a dead structure, an elaborate
tool made ready to work—and made to perfection—by the semi-fluid
living substance of the cell. Here is something to wonder at, for it
looks as if it were designed.

Dozens of these nematocysts may be packed close together in batter-
ies. They are not all of this penetrant type; some, as they come out,
curl round to entangle the prey, and others are sticky to hold it fast.
They are often arranged in groups with the sticky and entangling ones
to the outside; these have longer triggers and first secure the victim
and then the penetrants in the centre with shorter triggers deliver the
coup de grace. If the creature stung is small, it is quickly overcome and
carried by the tentacles towards the mouth; if it is much larger—
perhaps a would-be predator—it will feel the painful thrusts and so
tend to give the stinger a wide berth, as we would do if stung by a
jelly-fish when bathing. Those who wish to see something of the great
variety of form exhibited by nematocysts of different coelenterate
species should look at the beautiful studies by F. S. Russell (1938-40);
many of the threads are armed with little spines arranged in spiral

patterns along the greater part of their length (see Fig. 24, E and F)
Until recently it was difficult to understand how this long and very
slender hollow thread could be made to turn inside out when it would
seem that there must be so much friction between the inner and
outer parts as the one moves along inside the other. Some workers
have tried to make out that the appearance of the tube being turned
inside out is an illusion and that the thread is formed from a spinaret
like that of a spider; this is conclusively confounded by the beautiful
studies of L. E. R. Picken (1953) and Miss E. A. Robson (1953) of
Cambridge. They not only show in photographs the tube being
everted, but remove the problem of friction by discovering that the tube
swells as it comes out and so gives more space for the movement of
the inner part as it travels to the outside.

The simplest Coelenterate, like that just caricatured as a bottle,
would be a single polyp such as the fresh-water *Hydra* or one of a few
solitary marine ones; these belong to the class Hydrozoa (Gk. *hydor*,
water). As is well known, *Hydra* can reproduce itself simply by budding
from its side a new individual which then breaks off and walks away.
The majority of marine hydroids can reproduce in the same way,
but the new polyps so formed remain attached to their parent so that
little branching colonies are built up of very many polyps all connected
with each other. They occur in great numbers, and there are many
different species, growing on seaweeds, rocks and piers along our
coasts and on the bottom of our shallow seas: some look like miniature
ferns, others more like encrusting mosses spreading over the surface
of some stone. In each colony there are some individuals which can
reproduce sexually, but these, although budded from the colony, do
not usually remain attached to it; they are not ordinary polyps, they
are the small medusae which are typically set free to float and drift
in the plankton and there produce their eggs and sperm, so spreading
the species far and wide. There is in fact an alternation of generations:
non-sexual budding polyps and sexual medusae. The medusae are the
tiny jelly-fish which we referred to at the beginning of the chapter;
they are among the most characteristic and beautiful objects of our
plankton.

While they occur in great variety, two principal types can b
recognised: the Anthomedusae (Gk. *anthos,* a flower) and the Leptomedusae (Gk. *leptos,* thin); these are the medusae corresponding
respectively to two different orders into which the hydroids are divided:
the Gymnoblastea (Gk. *gymnos,* naked; *blastos,* germ, bud or branch)
and Calyptoblastea (Gk. *kalyptos,* covered or hidden) according to

whether their polyps are freely exposed or enclosed within a cup-like protective outer casing.

I will first describe a typical Leptomedusan—to begin with, its structure is more easily explained—and at the same time try to show how a medusa, although looking so different, is in origin a modified polyp. Let us return to the simile of the bottle. Instead of a normal upright bottle like the polyp, imagine a very flat one like a rubber hotwater-bottle that is circular instead of oblong and has its neck and mouth, not at its edge, but opening through the very centre of one of its flat circular surfaces. Its mouth is still surrounded by a ring of tentacles but they are now well away from it round the edge of the circle; in other words it is our old bottle but extraordinarily flattened in a vertical direction and drawn out sideways. Remembering that it has an outer covering and an inner layer, we can imagine the two wide circular surfaces (really its drawn-out top and bottom) pressed together so that their inner linings actually fuse with one another to form a single layer except at certain places; these are a main cavity in the middle into which opens the mouth, a circular tract right round the margin against the tentacles and four radial channels connecting the central cavity with this circular canal. These channels carry the food-supply from the central stomach to the periphery of the animal. As the little medusa grows on the parent colony, before it is set free, these inner canals and layers are indeed formed just as I have described. Now if we turn the whole thing over so that the mouth and the neck hang down from the centre of the underside like the handle of an umbrella and the tentacles hang round the edge like a fringe—then we have a little medusa in its simplest form. The medusa of the hydroid *Obelia* is shown in Fig. 25 (p. 98) where it is compared with the polyp; it is also shown in a photograph on Plate V (p. 102). It is indeed like an umbrella because, instead of being actually flat, it is curved in just the same way; in technical language the outer and under surfaces are respectively always called the ex- and sub-umbrella surfaces and the mouth with its neck is known as the manubrium (handle). Not only can it float like a parachute but it can swim upwards or sideways by a rhythmic pulsation of the umbrella, opening gently and contracting more vigorously. Its prey—an active creature such as a small copepod (crustacean) or a young fish—having been caught and paralysed by the tentacles with the aid of their stinging cells, is conveyed by them towards the manubrium, which usually curves over to receive it like an ele-

---

Plate 3.   LITTLE JELLY-FISH: MEDUSAE OF THE HYDROZOA

ANTHOMEDUSAE:
   1. *Turritopsis nutricula* (× 10) Southampton, August 1954
      a. actively swimming, tentacles contracted, velum producing jet
      b. at rest with tentacles spread out in search of food

LEPTOMEDUSAE:
   2. *Obelia sp.* (× 10), Bangor, August 1954
      a. side-view, umbrella inside-out
      b. seen from below
   3. †*Phialidium hemisphaericum* (× 10), 32 tentacle form, Bangor, August 1954
   4. *Eutonina indicans* (× 5), Cullercoats, September 1954
   5. †*Phiadidium hemisphaericum* (× 10) 16 tentacle form, Southampton, August 1954
   6. *Lovenella clausa* (× 10), Bangor, August 1954
      †These two forms may be distinct species, see Russell (1954)

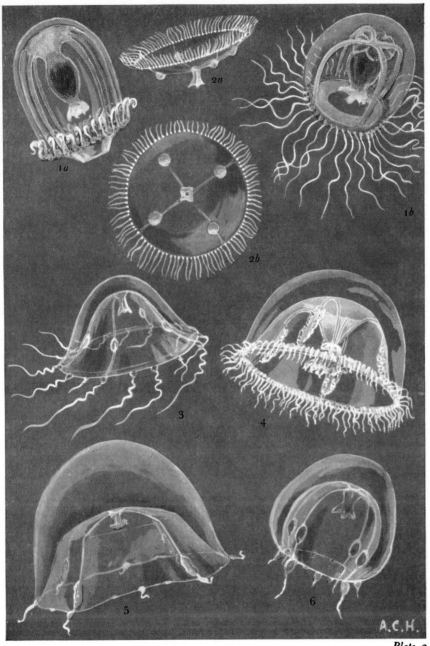

Plate 3

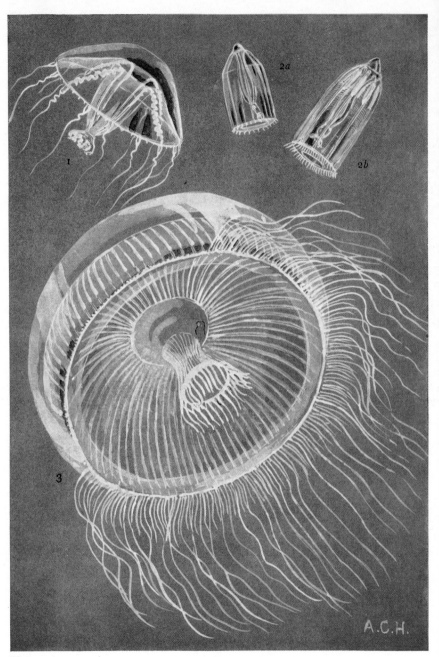

Plate 4

hant's trunk reaching for a bun. At the base of each tentacle is a patch of pigmented cells sensitive to light and at eight points round the edge of the umbrella
re special organs of balance which tell the animal which way up it is and allow it
o correct its position if it is tilting in the wrong direction. Each of these stability
evices (statocysts) consist of a hollow sphere lined by cells provided with little hair-
ke sensory processes; inside this is a small calcareous nodule which rolls about like
ball in a bowl and so stimulates the sensory 'hairs' it comes in contact with. If the
nimal tips over to one side each little ball will roll against the corresponding side of
ts sphere; at once nerve-impulses are sent from the sensory processes to the sub-
umbrellar muscle-fibres of that side of the medusa; accordingly they are now stimu-
ated to more violent contraction than the rest, with the result that the medusa
ilts back to its original position. It is likely that these little devices also serve for
he detection of vibrations in the water and, since they are placed at intervals from
ne another round the edge of the umbrella, may enable the medusa to locate the
osition of small prey swimming near it. I have seen a small medusa, an ephyra
not a Leptomedusan) when swimming upwards in an aquarium, come suddenly
upon a group of tiny *Balanus* nauplii (the larvae of the acorn barnacle); by accurate
left twists of the manubrium it picked up one after another of the nauplii, actually
apturing them with its mouth. The ephyra, which is the young stage of a large
elly-fish like *Aurelia*, has as yet no tentacles and must surely capture its active food
y using somewhat similar sense-organs to guide its mouth in the right direction.

While the medusa I have described, before this little digression on feeding, was
Obelia, the account would do just as well for most Leptomedusae, for they vary only
n such details as the number of tentacles and radial canals. They may be recognised
at once because the reproductive organs, the female ovaries or the male testes, are
always situated on the radial canals and never on the manubrium as they are in
he other group, the Anthomedusae. They are either male or female, never hermaphrodite. When newly budded from a hydroid colony they are very small and often
appear with the umbrella inside out; then they gradually grow bigger as the ovaries
or testes develop, but, with two important exceptions about to be mentioned, rarely
exceed half an inch or an inch in diameter according to their species. Each egg
when fertilised drops into the water, segments and forms a little oval ciliated larva
called a planula, which eventually settles on the bottom and becomes the first hydroid
person of a new colony. Several different kinds of Leptomedusae are shown on Plate
3 (p. 96). The Anthomedusan—also shown diagrammatically in Fig. 25—is
built on the same general plan as the Leptomedusan just described, but may be
distinguished from it by the following characters in addition to the position of the
reproductive organs already mentioned: it is always more rounded, more like a bell
than an umbrella, and round the inner margin of this bell is a wide flange, the velum,

---

Plate 4. MORE HYDROZOAN JELLY-FISH

1. *Tima bairdi* (nat. size) Cullercoats, 1952
2. *Aglantha digitale* (×2)
   a. young specimen, Cullercoats, November 1954
   b. adult, drawn from preserved specimen from Aberdeen, showing characteris-
      tic irridescence
3. *Aequorea vitrina* (×½), Lowestoft, August 1954. Marginal tentacles omitted from
      one side so as not to obscure view of sub-umbrella surface

projecting inwards and so considerably reducing the size of the opening to the bell. A few Leptomedusae are bell-like rather than umbrella-like, but they never have a wide velum. The Leptomedusae usually have many tentacles, but the Anthomedusae have rarely more than eight, often four or even two, and in a few species only one; these tentacles however are usually longer in relation to the size of the bell. There are no balance organs, but the light receptors are usually more developed and are often furnished with a lens to form a primitive eye. These Anthomedusae, while usually smaller than the Leptomedusae, are relatively more powerful swimmers. The water inside the bell is more enclosed and is still further confined by the velum; when the bell contracts the water is shot out as a jet through the narrow opening of the velum— they are in fact jet-propelled. Four different species are photographed in Plate IV

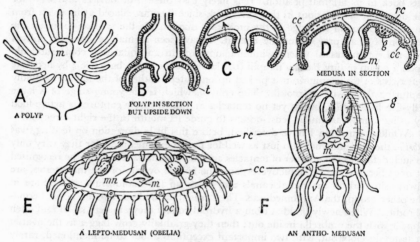

FIG. 25

Diagrams comparing the polyp (A) and medusae (E and F) of hydrozoan animals. B and D represent sections of a polyp and medusa respectively, while C shows how the latter was evolved from the former; the outer layer (ectoderm) is hatched, the inner layer (endoderm) is white, and the heavy black line between represents the mesogloea. Other lettering as follows: cc, circular canal; g, gonad (ovary or testis); m, mouth; mn, manubrium; oc, otocyst (or statocyst), re, radial canal; t, tentacle; v, velum.

and another drawn in colour is included in Plate 3 (p. 96), where it is shown in two positions: actively swimming with its tentacles contracted and floating with its tentacles stretching out for food. Those who wish to make a special study of these beautiful little jelly-fish should consult the magnificent monograph which Mr. F. S. Russell has recently produced (1954); he describes and illustrates all the known British species.

Two Leptomedusae deserve special mention as reaching an unusual size: *Tima bairdi* and *Aequorea* (several species); they are shown in Plate 4 (p. 97). *Tima* is a North Sea species occurring off the north-east coast in autumn and early winter; it grows to 2½ inches in diameter and is characterised by the very long reproductive

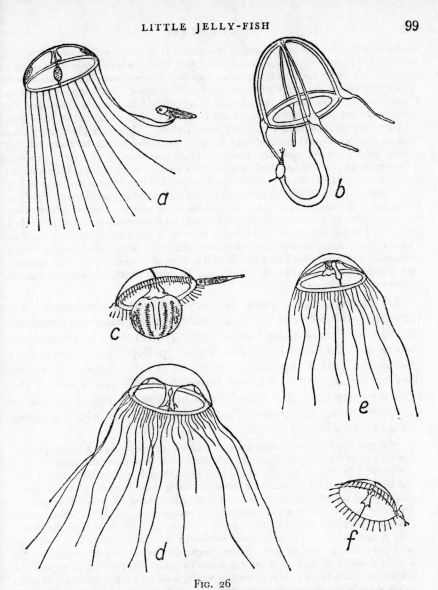

FIG. 26

Small medusae capturing prey drawn by Dr. Marie Lebour (1922, 1923) from specimens kept in a plunger-jar: *a, Phialidium* catching a young *Cottus* (×2); *b, Sarsia tubulosa*, passing a captured copepod from its tentacle to its mouth (×4); *c, Cosmetira pilosella* eating a Pleurobrachia and holding young *Labrus* (×1½); *d,* the same *Cosmetira* with tentacles extended ready to feed; *e, Laodicea undulata* with its stomach full of young blennies (×2); *f, Obelia,* catching a copepod (×4).

bodies coiled along the radial canals, the stomach hanging well down below the
bell and the lips of the mouth drawn out into a beautiful frill. *Aequorea* is quite a
surprise; it is so big for a hydromedusan—up to 7 inches across—that it might well
be taken for one of the larger jelly-fish, the Scyphomedusae. Yet it is produced from
a tiny hydroid: *Campanulina*. I was very lucky when I had nearly completed the
drawings for my plates to come across this beautiful medusa at Lowestoft (in August
1954); it was the first time I had ever seen it alive. Altogether there must have
been at least a hundred specimens of *A. vitrina* swimming in the harbour—all full
grown. As I drew the specimen, shown in Plate 4, swimming in a large glass jar,
it frequently lengthened and shortened its tentacles to twice or half the length shown
in my drawing. Occasionally it enormously expanded its mouth so that one looked
into a large cavity as if into an open but inverted sponge-bag; and then as if pulling
the strings of the bag it would contract the opening to a narrow funnel. My drawing
was made from an actively swimming specimen; on a closer examination of it when
preserved I found that there should have been even more marginal tentacles than
I have shown—approximately four to every radial canal. It is a characteristic which
distinguishes this from the other two British species, *A. forskalea* which has approximately as many tentacles as radial canals, and *A. pensilis* which only has one tentacle
to every three or four canals.

All these medusae, from the very smallest to the largest, are
voracious carnivores. Dr. Marie Lebour (1922–23) has made a wonderful study of their feeding habits, keeping them alive in plunger-jars
such as were described in the last chapter (p. 82). In Figs. 26 and 27
(pp. 99 and 101) I reproduce a selection of her delightful drawings
showing the capture of fish and copepods by different species. The
following quotations give some idea of her observations of just one of
the many species she studied: *Phialidium hemisphericum* which is shown
in Fig. 26 and Plate 3 (p. 96).

"Many *Phialidium* were seen to catch and eat the young fishes. In the plunger
jar they throve well, growing to a large size. The medusa would float about in
the jar with its tentacles outstretched to the finest threads, several times longer than
the diameter of the bell, and wait for some living thing to come along. Directly
this touched the tentacle it reacted and presumably stung the prey. Several more
tentacles then came into use and together entangled the fish, or whatever the
food caught might be. The fish would struggle, and the long tentacles would
play it until it was exhausted. Very rarely the fish would escape and run away
with the tentacle. Usually it would be caught and killed in a few minutes. The
tentacles would then, helped by the umbrella folded in, proceed to deposit the
fish in the manubrium. Sometimes this would only take a few minutes; at other
times more difficulty would be encountered, and after twenty minutes or more
the umbrella would turn upside down and the fish would be dropped into the
manubrium. From a few hours to half a day or, rarely, more was taken to digest
the fish, sometimes the head part being disgorged."

She then gives the following short diary of the feeding of one individual which she kept from February 17th when it measured 6 mm
across until March 15th during which it grew to about 12 mm; it

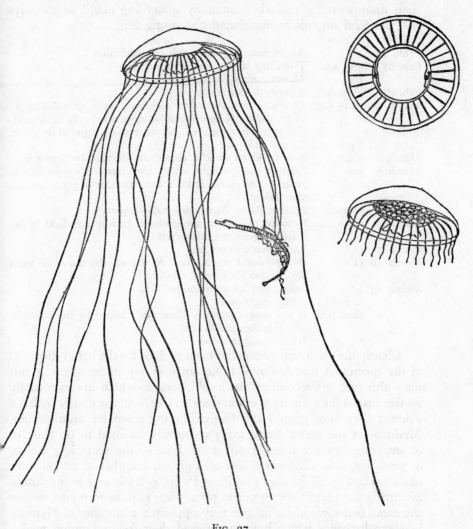

FIG. 27

Young *Aequorea* feeding in a plunger-jar, drawn by Dr. Marie Lebour (1923).
Catching pipe-fish, with two pipe-fishes coiled round in its stomach, and with
a stomach full of small blennies natural size. A large *Aequorea* is shown in
Plate 4 (p. 97).

then disappeared, "probably eaten by something else," as she says. (All the food organisms mentioned are young fish).

Date	Time	Food
"Feb. 22		A solid mass, probably a fish, in manubrium.
Feb. 27	9.15 a.m.	Two *Cottus bubalis*, one half digested.
	4.30	Almost completely digested.
Feb. 28	9 a.m.	A partly digested *Cottus*.
March 1	10.15 a.m.	Caught a live *Ammodytes tobianus*. Inside manubrium by 10.30, within an hour it is an opaque mass, only distinguishable as a fish by its eyes. Almost entirely digested by 3 p.m.
	6 p.m.	Caught a *Cottus*.
March 2	9 a.m.	Another *Cottus* caught, completely digested by 6.30 p.m.
March 3	9.20	Another *Cottus* caught, which after frantic struggles dies in about ten minutes. It takes forty minutes to get it into the manubrium.
March 4	12 noon	Another *Cottus*. Nearly digested at 6.30 p.m.
March 5	11.30 a.m.	Remains of food in manubrium. Hardly any food in jar. A small *Agonus cataphractus* put in.
March 7–9		Phialidium does not eat.
March 10	11 a.m.	It has caught the *Agonus*. At 6 p.m. the tail and trunk are digested, the rest got rid of.
March 13		Some small *Gonius minutus* put in.
	12 noon	It has caught one.
	12.15 p.m.	It has caught another. These both helped by the umbrella edge into the manubrium.
	3 p.m.	It has caught two more."

Clearly the medusoid person has been evolved for the better dispersal of the species. A few Anthomedusae not only reproduce sexually but may also bud off secondary medusoid persons which are eventually set free just as the primary one was once set free from the earlier hydroid colony; they thus extend the free-swimming stage for another lap. Medusae of the genus *Sarsia* are particularly inclined to do this; in *S. prolifera*, they are budded off at the base of the tentacles, and in *S. gemmifera*, from different points on a greatly lengthened manubrium as shown in Fig. 28 (p. 104) and Plate IV (p. 49). Here we see evolution favouring the planktonic dispersal phase; but curious to relate, within the same order we also find the very opposite: a number of Gymnoblastean hydroids which have suppressed their free-swimming medusoid generation. There are some Calyptoblastean ones which have done the same. These hydroids produce medusa-buds but do not liberate them; in different species we see them in all stages of degeneration from a complete medusa with manubrium and radial canals to just a globular sheath from which the sexual cells—or sometimes developing larvae—are finally allowed to escape. Evidently in these forms the medusoid phase has led to some disadvantage and selection

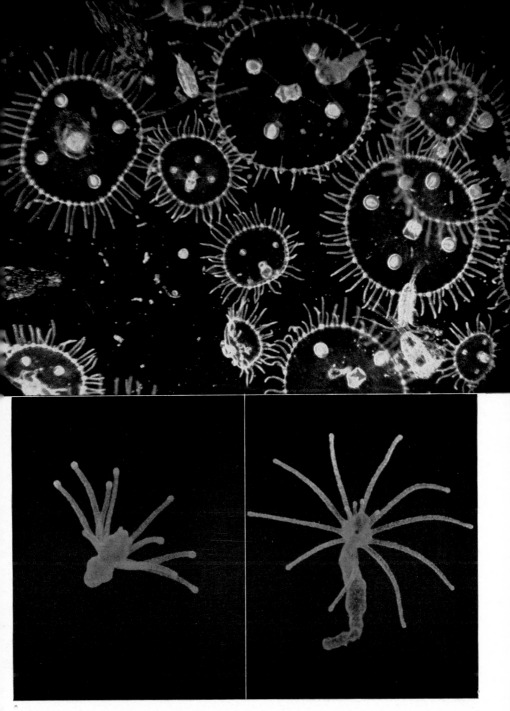

Plate V (*top*) *a.* Medusae of *Obelia* of various ages, x 17. (*bottom left*) *b.* An actinula larva of the hydroid *Tubularia* soon after being liberated, x 25. (*bottom right*) *c.* Young *Tubularia* after settlement of the free swimming larva, seen in *b*, x 25. (*a* only by electronic flash).

(*Douglas Wilson*)

Plate VI. The Portuguese Man-of-war *Physalia physalis* eating a fish (a goldsinny, *Ctenolabrus rupestris*). The gastrozoids are attached to the fish with their mouths widely spread over its surface, while the dactylozooid tentacles are detached and trailing free. The float is photographed in the act of twisting over to wet its surface: a periodic action. (Normally the crest of the float is erect as seen in Plate 5). x $\frac{3}{4}$ approx.

(*Douglas Wilson*)

has tended to preserve those variations in which it failed to develop
to its original complete form; perhaps the factor most likely to operate
in this way would be a change in the current systems so that now the
moving waters tended to carry the majority of the medusae to destruction, to some part of the sea where conditions would be unsuitable for
the next generation. There is good reason to believe that our common
fresh-water *Hydra*, and its allied genera, do not represent primitive
forms which have not yet developed medusae, but are at the extreme
end of such a process of medusa-reduction so that all traces of it have
gone leaving only the production of eggs and sperm by the polyp.
Indeed if *Hydra* had had free-swimming medusae as the reproductive
persons it is very unlikely that it would ever have been able to invade
the rivers from the sea; if the polyp crawled upstream little by little
all the ground gained would be lost by a free-swimming medusa
which would be swept downstream into the sea again. There are a
few fresh-water hydroids with medusae, but these are either only
produced very occasionally or are usually forms which inhabit such
large sea-like lakes as those of Victoria Nyanza and Tanganyika in Africa.

What would happen if, after the medusoid form had been reduced to
a mere sac in the manner just described, conditions changed and once
again planktonic dispersal would be an advantage to the race? Would
the medusa be redeveloped, over the course of many generations, to
be once more set free? This is not just an idle speculation as we shall
see; for some species the environmental conditions do appear to have
changed back again to favour once more a better means of dispersal.
Evolution however hardly ever retraces its steps and goes into reverse;
if a structure is lost and then wanted again, a substitute structure is
usually evolved to take its place and not a resurrection of an old one.

In the plankton of our coastal regions we may come across a very
delicate object floating like a medusa—yet *not* a medusa; it is in fact a
solitary hydroid polyp with greatly elongated tentacles spread out like
a star so as to offer, parachute fashion, as great a resistance to the water
as possible. It is the *actinula* larva: the young stage of the gymnoblastean hydroid *Tubularia*. This is one of the most beautiful of our
hydroids, which may not uncommonly be found in crevices or hollows
under overhanging rocks at very low tide. It produces clusters of
reduced medusae—gonophores as they are called—which hang like
little bunches of grapes around their rose-coloured flower-like heads;
the eggs within each of these, after being fertilized, develop one at a
time into the little planulae, which are however kept inside the gonophore until they have develped their mouths and long tentacles. Then

they are set free—to drift like medusae in the plankton for a time before they settle down, each to establish a new colony, perhaps far from its parental home. The production of such curious pelagic polyps is seen in the next chapter to have quite an extraordinary sequel; similar forms of the past are now thought to provide the key to the understanding of those puzzles of the animal kingdom: the composite Siphonophora which, although of undoubted hydrozoan affinities, will be dealt with among the larger jelly-fish. Photographs of the actinula larva and its development are shown on Plate V (p. 102).

You may have to take very many tow-nettings before you are lucky enough to catch one of these beautiful little actinula larvae. The best

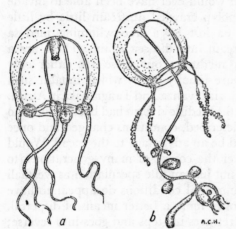

FIG. 28

Medusae which produce more medusae as buds, either as in (a) *Sarsia prolifera* at the base of the tentacles, or as in (b) *S. gemnifera* from a greatly lengthened manubrium. Drawn from specimens beautifully preserved by the late Mr. E. T. Browne and now in the British Museum. A photograph of *S. gemnifera* is shown in Plate IV, but the one drawn here is of special interest in that one of the secondary medusae is producing yet another generation of buds.

way to see them is to collect some of the colonies of the Tubularian hydroids along a rocky coast at extreme low tide in the late summer; if they are kept in a small aquarium it may not be long before they are seen, if watched with a lens, to be giving off these beautiful little living stars from their drooping gonophores (Fig. 29B). The body of the actinula emerges first and it hangs for a time suspended from the gonophore by the tips of its long tentacles which are still held together; finally it is liberated and, with its tentacles at once opening out to give it support, it floats away. While you are about it, I would also recommend you to try to see the launching of an *Obelia* medusa; it is quite an exciting event and not at all difficult to witness. At low spring tides colonies of *Obelia* can usually be found growing on the wide ribbon-like fronds of the Laminarian seaweeds which then come just within reach; at a first glance they appear as little white hairy

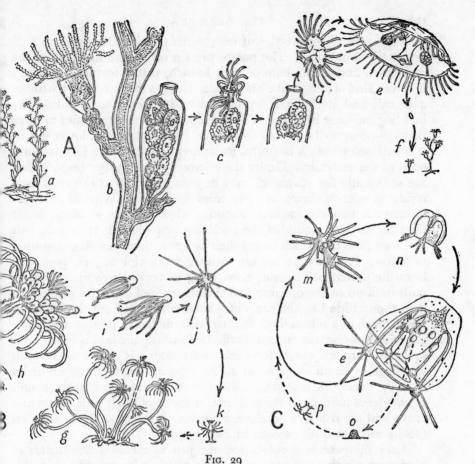

FIG. 29

Three different kinds of hydrozoan life history. A, the typical alternation of polyp and medusoid generations as seen in *Obelia; a,* a colony of polyps natural size; *b,* a small part of the colony highly magnified to show the medusoid buds within their urn-like protective casing; *c* and *d,* a young medusa escaping; *e,* the mature medusa with clusters of eggs; and *f,* the small larva derived from the egg settling on the bottom to form a new polyp colony. B, *Tubularia* in which the medusoid phase is surpressed, but replaced by a floating larval polyp, the actinula: *g,* polyp colony natural size; *h,* a single polyp magnified to show its medusoid buds which grow no further but whose eggs remain inside till they develop into the actinula which is shown being liberated in *i* and *j; k,* shows the actinula settling down to form a new polyp colony (as shown in Plate V). C, *Margelopsis haeckeli* in which the fixed hydroid colony is surpressed and the medusa is the predominate phase; *l,* the medusa with eggs which during summer develop into floating polyps, *m,* which then, without settling down, bud off another generation of medusae, *n;* as winter approaches larger eggs are formed which fall to the bottom and form a resting stage, *o,* from which a new floating polyp, *p,* arises in the spring. A and B are drawn from living specimens; C *l, m* and *n* are from Russell (1954—after Hartlaub), and C *o* and *p* from Werner (1955).

growths upon the seaweed, but under a lens each stalk is seen to be a delicate hydroid stem. The polyps branch off right and left from the main stem and at the base of each branch, if it is mature, appears a different kind of person: the blastostyle. This is a modified polyp without a mouth and its body is cylindrical and covered with button-like budding medusae in all stages of development from larger ones towards the top—soon to be freed—to smaller, less mature ones, lower down; it is enclosed within a beautiful transparent vessel, like a Grecian urn, open at the extremity. Under the microscope, the larger buds, if they are sufficiently far advanced, may be seen to twitch at intervals as if trying to wrench themselves free from the parent. What is so extra-ordinary is that the medusae awaiting liberation are so much wider than the narrow mouth of the enclosing urn which is their only exit to the sea; it is this which makes their escape such an amusing operation to follow. Presently we see one, with a more violent jerk, tear itself from the parent polyp-stem; now, using its tentacles as little arms, it pulls itself up to the opening and thrusts some of its tentacles through it to the outside like the arms of a man emerging from the conning-tower hatch of a submarine. Pulling with the 'arms' outside and push-ing with those within, the umbrella-body of the medusa is folded and squeezed through the bottle-neck to freedom. It swims away with violent jerky rhythmic pulsations. As often as not, the little umbrella in its very early life is seen to be inside out, as shown in Plate 3 (p. 96). It now feeds hard upon the still smaller forms of life it runs across and grows and nourishes its developing gonads. A sketch of the birth of an Obelia medusa is also shown in Fig. 29 (p. 105).

Lack of space will only allow me just to mention two orders of Hydrozoa which have become entirely planktonic and exist as adults only in the medusa phase. They do this either by their larvae budding off new medusae at once and so reducing the polyp stage to a mere transitory vestige, or, in some forms, by the larvae having a longer phase of development as parasites inside other larger medusae before giving rise to their own new generation. They are the Trachomedusae and Narcomedusae.

The former are represented by *Aglantha digitale* which comes into the northern North Sea in large numbers in the influx of Atlantic water which flows to the south after curving round the Shetland Islands. A true definition of its diagnostic charac-ters would involve us in zoological technicalities beyond the scope of this book. While the most radical difference between it and the foregoing forms is in the mode of formation of the sense organs, it can easily be recognised by its characteristic form which is shown in Plates 4 and XI (p. 94 and p. 140); it is often beautifully iridescent. A feature seen in this animal and common in the Trachomedusae—although not

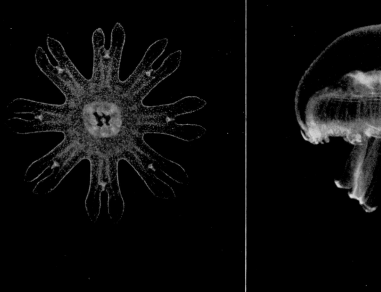

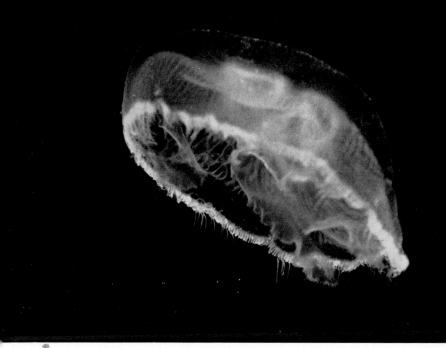

Plate VII (*top left*) *a.* Ephyra larva of the jellyfish *Aurelia*, showing mouth in centre, seen from below; note the 8 tentaculocysts, one on each lappet, x 12. (*top right*) *b.* Young *Aurelia*, bell contracting, lateral view, x 8. (*bottom*) *c.* Adult *Aurelia* pulsating, photographed at the end of contraction, x 1 approx. (Compare with Plate 6). (*b* only by electronic flash).

(*Douglas Wilson*)

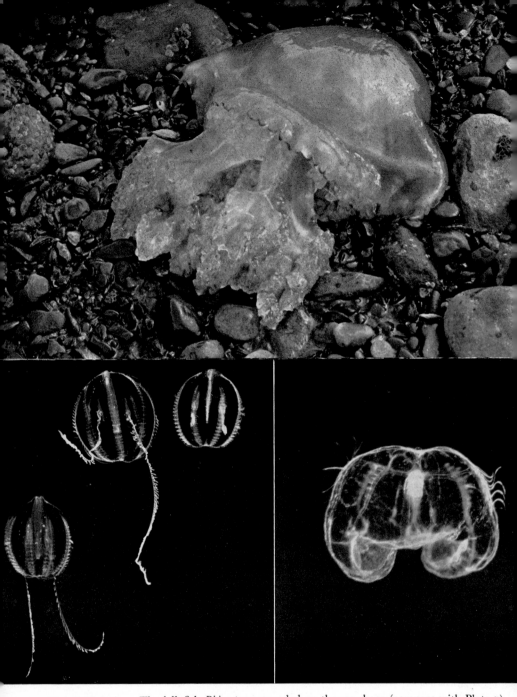

Plate VIII (*top*) *a*. The jellyfish *Rhizostoma* stranded on the sea shore (compare with Plate 7). (*bottom left*) *b*. The ctenophore ('comb-bearer') *Pleurobrachia* swimming mouth upwards, x $1\frac{1}{2}$ approx.; the rows of comb-plates are clearly seen and the tentacles of the specimen on the right are contracted into their pockets. (*bottom right*) *c*. A lobate ctenophore *Bolinopsis*, a young specimen swimming mouth downwards; the balance organ can just be seen on top. (By electronic flash) x 15. (*Douglas Wilson*)

invariably found in, nor confined to, members of this group—is the long so-called false or pseudo-manubrium. If we look at it carefully we see that actually the manubrium proper is quite short and leads as usual into the main gastric cavity; it is the position of the latter, however, which is extraordinary, for it hangs at the end of a long trunk-like structure down the length of which run the four radial canals conveying the digested food up to the umbrella body above and around it. It is usually a prehensile device for the more ready capture of prey, but in some it has become an organ of temporary attachment for parasitic feeding upon other medusae. A large pseudo-manubrium is also seen in a common North Sea Lepto-medusan *Tima bairdi* also figured in Plate 4, and to which we have already referred (p. 98).

The Narcomedusae are only represented in our seas by a few species which are confined to the deep water beyond the continental shelf. The one most frequently met with in tow-nets fished below some 200 fathoms is *Aegina citrea* which is shown in Plate 8 (p. 129); by a looping up of the margin, the tentacles appear to be high up on the upper surface of the bell. I have seen more than half-a-dozen of these beautiful orange and yellow medusae drawn up from the depths and each time the long tentacles, always only four in number, have been stretching upwards as shown in the Plate. When put in a glass vessel they hang motionless in the water in this position. I have only seen them make the slightest movements.

Having referred to these two entirely planktonic groups I will now go back to draw attention to one very interesting member of the Anthomedusae which has gone the same way. This is the remarkable *Margelopsis haeckeli*, illustrated in Fig. 29c (p. 105); it may occasionally be abundant in parts of the southern North Sea and eastern end of the English Channel (Russell 1954). The eggs are not discharged by the parent medusa but develop within its bell cavity into the little polyps which are carried along—under their mother's wing, as it were— until their tentacles have grown long enough to give them sufficient support to enable them to float on their own. These little larvae, when set free, remind us of the actinula larvae of *Tubularia* which we have just discussed, except that they are not produced from a fixed medusoid bud on a hydroid; and further they do not themselves settle down to form a bottom-living or coastal hydroid but continue to float until they have budded off and liberated the next generation of medusae. Quite recently Werner (1955) has discovered that, at the approach of winter, a young developing egg drops to the bottom as a resting stage, to float up again in the spring. The sedentary hydroid stage has been eliminated, just as it has in the Tracho- and Narco-medusae. We see many examples of evolution repeating its pattern in parallel directions.

In the beginning of this chapter we considered the possible evolution of the Protozoa from single-celled plants; for we saw that there must have been plants before there were animals at all. After dealing with

the Protozoa we have passed to consider some of the simplest of the
Metazoa. The problem of origins—even if speculative—is always
interesting; it may be fitting then to end the chapter with a brief
consideration of the possible way in which these many-celled animals
could have been derived from simpler protozoan ancestors. As Dr.
Baker has pointed out in the article (1948) to which I have already
referred, it must be quite a different matter from that of the evolution
of the many-celled plants from the single-celled ones. Botanists are
familiar with several lines of many-celled forms among the lower
algae, which must have evolved independently from the protophytes.
As the plants are feeding upon the dissolved chemicals in the water
surrounding them and trapping the energy of the sunlight, they can
take up food almost as easily if they are made up of many cells—especially if they have a linear or plate-like arrangement—as they can if
consisting of only one. It is much more difficult to imagine how the
step from one to many cells could have been taken by the animals—
for the animal's method of feeding is so different; it is actively hunting
for its food. There are two theories as to how the metazoa have arisen
which have been put forward in the past. One is that the products of
cell-division remained connected with one another as is seen in some
colonial forms of protozoa, and then gradually became differentiated
so that some were more concerned with capturing food and others
with protection, reproduction, etc., until eventually the aggregation
took on a new individuality of its own. It is most likely that the
Parazoa or sponges—much more loosely integrated cellular animals
than the Metazoa proper—have been evolved from the Protozoa in
such a way. It is more difficult, however, to see how the polyp-like
coelenterates were evolved from a group of actively feeding animal
cells. A second hypothesis considers the metazoan to have arisen by
the division of the body of a multinucleate protozoan into a number
of cells by partition rather than by the aggregation of a number of
separate animal cells into a colony. This view was originally put
forward by the protozoologist Dobell (1911) and it has been revived
more recently by Hadzi who considers that the coelenterates are not
the most primitive of the Metazoa. De Beer (1954) has given a summary in English of Hadzi's views. He believes that the first Metazoa
were simple flatworm-like creatures derived by partition from a multinucleate ciliate protozoan and that the coelenterates have been evolved
from such flatworm ancestors. There appear to me to be serious difficulties in the way of this view: notably the nerve-net of the Coelenterates is in a much more primitive condition than that of any such

system known in the flatworms. Although it is just as speculative, I have recently (1953) suggested yet a third hypothesis which I believe does not present the same difficulties as the other two: that the metazoa are not in fact derived from protozoa at all, but from relatively simple metaphytes. We have seen that among the single-celled flagellate organisms there are several groups having both representatives which are feeding as plants by photosynthesis and others which are feeding as animals. Here without doubt animals have been evolved from plants quite independently at several different times. We have also seen that some of the higher plants have taken to capturing insects to supplement their supply of nitrates and phosphates : to this extent they are becoming animal in nature. One of these insectivorous plants is particularly interesting—the aquatic bladder-wort *Utricularia* which on its underwater stems has developed little vesicles into which water-fleas are attracted; once inside they find themselves in a trap from which they cannot escape. They are killed and slowly digested. Let us suppose that one of the very primitive plants—perhaps just a sphere of green photosynthetic cells—were to be in water short of phosphates and nitrates, and that a variation occurred which formed a hollow pocket at one side into which small protozoa might enter; perhaps first they might do so for shelter, but then be trapped for food—just as today we see water-fleas trapped by the bladderwort. It might have arisen in the first place as a sort of symbiosis. We have seen some animals harbouring little plants; these many-celled plants may instead have harboured little animals without killing them: giving them shelter, oxygen and perhaps some carbohydrate food, and in return getting their highly nitrogenous excretory products. It would only be through a series of steps, no greater than so many evolutionary series we have seen, for the host-plant by variation and selection to produce, first, a means of attracting animals, and then devices for capturing them and pushing them into the now digestive cavity; the insect-eating sundew *Drosera*, which has developed tentacles on its leaves for the capture of flies, shows us what plants can do in this direction. With the success of such an evolutionary line we can well imagine the subsequent loss of photosynthesis as a means of feeding, just as we have seen more than once a similar loss among members of the flagellates which too have turned from a plant to an animal mode of nutrition. Such a sequence of events seems, to me at any rate, to be as likely an origin of the metazoa as the other possibilities suggested.

CHAPTER 7

SIPHONOPHORES AND THE LARGER
JELLY-FISH

IN THE LATE SUMMER of some years, along the shores of Devon and
Cornwall, or those of southern Ireland, we may find washed up by
the waves, large numbers of little objects which at first sight look
more like manufactured toys of plastic than things of nature. Oblong
in shape, each is like a miniature raft measuring up to 2½ inches long
by 1½ inches broad; it has a triangular fin set like a sail diagonally
across it. Indeed that is just what the fin is: a sail set to catch the wind,
and drive the animal along the surface of the sea like a small model
of a sailing barge. Below it hang the polyp persons and tentacles which
no doubt take the place of a keel; it would not surprise me if the
tentacles stretched out astern acted like a rudder to keep the craft
sailing somewhat obliquely to starboard and so present a greater
frontage of water to the other tentacles in search of prey. 'Jack Sail-
by-the-Wind', the old sailors used to call them; today I do not suppose
the steamship-men, travelling fast, ever notice them, but in the old
days of sail our seamen, when becalmed, might often see the surface
of the tropical seas dotted with them. *Velella velella*[1] is their scientific
name and every few years, due to the vagaries of the current system and
the wind (as already noted in Chapter 2), numbers may reach our
coasts, together with other wanderers, as strays from warmer areas of
the Atlantic. They are members of that remarkable group of animals
the Siphonophora (Gk. *sipho,* a siphon or tube) which are really very
unusual Hydrozoa related to the little hydroids and medusae, dealt
with in the last chapter; we have reserved them for discussion here,
with the larger jelly-fish, because they *are* so very different: both
larger and so much more complex than the forms already considered.
We shall be very lucky if we see *Velella* actually drifting in to the shores
of England; we would stand a better chance of meeting it still afloat
if we were staying at the right time on one of the lovely arms of the

[1] = *V. spirans.*

sea that carry the Atlantic water deep into the coasts of Counties Cork and Kerry. Occasionally they have been picked up on the coast of the Isle of Man (Williamson, 1956). They are coloured like the ocean itself, deep blue, and made to float by a series of air-filled buoyancy chambers, made of a horny material and arranged in a concentric manner; as they grow bigger additional outer chambers are added round the edge from time to time. The specimen illustrated in Plate 5 (p. 112), was drawn partly from a specimen actually stranded in Cornwall at Porthnanven, about 4 miles north of Land's End, in August

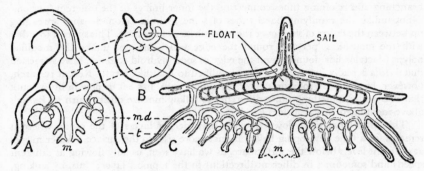

FIG. 30

Diagrammatic sections comparing a hydroid polyp such as *Corymorpha* (A) with the young and adult stages of *Velella* (B and C). Broken lines connect the corresponding parts of A with those of B or C; continuous lines connect the parts of B with what they become in C; *m*, mouth; *md*, medusoid buds; *t*, tentacles. The drawing of B (the Conaria larva) is simplified from Garstang (1946) after Woltereck.

1935, and picked up and sent to me with many others by the Rev. Peter Hartley, and partly from some quite recently collected by the R.R.S. *Discovery II*. It was not coloured from life but from my memory of having seen it alive, refreshed by colour notes of the living animal kindly supplied by Dr. H. G. Vevers and from a colour sketch made by Mr. W. Fry on board the *Discovery II*.

They have usually been thought of as a complex colony made up of a large number of polyp 'persons' crowding the *under surface* of the raft and hanging 'mouth downwards'. In the middle is a large central 'person' with a big mouth—clearly the main feeding member; round the edge there spreads out a fringe of 'persons' without mouths but heavily charged with stinging cells both for the capture of food and the protection of the whole; in between these and the central mouth are a large number of smaller polyp-like 'persons' with mouths, but also serving as the 'blastostyles' of a hydroid colony, for they bud off small male and female medusae which are eventually set free. These will produce eggs and sperms which will start a new generation by the fertilised egg developing through a planula into a little larval form: the *Conaria*.

Velella and some other closely related but circular forms (*Porpita, Porpalia,* etc.) belong to one very distinct division of the Siphonophora—the Disconantha; they are clearly very different from the rest. Some zoologists have thought them to be just modified medusae with the central 'person' as the manubrium, the outer 'persons' being the tentacles and the 'blastostyles' extra budding devices; but the remarkable horny float, which on microscopic investigation is shown to be produced as a secretion inside a pocket of the outer skin which has been folded inwards, is something quite unlike anything in a medusa. The microscope further reveals something which is not usually described in the more elementary books: that there is a remarkable series of branching and rejoining tubes connecting the inner linings of the different 'persons' —not unlike the ramifying basal tubes of some hydroid colonies—and spreading up between the layers of skin, over the *upper* surface of the float. This system, together with the mouthless 'persons' round the edge which are not unlike the modified polyps (dactylozoids) found round the edge of some hydroid colonies, have suggested that *Velella* is really just such a colony modified to float like a raft. Recent research, however, by the late Professor Walter Garstang (1946) has left us now in little doubt that it is really something quite different; to follow him we must return to *Velella's* above-mentioned larval form: *Conaria*.

This little larva sinks down into the dark deeper layers of the ocean where it remains for a time developing before it once more rises to the surface. Water masses at different levels below the surface are, as we have seen, usually flowing at different speeds and sometimes in different directions to the topmost layer; thus, by sinking, *Velella's* little daughter can best be carried furthest from its parent in order to spread the species as widely as possible: a pretty example of a reversal of what is usual in animals living on the sea floor which send their larvae *upwards* for the same reason. The *Conaria* larva is in fact a floating polyp just like the *Actinula* larva of the hydroid *Tubularia,* which we saw in the last chapter was adapted to float for a time before settling down. *Conaria* never settles *down*; with the development of a float it rises to settle upside-down on the surface. Garstang makes a striking and detailed comparison between the growing structure of the *Conaria* and the development of certain large solitary hydroids (*e.g. Corymorpha*) and comes to the conclusion that *Velella* is not indeed a colony at all in the ordinary sense of the word, but a very large and inverted hydroid head, like the single head of a *Tubularia* or *Corymorpha*. The so-called central

---

Plate 5. RARE VISITORS FROM THE SOUTH-WEST

1. The 'Portuguese man-of-war', *Physalia physalis* ($\times \frac{1}{2}$), drawn on the R.R.S. *Discovery II*, south-west of Ireland, September 1954
   a. whole animal floating with crest erect
   b. enlarged view of part of one of the long tentacles. (See photo. of *Physalia* eating fish in Plate VI)
2. *Velella velella* (=*spirans*), (natural size) drawn from a preserved specimen stranded in Cornwall but coloured from notes of the living animal (see p. 111)
3. The planktonic snail *Ianthina janthina* clinging to its bubble raft (natural size)

Note: The blue of *Physalia's* tentacles should be much more intense, and the shell of *Ianthina* a more lavender blue; in drawing attention to this failure in reproduction I wish to state that in nearly every other plate the colour values are excellent.

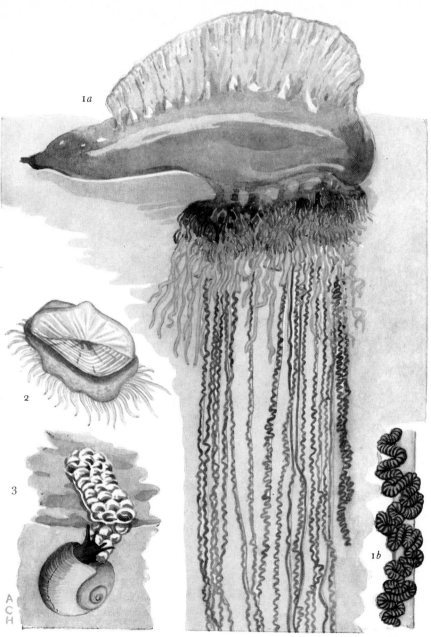

2

3

A
C
H

1b

Plate 5

Plate 6

'person' is its mouth and the so-called dactylozoids round the margin are its main tentacles, whereas the blastostyles correspond, except in their small open mouths, to those found hanging round the heads of such large hydroids, in exactly the same position, between the mouth and tentacles. The comparison is shown in Fig. 30 (p. 111). Moreover, the ramifying system of little inner lining tubes is very similar to such a system found in the *stalk* of the solitary above-mentioned *Corymorpha*. The final proof of the relationship is coming from the as yet unpublished studies being made by Mr. George Mackie in our department at Oxford. He finds a very close resemblance between the microscopial structure of the tissues of *Porpita* (a close relative of *Velella*) and those of the hydroid *Corymorpha*, and also a striking similarity in their behaviour patterns—feeding reactions, movement of tentacles, etc.—as seen by him in living specimens of *Porpita* in the Canary Islands. *Velella* has been evolved from a floating actinula of a *Corymorpha*-like hydroid in the distant past: an actinula which, by developing a float, has led to a new race of pelagic beings.[1]

Before leaving *Velella* I must mention the remarkable pelagic snail *Ianthina* which has a delicate violet-blue shell and like *Velella* floats on the surface of the ocean, but does so by making a bubble raft for its support. While it will be referred to more fully in a later chapter (p. 148), I introduce it here, and illustrate it on Plate 5 (p. 112), not simply because it may also be carried to our shores as a rare visitor, but because it preys upon *Velella* with unusual results. Mr. Peter David, zoologist on the staff of the National Institute of Oceanography, has told me how on board the R.R.S. *Discovery II* in the early summer of this year (1954) they came across large numbers of these two animals in the Atlantic some 500 miles west of Cape Finisterre, and that many of the *Velella* had small *Ianthina* attached to their under surface. Keeping them alive in an aquarium he saw the *Ianthina* browsing on the *Velella* and gradually clearing the under side of tentacles and blasto-styles. He noticed that while the *Ianthina* were feeding they periodically exuded a purple dye. He thinks it possible that this may be used to anaesthetize the *Velella*, because while such specimens appeared life-less, they did not shed their tentacles as they usually do when they die; also such *Velella*, if put into formalin to preserve them, do not contract their tentacles anything like so much as those not attacked by *Ianthina*. Eventually he observed that nothing remained of the *Velella* except the horny float which is left to drift; such dead floats often become a raft upon which the larvae of goose barnacles (*Lepas anatifera*) will settle to form strange planktonic clusters of these long pendant

[1] In the Southern Ocean there is another remarkable floating hydroid form, *Pelagohydra*, which has a different, more balloon-like float—no doubt evolved quite independently but in the same fashion.

Plate 6.   THE COMMON JELLY-FISH *Aurelia aurita* ($\times\frac{1}{2}$), looked at from below and from above. Lowestoft, August 1954

TOS—I

barnacles which more usually become attached to the bottoms of ships. These should not be confused with groups of another species of barnacle of the same genus, *L. fascicularis*, which, like *Ianthina*, are buoyed up by gas-bubble floats which they themselves secrete; they too may drift to our coasts from the more open ocean along with *Physalia* and *Velella* (Wilson, 1947).

The remaining Siphonophores, in the division Siphonatha, are much more of a puzzle than *Velella*. They certainly are *composite* animals: made up of a large number of persons, some of which are polyp, some medusoid in nature—and some which it is impossible yet to be sure what they are. Although we have seen that there is a differentiation of individuals in some ordinary hydroid colonies, nowhere else in the animal kingdom do we get the extraordinary integration and co-ordination of persons that we find in these composite Siphonophores. The whole colony now appears to act as one, as if it was endowed with some new and higher individuality greater than that of the component persons. Here indeed is a problem. Every kind of person about to be mentioned may not be found in any one species; different types of Siphonophore are characterised by their own particular assortment. Of polyps there are the three kinds already mentioned as being sometimes found in the more ordinary hydroid colonies: *gastrozooids* or feeding polyps which are of course always present, *dactylozooids* for protection and capture of prey, and *blastostyles* which give rise to medusa-buds. Then there are the medusoid persons. Fully formed sexual medusae to be set free like those of *Velella*, are found only in *Physalia*, the 'Portuguese man-of-war', and then it is only the female ones which are liberated; in all the rest they are *gonophores* or medusoid persons in various stages of reduction. In many forms there are special medusae developed as locomotor organs: powerful pulsating *nectophores* or swimming bells using jet-propulsion. Then there are bodies about whose nature we are much more doubtful: the *hydrophyllia* or bracts, which are usually leaf-like shields hanging down to protect the more important gonophores and gastroozoids. A float or *pneumatophore* is sometimes present and has been regarded by some as a medusa, but it seems more likely that it is formed simply as a hollow pocket by an infolding of the outer layer at the base of the primary polyp which grows from the initial planula larva.[1]

[1] In my first edition I had a footnote stating that *Stephalia* has a gas-producing organ formed from a modified medusa; Mr. Totton informs me that this has now been proved incorrect. He also tells me that he has recently received from the University Museum of Dundee a fine specimen of *Stephalia* taken in Scottish waters in 1907; I had said that it never reached our seas.

There are three main types of siphonophore in this second division. It is fortunate that at least one of each may occasionally be found in our waters. One type has both a float and swimming bells; another has the float enormously enlarged but has no locomotive organs; and the third is just the opposite, being without a float, but having the swimming bells highly developed to give a powerful form of jet-propulsion. We seem to see two main lines of evolution: one towards a more passive and the other to a more active mode of life.

We will deal first with the type that has both methods of support moderately developed. *Physophora hydrostatica*, which is an excellent example, is not infrequently met with in the North Atlantic drift off the western Isles of Scotland and may occasionally enter the northern North Sea. It is one of the most beautiful and delicate of all pelagic organisms. If caught in a tow-net it will most likely break into many fragments; it should if possible be gently lifted from the sea in a bucket. It was one of the great joys of going south in the old R.R.S. *Discovery*, to be able to lift these exquisite creatures from the ocean surface and watch them alive in aquaria on deck as we proceeded through the tropics: I refer to the 1925–27 expedition to the Antarctic, using Captain Scott's famous old ship, on which I had the honour to serve as a zoologist. Having only auxiliary steam and a very limited supply of coal, which we had to conserve as much as possible, we sailed whenever we could and at times were almost becalmed. I shall never forget one glorious day when we bore slowly forward upon a calm and intense blue sea with a very gentle swell. I was lowered down on a 'bo'sun's chair' below the jib-boom, till my bare feet just splashed in the warm water, and there I sat with a bucket of water and a hand-net ready for what came along. Specimens of *Physophora* with their rose- and salmon-pink bracts, looked like exotic water-lilies floating just below the surface; usually they were just out of reach, but one I got into my bucket and so made the colour sketch which is reproduced in Plate 9 (p. 144). At the top of a long stem is a pear-shaped transparent float, and below it on two sides of the stem are a series of nectophores or pulsating bells which are also of a glass-like transparency; a little way below these is the circle of pink bracts which curve out and downwards just like the petals of a pendant flower. Why should they advertise themselves so brightly? It seems probable, since they are provided with powerful stinging cells, that they have developed a warning colouration analogous to that so often displayed by stinging insects or venomous snakes: in this case a warning to fish to give them

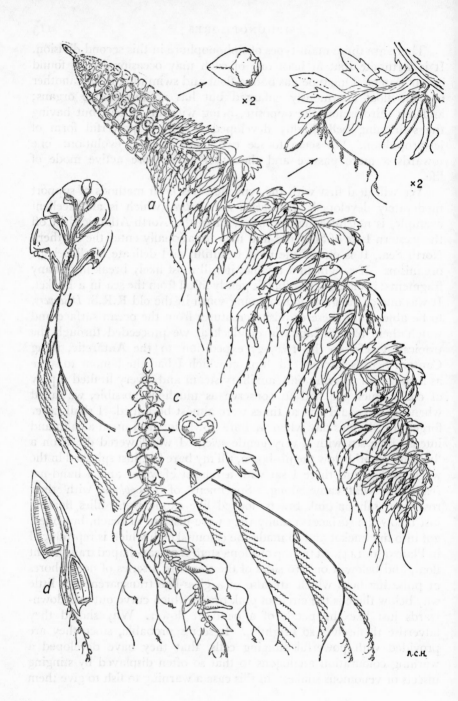

×2

×2

a

b

c

d

A.C.H.

a wide berth. Within the circle of bracts, and just visible between them, are the gastrozooids, which let out long dangling tentacles to capture prey in the waters beneath, and the gonophores which hang down like clusters of grapes. Occasionally you will see one of the tentacles jerk upwards, like contracting elastic, to convey some small crustacean to its appropriate mouth. I have drawn *Physophora* placidly floating, but at times it will turn over on its side and swim actively with its main axis in a horizontal instead of vertical position.

Other not dissimilar members of this group which may be met with in our western waters are *Agalma elegans* and *Forskalia* (Fig. 31, opposite) with its vast array of swimming bells. The drawing I have made of *Forskalia* is partly from a preserved specimen sent me by Mr. A. K. Totton from Villefranche where he has been making a special study of the siphonophores and partly from one of his splendid photographs of the living animal. He tells me that swimming bells of *Forskalia* have definitely been taken in tow-nets off the western end of the English Channel. Very occasionally the delicate and slender *Nanomia* (*Stephanomia*) *bijunga* may enter the Channel and in May 1929 was found in large numbers in the Salcombe Estuary (Berrill 1930); or the allied *N. cara* is taken off the southwest of Ireland, as that illustrated in Fig. 31.

Mr. G. Mackie, to whom I have already referred, is making a special study of the remarkable integration of activity of the different persons in the siphonophore colony which gives rise to its surprising new level of individuality. In particular he is studying the nervous connections between the different component parts. In the species *Forskalia*, just mentioned, which he has studied at the Stazione Zoologica at Naples, he has observed that the colony can swim either forwards, i.e. with the float as the spearhead, or backwards; although all the swimming bells do not keep time with one another, they stop and start together and, by an alteration in the contraction of their openings, they can, in unison, project their jets of water either backwards or forwards. If he touches the float the whole animal will retreat and if he tickles its 'tail end' it goes forward: just as if it were really a true individual. When a single swimming bell is shed from the colony—if we can continue to use the word colony—it immediately starts pulsating and goes on doing so for a long time "as if freed from restraint" as Mr. Mackie says. It goes on pulsating like the vertebrate heart often does when its owner is dead and it has been removed from the body;

← FIG. 31

Some rare siphonophores from British waters: *a*, *Forskalia edwardsi*, a drawing (natural size) made in part from a preserved specimen and in part from several photographs of the living animal taken by Mr. A. K. Totton. Many of the leaf-like bracts have been omitted in the main drawing for clarity, but are shown complete in the inset enlarged (× 2) view of a small part; *b*, *Hippopodius hippopus*, drawn in part from a specimen taken by Dr. J. H. Fraser 60 miles N.W. of Cape Roth and in part from photographs of the living animal taken by Mr. George Mackie at Naples, natural size; *c*, *Nanomia cara* (formerly *Stephanomia*) drawn from a British Museum specimen caught at Valentia Island, Kerry (× 2); *d*, *Chelophyes appendiculata*, drawn (× 1½) from a specimen caught off the north of Scotland by Dr. J. H. Fraser.

and just as the rhythmic contractions of the heart muscles are under the nervous control of the whole animal when it is healthily alive, so it appears that the individual swimming bells of the so-called colony are under the nervous direction of a higher integrating system. He notices that the very young swimming bells appear to pulsate independently of the older ones (which may at the time be quiescent) as if they were not yet under control; again this recalls the vertebrate heart, which, as we can see in the embryo chick, starts to beat before it has any nerve connections to control it. "In evolution", he says, "one can imagine the gradual taming of the medusoid members along with other modifications, so that they only swim when the colonial restraint is lifted."

Physalia, the 'Portuguese man-of-war', is the supreme example of the next type in which there are no swimming bells but an enormous float; this when fully grown may measure nearly a foot in length. It is generally recognised now that there is only one species, *Physalia physalis*, which has a world-wide distribution. Like *Velella* it drifts on the very surface of the sea and its bladder has a crest along the top which serves the same purpose as *Velella's* sail; like *Velella* too, it is occasionally but much more rarely brought to the mouth of the Channel and stranded on the shores of Devon or Cornwall. The last big invasion of *Physalia* into our waters was in the late summer and autumn of 1945[1] when Dr. Wilson of the Plymouth Laboratory made a study of its progress month by month, and wrote a very interesting account of it (1947) together with notes of all the earlier records; apart from isolated specimens, there were former invasions in the years of 1862, 1912, 1919, 1921 and 1934. The maps in Fig. 32 opposite, showing the strandings on the south and west coasts in 1945, are compiled from his several monthly charts; and the magnificent photograph in Plate VI (p. 103) was taken by him. On Plate 5, I show a water-colour study which I recently made from a living specimen on board the R.R.S. *Discovery II*. The large float in the sunlight is of an almost iridescent blue merging into mauve and pink at the top of the crest where sometimes there is a touch of orange.

In development this huge bladder has been formed in the same way as the smaller float of *Physophora*, by an infolding of the outer layer to form a hollow pocket which has then greatly enlarged; the lower part of its lining is developed as a gland for the production of gas and at one end there is a closed tube to the outside which represents the original infolding and has been thought by some to act as a valve for the release of gas if necessary. Mr. G. Mackie, who recently joined Mr. Totton at the Canary Islands for a special study of these animals, tells me that he has made a fresh analysis of the gas and finds it to be a mixture of oxygen and nitrogen almost the equivalent of air; he doubts if the closed tube I have just mentioned does in fact serve as a valve. Along the underside of the float hang clusters of persons: gastrozooids, blastostyles bearing not only numbers of medusoid gonophores but protective

[1] Since writing this there has been another big invasion in 1954.

FIG. 32

Charts showing positions at which the 'Portuguese Man-of-war' *Physalia* was seen from July to October, 1954. The symbols represent the expressions o quantity reported (see key) and show July and September records by circles and August and October records by squares. Summarised by permission from the more detailed charts of Wilson (1947).

bodies, like small dactylozooids, called *gonopalpons*, and then the much larger dactylozooids proper from which hang very long trailing tentacles of the most vivid blue; when fully extended these may be some 30 feet or more in length.

As *Physalia* is driven forward sailing at an angle to the wind the long tentacles stream out behind fishing for prey. They are charged with the most powerful stinging cells of any known jelly-fish (Fig. 24, p. 93) and can paralyse quite large fish, which on being encoiled are pulled up to the gastrozooids. These, as Wilson has so well shown in his account, are remarkably distensible; they expand their mouths enormously and together completely cover the prey which now hangs as in a bag, bathed in the digestive juices. Fig. 33 (p. 120) is a sketch of a fish taken from a *Physalia* at the Canary Islands by Mr. Mackie; it is covered with a mosaic of the expanded mouths of the gastrozooids, the basal parts of which have been cut and now appear as contracted stalks arising from the centre of each. Here a 'crowd' of so-called 'persons of the colony' come together to form one common stomach. Anyone who has had the misfortune to get part of the tentacle of *Physalia* across his hand or arm, when pulling in some large tow-net, will not forget the intense burning pain. I have only experienced a sting of a few inches across the wrist and that was unpleasant enough. Sometimes the tentacles may get wound round ropes being hauled inboard by bare-armed sailors, and there have been cases, so it is said, of arms being paralysed for a time. In more tropical seas there may often be found in among the tentacles one or two specimens of the little fish *Nomeus gronovii* which has silver sides banded with vertical stripes of the same intense blue as the tentacles.

It seems as if it must be a remarkable case of mimicry, for the intense blue tentacles must be avoided like the plague by most predatory fish. Are these little fish immune to the paralysing poison of the sting or do they deftly avoid the tentacles? Are they engaged in some subtle partnership with their overlord? It is one of the many little puzzles of pelagic natural history yet unsolved.

Physalia shows considerable variation in the colour of its float. The one I have drawn shows the full range of hue. Some, however, have no trace of orange on the crest, others may have little of the peacock-green at the base of the float, and occasionally, except for the pink on the crest, the float may be a milky-white instead of blue. The float is not just an inert bladder incapable of movement; at regular intervals its whole body twists over to wet itself first on one side and then on the other, and then comes up again into its normal erect position as shown in the drawing. I was

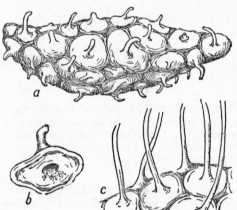

FIG. 33

a—a partly digested fish entirely enclosed by a large number of expanded feeding polyps of *Physalia* which have come together to form one large communal stomach; it is drawn from a specimen obtained by Mr. G. Mackie; the little stalks represent the cut stems of the polyps. *b*, one of the polyps detached. *c*, a sketch of the long connecting stems of such polyps as seen in a living specimen—and as photographed by Dr. Wilson in Plate VI (p. 103).

quite unaware that it possessed such remarkable powers of coordinated movement until I sat sketching it floating in a large enamel bin on the deck of the *Discovery II.* Where I first sat drawing there was rather a breeze and I noticed that my *Physalia* did its twisting-and-watering-the-float-act about every 8 or 9 minutes; then when I moved to a more sheltered spot, where evidently the drying effect of the wind was less, it only turned over to wet itself about every quarter of an hour. Wilson's photograph in Plate VI shows *Physalia*, in addition to feeding, caught in the act of bathing one side of its float.

Some observations have been made which suggest a very interesting adaptation in *Physalia*. We have already noted that the crest of its float is placed diagonally across it as is the sail of *Velella*. In examining a number of specimens caught from both north and south of the equator Dr. A. H. Woodcock, of the American Oceanographic Institution at Woods Hole, found (1944) that the majority of those from the north are mirror images of the majority of those from the south; he believes that their sails are 'set' to overcome more easily certain complex effects produced by wind action on the surface of the sea—effects which are exactly the reverse of one another in the two hemispheres. A sufficient number of specimens have not yet been examined to establish beyond doubt whether there are always more of one kind in the one hemisphere; if, however, it should be confirmed it will be another remarkable example of the power of natural selection : the evolution of unconscious navigation. Mr. Mackie, however, who has just returned from the Canaries tells me that there he and Mr. Trotton came across right- and left-handed *Physalia* in shoals according to different winds. Placing one of each kind together in the water, he observed that they moved apart at an angle of about 90°. "This", he writes,[1] "explains the preponderance of one or other kind in single catches. Assuming that the two sorts are produced in roughly equal numbers in each brood, one can envisage occasions when, for the race as a whole, this would be an advantage: for every swarm blown onto a beach by the wind, another swarm of equal numbers would be blown away from the shore." Nevertheless one form may be at a greater advantage in the north than in the south and *vice versa*. I have noticed that the southern hemisphere specimens of *Velella* shown in the photographs in Dakin's *Australian Sea Shore* are mirror images of our typical North Atlantic forms; and Totton (1954), who distinguishes the two types of *Velella* as N.W. and S.W. (according to the direction of the sail in relation to the long axis considered as north), mentions that in the *Discovery* collections there is a series of S.W. forms from the South Atlantic and a series of N.W. ones from the western end of the Mediterranean. Clearly *Velella* will also be worth investigating in this respect.

We now come to the Siphonophores which have developed in the opposite direction to the sailing *Physalia* and are rapid swimmers without a float; these belong to the sub-order Calycophorida (Gk. *kalyx*, a cup). They are the most frequent visitors to our waters and one species, *Muggiaea atlantica* (Plate 9, p. 144), turns up in the plankton in most summers, if not actually off Plymouth, then in the region off the Channel mouth. They are smaller than the other Siphonophores and are all built on very much the same plan. In most species there are two swimming bells at the head of the colony; although sometimes, as in *Muggiaea*, there is only one; then there extends a long trailing stem which may be a foot or more in length and along this, at intervals, are little groups of persons usually consisting of a gastrozooid, a medusoid gonophore and a protective bract. There may be a dozen or more of these little groups strung out along the stem. It is drawn rapidly along in an undulating fashion by the powerful jet-propulsion of the swimming bells in front. It is like a train with the locomotive pulling a long line of carriages behind it, carriages from which hang long fishing lines armed, not with hooks, but with paralysing darts for the capture of small crustaceans and other members of the plankton. In time of danger the whole stem can be quickly drawn up into the shelter of a protective hood at one side of the swimming bells. While the little medusoid persons are not themselves freed from their accompanying zooids, a curious thing happens when they are mature the covering bracts become enlarged and each little group of medusa,

[1] In a personal communication.

gastrozooid and bract, breaks away from the parent stem, so that they scatter—drifting apart, with the bract now playing the part of supporting parachute.

These Calycophorid siphonophores are really much more abundant to the west and north of our islands than is generally supposed. I have already drawn attention (p. 35) to Dr. J. H. Fraser's important observations on what he calls the Lusitanian plankton carried to these regions. In additon to the more typical forms with one or two swimming bells (such as *Praya cymbiformis, Nectopyramis diomedeae, N. thetis* and *Bassia bassensis*), he has frequently recorded the beautiful and much larger *Hippopodius hippopus*, which has many swimming bells. My sketch of this latter species, in Fig. 31 (p. 116), is made from one of his western Scottish specimens but with the trailing stem drawn extended from a photograph of the living animal taken by Mr. Mackie. In the same figure I have included a drawing of another of his northern Scottish specimens *Chelophyes appendiculata* which, however, is not one of the strictly Lusitanian forms, but a more cosmopolitan species.

The origin of the whole group of these compound and complex Siphonophores of the second main division is more of a puzzle than that of *Velella* and its allies. Some naturalists think of them as being derived from a medusa by repeated budding similar to that seen in the genus *Sarsia* (p. 102); others regard them as hydroid colonies which once had a fixed existence but then, by the development of a float, took to a drifting pelagic life and subsequently, by retaining some of the medusae to become specialized for locomotion, took to active swimming. Professor Garstang, although not producing such strong evidence as in the case of the *Velella* group, believed that the rest of the Siphonophora also owe their origin to an 'actinula-like' larva which, instead of settling down, remained afloat to form a drifting planktonic colony. We must not pursue these speculations further; much yet remains to be made out both of the natural history and the physiology of these very curious animals, whose composite nature presents us with so great a puzzle concerning their individuality.

We must pass now to the true jelly-fish which are placed in the second great coelenterate class of the Scyphozoa or Scyphomedusae (Gk. *scyphos*, a cup). Few animals exemplify a drifting life more completely than do these large jelly-fish, which we often see in summer months floating past our boat or pier. Although they are capable of a slow progression or an upward movement by a rhythmic pulsation of their umbrella-like bodies, they are nevertheless borne along by ocean currents just as parachutes are carried in the wind. Because they are drifting with the waters, they are, in spite of their large size, as much members of the plankton as the more numerous and much smaller forms. They are Coelenterates in which the sexual medusan phase predominates and grows big, while the polyp generation is reduced to a small sedentary phase, for passing the winter attached to the sea-bed

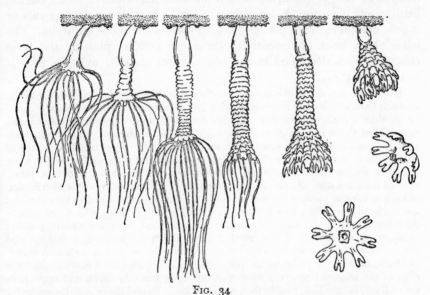

FIG. 34

The polyp (scyphistoma) stage of the common jelly-fish *Aurelia* shown producing young (ephyra stage) medusae by transverse budding (strobilation), sketched from successive photographs of living specimens taken by Dr. D. P. Wilson in the Plymouth Aquarium. It is hanging downwards from a projecting rockface.

or to rocks along the coasts. In the spring this small polyp, or *scyphistoma* as it is called, undergoes an extraordinary mode of budding, known technically as strobilation (Fig. 34, above); it becomes divided by a series of transverse grooves round the body until it looks like a pile of plates. That is very nearly what it is: a pile of small saucer-shaped medusae, which, one after the other, will break away from the top and swim off to grow in due course into large jelly-fish; actually the *scyphistoma*, as illustrated, often hangs downwards from a rock face.

There are only four kinds of these large jelly-fish which may be considered really common in our seas and they may all at times be quite numerous. *Aurelia aurita*, which is shown in Plates VII and 6 (p. 106 and 113), is the one most often seen, so we will describe this species and then note the main points in which the others differ from it. It may reach the size of nearly one and a half feet in diameter and is at once recognised by its bright purple- to pale lilac-coloured ovaries or testes which, because they line the walls of four pouches leading off the central stomach, appear as four conspicuous ovals or nearly complete ovals, grouped round the centre of the umbrella. The jelly-fish is built on essentially the same general plan as the little *Obelia* medusa described in the last chapter (p. 96), only its jelly is so much thicker.

Its central mouth, at the end of a rather short manubrium, leads up into the stomach from which open the four pouches just mentioned; the manubrium appears longer than it really is because the lips are drawn out into four long processes sometimes called the oral arms. Around the slightly lobed margin of the umbrella runs a circular canal which is connected to the central cavity by a complex system of radial tubes: unbranched and branched canals radiate alternately. While these relatively simple animals have no blood-supply, this elaborate layout of tubes, derived from the digestive cavity, plays the part of such a system in the distribution of food to various parts of the body; their inner linings are ciliated so that currents are made to flow along them. The straight unbranched tubes, which are usually magenta coloured, are carrying already digested food from the gastric pouches outwards to the periphery like arteries; those which are normally colourless and receive branches from the circular canals are carrying fluid (and no doubt waste-products) back to the centre again like veins. In this species the tentacles round the edge of the umbrella are very numerous but comparatively short, and appear, as we shall see, to take little part in the capture of food. Round the edge of the umbrella are eight sense-organs, called tentaculocysts, situated in the slight indentations between the eight lobes. Each tentaculocyst works on the same principle as the otocyst of *Obelia*, but is built on a different plan : the little calcareous nodule is enclosed tightly in a sheath of cells with sensory processes and this hangs down like the clapper of a bell within a cavity which is an outgrowth of the inner canal system and not, as in *Obelia*, a pocket formed from the outer layer. If the animal alters its position the 'clapper' swings against one side of the 'bell' and those sensory cells on that side of the 'clapper' will indicate the side towards which the medusa is tilting and stimulate the appropriate muscles to contract more strongly to bring it up again on an even keel. Against each tentaculocyst there is a small patch of pigmented light-sensitive cells and the whole little sensory complex is covered by a protective hood. No doubt the tentaculocysts also record vibrations in the water. In the very young medusa, or *ephyra* as it is called (Plate VII, p. 106), these sense-organs are very conspicuous; it was a late *ephyra* stage that I described on p. 97 capturing the barnacle nauplii by quick accurate thrusts of the manubrium.

Dr. Marie Lebour (1922–23) observed that very small *Aurelia* catch young fish as other small medusae do, but that when they reach a size of a little over an inch across they would no longer do so; on examining the stomachs of older ones she only found the remains of small plankton animals such as copepods and crab larvae. Professor Orton in the same year (1922) made the surprising discovery that even the very large adult *Aurelia* are plankton feeders. He found that they catch the plankton—copepods, oyster and barnacle larvae, young polychaete worms etc.—all over both their upper and lower umbrella surfaces in streams of mucus like catching flies on a sticky fly-paper; here, however, the fly-papers are really moving bandings—conveyor belts—of mucus driven by cilia to carry the food towards the middle of each marginal lobe of the umbrella, where, on the very edge, they form a hanging blob of food which will be licked off by one of the long tongue-like extensions to the lips of the central mouth. Recently A. J. Southward (1955) has confirmed Orton's observations and given us a more detailed study of these remarkable ciliary currents.

Jelly-fish are either male or female and the sperm from a passing male will most likely be attracted by some chemical stimulus to enter the manubrium of a female, for fertilization of the eggs is internal. The eggs develop, as in the Hydrozoa, into little ciliated larvae, the planulae, which then pass out through the mouth and may often be found swarming in prodigious numbers over the surface of the 'oral arms' and giving their margin the yellow appearance shown in Plate 6. Each little planula, if it escapes the attacks of many predators, will usually settle to the sea floor to form the next little hydroid or scyphistoma generation; occasionally, however, it has been found that a planula my develop direct into a young medusa, so cutting out the sedentary phase.

Two other of our large jelly-fish are not unlike *Aurelia* in general plan; but in addition to having a more extensive development of the lips, they have long trailing tentacles: they are *Chrysaora* and *Cyanea*. These, as Southward (*loc. cit.*) has shown, have not the same ciliary currents on the umbrella surface but depend upon their tentacles for the capture of food. *Chrysaora hyoscella*, figured in Plate 7 (p. 128), has an umbrella or disc somewhat like an inverted soup-plate, with twenty-four long tentacles dangling from its margin and four 'oral arms' of about the same length (equalling a little more than twice the diameter of the disc); the colour is white to yellow with often a reddish-brown ring round the centre of the disc and sixteen radial streaks of the same colour surrounding the ring like the spokes of a wheel surrounding its hub. It rarely exceeds 8 inches in diameter.

I will now quote Dr. Lebour's account (1923) of the feeding of a young *Chrysaora* at Plymouth; and I reproduce three of her sketches in Fig. 35 (p. 126).

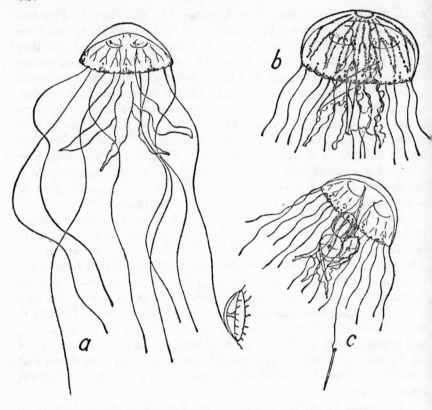

FIG. 35

Young *Chrysaora* drawn by Dr. Marie Lebour feeding in a plunger jar. *a*, catching another medusa, *Phialidium*; *b*, a larger specimen (× ½) with two young pollack enfolded in its oral arms; *c*, the same specimen as *a* but older, shown catching an **arrow** worm (*Sagitta*) and already containing *Sagitta*, *Phialidium* and *Pleurobrachia*.

"In May, 1923, a young medusa of this species was brought in from the Sound. It measured ca 25 mm. across, and was in the 8-tentacled stage, the rudiments only of the secondary tentacles being present. The tentacles and lappets that contained the sense organs were a bright chestnut brown; otherwise the medusa was colourless. This was placed in a plunger jar, and fed upon miscellaneous plankton. It would eat young fishes, medusae, Pleurobrachia, Sagitta, Tomopteris, and occasionally crustacea. Its chief food seemed to be Coelenterates and Sagitta, although it always took the small fishes when present. All these it would catch one by one with great rapidity with its tentacles, which at first would be greatly extended to their utmost capacity, so as to be many times the diameter of the

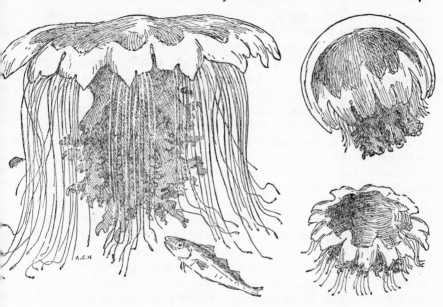

FIG. 36

Sketches showing different attitudes of *Cyanea capillata* as it swims by pulsation; the larger drawing shows an accompanying young whiting and two of the associated (probably parasitic) amphipod crustaceans *Hyperia galba*.

umbrella. As the food was caught the tentacles contracted and the lips swep off the food, which was collected in a temporary bag made by the lips below the stomach. Sometimes the food reached the stomach, sometimes not, and, as it was usually wholly or partly digested, it is possible that the digestive juices may be poured out on to it. Later, June 6th, another individual, ca 60 mm. across, also from the Sound, was introduced into the same plunger jar. These never attacked one another, although they would eat almost any other medusae. Two Aurelia, ca 55 mm. across, were put in and were both eaten by the larger Chrysaora; later two more of the same size and a few smaller were also eaten. Some Pollack (ca 25 mm.), Gunnel (ca 35 mm.), Gasterosteus (ca 30 mm.), Rockling (ca 25–35 mm.) were caught by both the Chrysaora. . ... At a slightly larger size these fishes were not eaten."

Cyanea is usually commoner than *Chrysaora* and grows very much bigger, being the largest jelly-fish known. Specimens 6 feet in diameter have been recorded from the Arctic Ocean, but in our own waters one with a diameter of 2½ feet would be a very large one and ones of 1 to 1½ feet in diameter are more usual; of course younger ones are of all sizes up to this. *Cyanea* shows a great range of colour, but there are two main kinds which may indeed be distinct species—although about this the authorities are not yet agreed. The form more usually met with is *capillata* which is drawn in its different swimming attitudes in Fig. 36 (p. 127); it varies through shades of brown to bright orange or deep red. The other kind, *lamarcki*, which is only common in the southern North Sea, ranges from deep blue to purple; it is shown in a colour drawing in Plate 7. The following general description will apply to either form. The disc is quite flat when relaxed, but inverted-basin-shaped when contracted. From the mouth the lips hang, not in arm-like processes, but to form a voluminous pleated curtain which is longer at the four corners of the mouth than in between; this is continually being raised and lowered in graceful undulations. Outside this hang a vast number of thread-like tentacles which may extend for a distance equal to four or five times the diameter of the disc. As *Cyanea* drifts in the water its tentacles, heavily charged with stinging cells, spread out, forming a wide fringe of threads for the capture of prey, almost like a giant spider-web. Sometimes a tow-net rope may get covered with these tentacles broken off from a specimen which has become temporarily caught on it; and these can cause quite a sharp sting to arms and backs of hands. When occurring in large numbers these jelly-fish can be a great nuisance to plankton workers as nearly every tow-net put out may come up containing one or more of them to spoil the rest of the catch—as I have often found in the Clyde sea area towards the end of July. There is an interesting association between young whiting and this jelly-fish as shown in Fig. 36; two or three of them may often be found sheltering underneath it and coming to no harm from it. Also associated with all four species of our large jelly-fish are to be found the amphipods (Crustacea), *Hyperia galba* (Plate 14, p. 167), with their very round bodies and large bulbous composite eyes; it is likely that these are there to steal the planktonic food of their host. I once collected as many as fourteen from one *Cyanea* landed in a tow-net and I am sure a great many more were lost in getting the jelly-fish out of the net. As suggested by Orton (*loc. cit*)

Plate 7.   MORE JELLY-FISH FROM OUR SURFACE WATERS

1. *Cyanea lamarcki*, a very dark blue specimen ($\times \frac{1}{3}$), Palling, Norfolk, August, 1954 (note *C.capillata* is similar but red-brown in colour)
2. *Pelagia noctiluca*, a visitor to our seas from the south-west ($\times \frac{3}{4}$), drawn from a preserved specimen, but after seeing it alive on the R.R.S. *Discovery*
3. *Chrysaora hyoscella* ($\times \frac{1}{2}$) Cromer, August 1954
4. *Rhizostoma octopus* ($\times \frac{1}{5}$) Millport, Firth of Clyde, October 1954

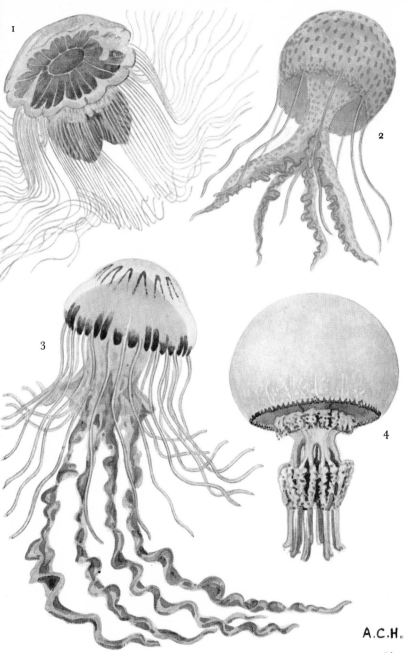

A.C.H.

Plate 7

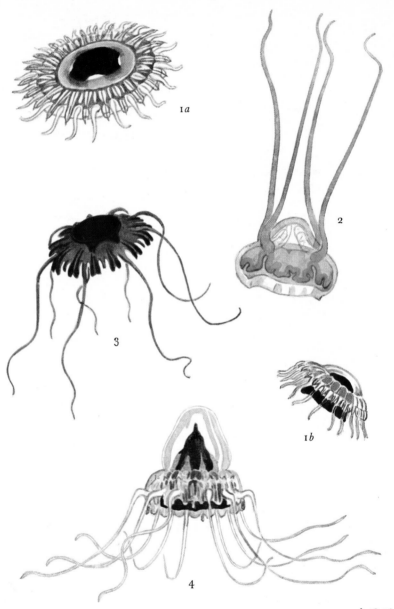

1a

2

3

1b

4

A.C.H.

Plate 8

"it is a reasonable deduction that these crustaceans may be feeding on the food collected by the jelly-fish." I had wondered if perhaps the young whiting came to the *Cyanea* to pick up these amphipods as food, but this can hardly be so, for they have been observed swimming round the medusa and when alarmed darting underneath for shelter. My friend Dr. Richard Bainbridge, on one of his underwater tours of observation in a frogman's suit, has recorded seeing a small 'gadoid fish' swimming round *Cyanea* but generally hanging in a vertical position above the bell. "It repeatedly sank towards the bell and then swam up again," he writes (1952), "the coelenterate was gently pulsating and the pair drifted obliquely upwards and eventually out of sight."

The fourth kind of large jelly-fish we may expect to encounter in our seas is very different from the other three; it is *Rhizostoma octopus*, which (together with its allied species) is unique in the animal kingdom —except for the sponges—in that it has a vast number of mouths: without exaggeration, thousands and thousands of them! It cannot be mistaken for any other jelly-fish. Above it presents a semi-opaque white dome—in shape like a young mushroom but in size like a football or bigger—with a deep purply-blue and slightly lobed margin entirely lacking tentacles; below this hangs a bunch of cauliflower-like objects which are really the greatly modified oral arms. The whole animal is shown in Plates VIII and 7 (pp. 107 and 128); but in Fig. 37 (p. 130) is a diagram to show how the margins of the oral arms have closed over and fused together during development to form a vast number of separate openings leading into little canals, which run together and join into main channels which are the original grooves of the oral arms running up into the manubrium. By a prodigious folding, twisting and fusing of the edges of these oral arms, the number of possible perforations has been enormously increased; this complex folding and the addition of a multitude of little sensory palps give the whole its cauliflower-like appearance. It used to be thought that these many mouths acted as suckers to cling on to larger prey—such as fish—and suck out nourishment; but it has since been shown that these little mouths and their canals are strongly ciliated and are often full of copepods and other small plankton animals: *Rhizostoma* is a plankton feeder, like *Aurelia*, but drawing in the water and its contained animals

---

Plate 8. JELLY-FISH FROM THE OCEAN DEPTHS
1. (*a* and *b*) *Atolla bairdii* (nat. size). 2. A narcomedusan: *Aegina citrea* (×1½).
3. *Nausithoe rubra*, (nat. size). 4. *Periphylla hyacinthina* (nat. size)

like a sponge. Closely related to *Rhizostoma* is a remarkable genus *Cassiopeia* whose species are all tropical. These have a flat disc instead of a domed top, and have developed this peculiar method of feeding to such an extent that they have become sluggish; they lazily lie on their flat backs on a shallow sandy bottom and, stretching out their shrubby sponge-like arms, suck in the plankton drifting past. From being planktonic they are becoming sedentary; their method of feeding has been well described by H. G. Smith (1936).

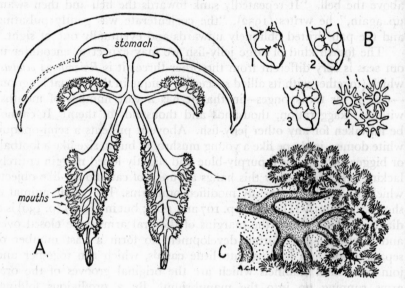

FIG. 37

A, a section through the jelly-fish *Rhizostoma* showing the system of canals running from the many mouths to the stomach. B, diagrams showing how these many mouths are formed by a complex folding and fusing together of the edges of the greatly extended lips of the original mouth. C, a drawing of a few of the mouths much magnified.

A few years ago I should scarcely have mentioned another jelly-fish because it was regarded as a great rarity in British waters: the beautifully coloured, and at night highly phosphorescent, *Pelagia noctiluca*, which is also shown in Plate 7 (p. 128). It is really a native of the subtropical seas but in recent years it has appeared on several occasions in the western Channel, and in December 1951 some sixty specimens were found along the shore of the estuary of the River Yealm in South Devon (Hunt, 1952). This very interesting record was due to Miss Agnes Russell of the yacht *Ardglass*; although not a biologist she noticed this unfamiliar jelly-fish while rowing ashore and took it to Mr. Hunt who at once searched for more and found this extensive invasion. Sir William Herdman (1923) records *Pelagia* at the Isle of Man as follows: "A small tankful of them once gave us a magnificent display in the dark at the Port Erin Biological Station, and when taken out in a bucket they looked like balls of fire, or rather incandescent metal, as the light is white and very intense. It was difficult to believe it would not

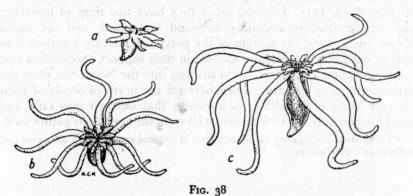

FIG. 38

Arachnactis, the planktonic larvae of the sea-anemone *Cerianthus*: *a*, a very young stage taken off Plymouth (× 10), *b*, and *c*, older stages from northern Scottish waters (× 2).

burn one's fingers when touched." Dr. Fraser (1955) has now shown the species to be at times an important element of the "Lusitanian" plankton carried to the north of Scotland as already described (p. 35). It was so in 1953. "*Pelagia noctiluca*", he writes, "increased in numbers from June onwards in this area and by early November was abundant enough to interfere with fishing operations west of Shetland. It replaced the usual jelly-fish *Aurelia*, which however was still present in those areas not affected by the inflow, for example the Moray Firth." When out on the *Discovery II* in late August 1952, we met with a great many in the region to the north east of the Azores where the specimen drawn in the plate was taken; at night we saw them passing along the side of the ship as globes of light below the surface. They gave off a phosphorescent slime when handled; after lifting specimens from a dip-net to put them in an aquarium we found our hands shining in the dark as if smeared with luminous paint. The beautiful dark red, purple and madder brown jelly-fish of the deep water beyond the continental shelf are shown in Plate 8 (p. 129) but will be referred to more particularly in discussing the life of the great depths (p. 227).

The last and third great class of the Coelenterates is the Anthozoa (Gk. *anthos*, a flower) consisting of the sea-anemones and corals. Since these exist only in the polyp form we should not expect to meet any in the plankton; nevertheless we sometimes find the young stages of the anemone *Cerianthus* in large numbers.[1] They are specially adapted for distribution by floating, much as are the actinula larvae of the hydroid *Tubularia*; they are given the name of *Arachanctis* and are shown in Fig. 38 (p. 131). Like the adult they have two rings of tentacles: the smaller ones immediately surround the mouth and the much larger ones, round the margin of the polyp, are greatly lengthened to spread out parachute-fashion to assist in their support. Sometimes vast swarms of these larvae are found drifting into the North Sea from the Atlantic in early summer; somewhere off the north of Scotland must lie vast banks of *Cerianthus*, an anemone that burrows into sand and mud and spreads out its long tentacles over the surface to gather food.

[1] There are a few floating adult anemones in tropical seas but they are never met with in our own waters.

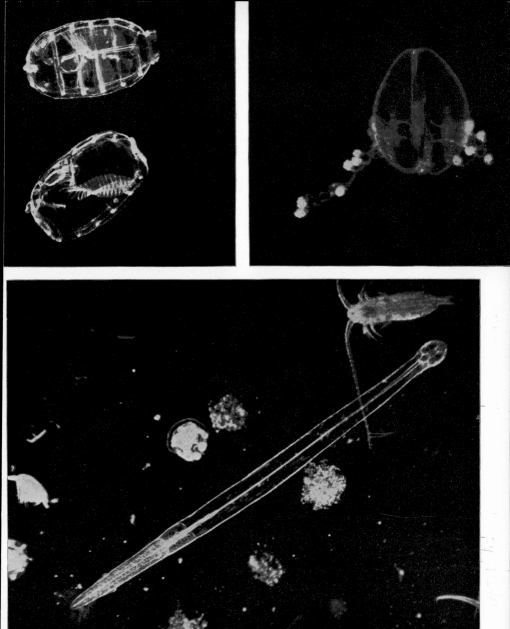

Plate IX (*top left*) *a*. Two specimens of the planktonic tunicate *Doliolum;* the upper one shows the circular muscle bands clearly, and the lower one shows the gill slits (side view), x 15. (*top right*) *b*. A very young specimen of the ctenophore *Pleurobrachia* with tentacles contracting, x 40. (*bottom*) *c*. A living arrow-worm *Sagitta setosa* in the plankton (head pointing upwards, near copepod *Calanus*), x 15. (All by electronic flash). (*Douglas Wilson*)

Plate X (*top*) *a.* The planktonic tunicate *Oikopleura*, with its tadpole-like tail, amid the needle-shaped diatoms *Rhizosolenia alata* (the white object on left is *Peridinium depressum*), x 50. (*bottom*) *b.* and *c.* The shelled pteropod *Limacina retroversa* in the act of swimming with its 'wings', x 60. (All by electronic flash). (*Douglas Wilson*)

CHAPTER 8

MORE ANIMALS OF THE PLANKTON
–BUT NOT THE CRUSTACEANS

PASSING ON from the true jelly-fish—the medusae—we shall look at the rest of the animals of the plankton group by group. In this chapter we shall deal with all those which are permanently members of this drifting community, except the Crustacea: they must have a chapter to themselves. We shall also reserve the great array of planktonic larval forms of bottom-living animals for separate treatment (in Chapter 10, p. 178) for they present some special and peculiar problems.

We take up our review with animals which might at first sight be regarded as just more jelly-fish and which are sometimes called comb jellies: the Ctenophora (Gk. *kteis, ktenos,* comb). In the older classifications they were indeed often placed as a fourth class of the Coelenterates on account of their superficial resemblance to them; actually they are so different in their organization and essential characters that they should certainly be placed in a separate main division or phylum. To emphasise this I have deliberately placed them in a different chapter; the only real character the two groups have in common is the production of a non-cellular gelatinous packing material for enlarging the size of the body without a corresponding increase in the amount of living tissue to be maintained.

The ctenophores may well be counted among the most beautiful and delicate of planktonic organisms, yet in truth, in spite of their innocent appearance, they are also to be considered among the most voracious. They sometimes occur in enormous numbers and must then seriously reduce the population of the animals upon which they prey, including the fry of many commercial fish as well as the small crustaceans upon which so many of the young fish depend for food; it has, I think, only recently been realised what a very important—not to say sinister—part these seemingly fragile creatures play in the general economy of the sea. As a result of the spring outburst of little floating plants which provide the necessary food, there follows the great increase in the animal plankton, particularly in the small crustaceans—the copepods—reaching a peak of abundance perhaps in mid-June; this is often followed by a very rapid decline which seems to coincide with a marked increase in the number of ctenophores. Their power of destruction is not surprising once we see their method of obtaining food.

I will describe the main features of the ctenophore most commonly met with: *Pleurobrachia*, or the 'sea-gooseberry' as the fishermen call it. Photographs of young and adult specimens are shown in Plates IX and VIII (pp. 132 and 107), and a diagrammatic sketch is given in Fig. 39 opposite. It will well repay study both in a small aquarium where its method of swimming and feeding may be watched, or in a small glass dish under a low-power lens where, being so transparent, the more important items of its structure may be seen quite easily. We shall then realise how radically it differs from a true jelly-fish. A full-grown *Pleurobrachia* is indeed just the shape and size of a large gooseberry. At its lower pole is the mouth, at its upper pole is a remarkable organ of balance,[1] and down the meridians, at equal distances apart, are the eight rows of beating comb-plates which give these animals their name of 'comb-bearers'. Each one of these little plates does indeed look just like a comb, as it is made up of a line of cilia fused together at their bases but free, like the teeth of a comb, at their outer ends. The comb-plates down each meridian beat rhythmically one after the other and so propel the animal forwards; if, as they beat, they catch the light, they refract it like little prisms, so that wave after wave of vibrating rainbow colours sweep in lines along the crystal-clear sides of the animal as it advances. An impression of this is shown in Plate 9, p. 144.

The balance-organ, which is enclosed within a little glass-like dome at the top of the animal, consists of a mass of tiny calcareous granules, secreted by cells and fused together into a ball; this is supported upon four little legs which are again made up of fused cilia, as indeed is the dome itself. From each of these little ciliary 'legs' run two bands of normal cilia which separate out to connect with the upper ends of the two lines of meridional comb-plates nearest to them. This little automatic stability device is so small that it can only be seen satisfactorily under the compound microscope. It is said to work thus: if the animal tilts over to one side the little ball presses upon one leg, or perhaps two, more heavily than upon the others and this causes a wave of more intensive ciliary action to pass along the bands to the corresponding comb-plates which in turn beat more violently and so bring the animal

[1] This is the usual morphological way of describing it; actually it may just as often be found swimming vertically upwards with its mouth uppermost as shown in Plate IX, p. 132.

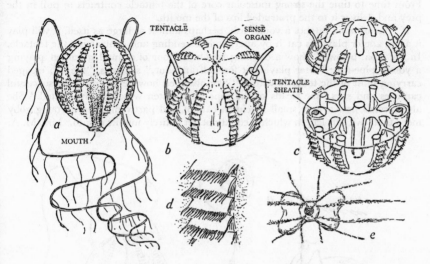

FIG. 39

The 'comb-jelly' *Pleurobrachia*. *a*, a sketch from the side; *b*, looking down on it at an angle; *c*, in same position but cut through the equator to show internal system of canals; *d*, an enlarged view of the 'comb-plates'; *e*, the organ of balance at the top of the animal seen from above (enlarged).

up on that side to correct the tilt. Actually I doubt if it is as simple as that: we do not see the animal always being brought back to a vertical position; it moves at all sorts of angles for considerable periods and no doubt the little ball-and-leg mechanism is providing its co-ordinating system with the necessary information regarding its position in relation to the vertical.

How different is the ctenophore's method of locomotion compared with that of a medusa: by ciliary plates instead of by muscular pulsation. Equally striking is the contrast in the method of capturing food. All coelenterates, as we have seen, are characterised by the possession of stinging cells; the ctenophores lack these, but have instead entirely different structures—the lasso-cells, which occur in enormous numbers on the long trailing tentacles. These tentacles which arise from little pockets on either side of the animal, stretch out to some ten times the length of its body and have an immense number of smaller branches hanging at right angles from them, as shown in Plate 9; they are heavily charged with the lasso-cells and sweep a great tract of water as the *Pleurobrachia* is propelled forwards like a paddle-steamer. Each lasso-cell consists essentially of an outer gland-cell covered with little adhesive knobs which exude a very sticky substance when touched, and two fibres which anchor it firmly to the basement membrane below the skin; one fibre is elastic and the other, coiled round it in a spiral, acts like a spring. Small crustacea, young fish and other members of the plankton are continually being caught on the adhesive cells which jerk out and spring back again as their victims try in vain to escape; as each struggles to be free it becomes more entangled with neighbouring lasso-cells

From time to time the strong muscular core of the tentacle contracts to pull in the prey and deliver it to the protruded lips of the mouth.

If *Pleurobrachia* catches a young fish, perhaps almost as large as itself, it will play it as an angler plays his catch by alternately extending and contracting the tentacle. In Fig. 40a, below, I show a drawing by Dr. Lebour of a small specimen playing a young pipe-fish; "after playing it for half an hour," she says, "the fish escaped carrying most of the tentacle with it." She observed however many other successful captures and says "the heads of the pipe-fishes eaten were usually ejected." She also records how a very small specimen caught and partly digested a young goby more than twice its length which it could not get entirely into its mouth. Mrs. M. A.

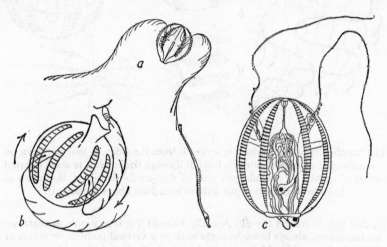

FIG. 40

Pleurobrachia feeding: *a*, capturing and 'playing' a young pipe-fish; *b*, swinging the prey (here a copeopd) towards the mouth by spinning round as shown by the arrows; *c*, a specimen with its 'stomach' full of young herrings. *a* and *c* are from drawings by Dr. Marie Lebour (1923) and *b* from a diagramatic sketch by Mrs. M. A. Connell.

Connell who has recently been making a special study of the ctenophores sends me an excellent account of how the food is conveyed to the mouth: "this is brought about", she writes, "by the rapid whirling of the Pleurobrachias towards the tentacle carrying the prey so that this is carried to the mouth in the current created by the comb-plates and then the prey is sucked off by the cilia lining the inside of the lips." She also provided me with the little sketch which I reproduce in Fig. 40b. Inside the mouth a widely distensible cavity leads to a central stomach, from which run a series of branching canals to carry food-supplies to different parts of the body: a system with quite a different arrangement of symmetry from that of the radial canals of a medusa. At times we may see the mouth-cavity filled with copepods awaiting digestion: the only conspicuous feature in the otherwise transparent and almost invisible being. In time of danger the whole tentacle may be withdrawn into the protection of the sheath at its base. The nature and origin of these tentacular muscles in the course of development are quite different from those of the coelenterates' system.

Dr. Marie Lebour, in addition to watching Pleurobrachia feed, has examined specimens from plankton samples and records that many contained several young herring at a time; one had 'at least' five in its stomach and I reproduce her sketch in Fig. 40c. In the autumn of some years, when the herring fry are swimming up into the plankton from the bottom where they were spawned off the Yorkshire coast, the waters may be swarming with these 'sea-gooseberries'; when serving as a fishery naturalist at the Lowestoft Laboratory I have at this season seen tow-nets hauled on board half-full and bulging with them. At such times these ctenophores must levy a very heavy toll upon the young herrings, and variations in their numbers from year to year must play a part in determining whether some broods of herring are more successful than others; it is possible that this may be one of the factors leading to the well-known fluctuation in the relative strength of different age-groups within a herring population. Man usually catches his herring in a great line of nets attached to a rope, the warp, let out like a tentacle trailing from his drifting vessel; it is amusing to reflect that nature, in the ctenophore, has long, long ago invented a similar method of fishing the herring—fishing however in miniature, when the fish are but half-an-inch or so in length.

Another but rarer ctenophore of our waters is *Hormiphora*, which is essentially similar to *Pleurobrachia* except that it is shaped like a pear instead of like a gooseberry; it is an inverted pear with its mouth at the narrow end. It occurs in more oceanic water. One of the most beautiful forms, and not at all uncommon, especially off the west coast of Scotland in May and June, is *Bolinopsis* (Plate VIII, p. 107), an example of the lobate ctenophores in which the body on two sides is drawn out into large lobes and the tentacles are replaced by (or more strictly, modified into) a fringe, following the margin of a ciliated groove running round on each side to the mouth. They capture the much smaller plankton, not by adhesive cells and trailing fishing lines, but by cilia-created currents along the wide surface of the lobes, which sweep in the suspended specks of life into the grooves leading mouthwards.

All ctenophores give off flashes of luminescence at night; but the lobate forms appear to be more brilliant than the rest. I think the most remarkable display of such luminescence that I have ever seen was caused by a tropical form closely related to our *Bolinopsis*. On the old *Discovery*, as we were returning in 1927 from the expedition to the Antarctic, we came back up the west coast of Africa; one night in the Gulf of Guinea, soon after leaving the island of Annobon, the whole surface of the sea was dotted with vivid flashes of light, almost as if rockets were being sent up from some under-water firework display to burst against the surface. Putting out tow-nets we collected the very broken remains of large ctenophores. We then stopped the ship and, by dipping buckets over the side, were eventually successful in securing specimens of a large and most delicate lobate ctenophore called *Deiopea*; a sudden shake of the bucket would produce a brilliant flash of light. This remarkable property of bioluminescence, common to so many members of the plankton, will be discussed in Chapter 13 (p. 248).

Last among the ctenophores to be mentioned is our largest— *Beroe*: it may be up to six inches in length. It is unfortunate that we cannot include the beautiful *Cestus* or Venus's Girdle, which is found only in warmer waters: a ctenophore in which the body is drawn out into a long transparent ribbon, a foot or more in length, with the beating comb-plates sending waves of iridescent colours chasing one another round its edges. *Beroe* (Plate 9, p. 144) which is shaped like a stout vegetable marrow, is of a delicate lavender colour, and has the characteristic eight rows of beating plates down its sides. To see these animals swimming and turning in the water is to be reminded of the manœuvring of the now almost forgotten airships—those sausage-like dirigible balloons of the early days of the century—and the lines down their sides also give them the appearance of Zeppelins. They have no tentacles and swim mouth forwards. The mouth opens into a large space, filling, indeed, almost the whole of the apparent mass of the animals, whose body turns out to be but a thin casing round this cavity; it is in fact a huge plankton-trap—huge, that is, from the point of view of its smaller victims. In our own waters it feeds almost entirely upon its smaller relatives *Pleurobrachia* and *Bolinopsis*. Dr. Lebour says "it was often seen to be full of either one or the other or both"; she also records copepods as sometimes being present as well, but these may have come from the *Pleurobrachia*. Mrs. Connell tells me it will also feed on Salps (p. 152) where they occur, as in the Mediterranean, and has given me another graphic account of its feeding. "On coming into contact with its prey", she writes, "a very rapid and wide extension of the mouth occurs followed by the almost instantaneous engulfing of the prey, rather like a vacuum cleaner. It is very spectacular and startling, especially if you have spent the previous half hour watching them swimming monotonously round the aquarium. I think the mechanism must work by a sudden reduction of pressure in the cavity caused by its rapid expansion and this is aided by the inward beating of the fused sickle-shaped cilia which line the inside of the mouth." She further tells me that she has noted an interesting succession of species, in the Clyde sea area at any rate, with *Beroe* appearing in numbers about a month after *Bolinopsis* and *Pleurobrachia*; "I think", she says, "that this is almost certainly associated with the fact thet *Beroe* feeds very largely on the other ctenophores and so does not appear until they are abundant."

The next great phylum to consider is that of the Platyhelminthes

(Gk. *platys*=platos, flat; *helmins, helminthos*, a worm). Very few of the free-living flatworms or Turbellaria are met with in the plankton; *Microstoma* however is not uncommon in the Clyde sea area. These little flat unsegmented worms glide about in all directions, swimming by means of their coating of ciliated cells and managing to keep up by offering to the water a relatively large surface area compared with their mass. Through the microscope, any tiny particles in the water near the live animal will be seen to be whirled round and round by the intensive ciliary action at its surface: hence the name of the group (L. *turba*, diminutive *turbella*, a disturbance). They are frequently seen to be undergoing asexual reproduction; at first a worm will become constricted by a transverse furrow half way down the body, and a new head-end and mouth will begin to be differentiated at the front end of the second half; then each of the two halves may be subdivided again, to give a little chain of four new individuals held together for a time. They may undergo still further subdivision, giving finally a chain of sixteen individuals; but eventually they separate to go their own divergent ways and finally proceed to sexual reproduction. The nemertine worms, which are figured in Plate 10 (p. 145), are only found in the plankton in the depths beyond the continental shelf and will be referred to in a later chapter (p. 227). Very much smaller unsegmented round-worms (Nematoda) may sometimes be found wriggling in the plankton of more coastal waters; these, and the rotifers which we may sometimes find in the sea-locks of the Western Highlands, are forms which we shall but mention, as they are not true inhabitants of the open sea.

We come now to some of the most characteristic animals of the marine plankton: the Chaetognatha (L. *chaeta*, a bristle; Gk. *gnathos*, jaw) or arrow-worms. Like the ctenophores, the chaetognaths are found nowhere else but in the plankton, and one species or another may be found in every ocean of the world. Regarding relationships, they are one of the most isolated groups in the whole animal kingdom; comparative anatomy and the study of development have so far failed to provide convincing evidence of their kinship with any of the main groups of animals, and the fossil record can do no better. The earliest known arrow-worms come from the rocks of the middle Cambrian which were laid down some 500 million years ago; yet these early forms differ only very slightly from those of today: seeing their imprint we have no hesitation in recognising them as chaetognaths. Next to the copepods, upon which they prey, the arrow-worms are the commonest of plankton animals. There are two species found immediately

Plate XI. Mixed zooplankton typical of the Atlantic influx into the northern
North Sea: part of a tow-net sample preserved in formalin, ×5. A key to the
different animals is given below. (*J. H. Fraser*)
a, anthomedusan *Hybocodon prolifera*. *b*, trachomedusan *Aglantha digitale*. *c*, arrow-
worm *Sagitta elegans*. *d*, polychaet worm *Tomopteris helgolandica*. *e*, copepod *Calanus
finmarchicus*. *f*, amphipod *Themisto gracilipes*. *g*, euphausiid *Meganyctiphanes norvegica*.
h, zoea larva of crab. *i*, larva of 'squat lobster' *Galathea*. *j*, appendicularian *Oikopleura
labradoriensis*. *k*, fish eggs (species not determined). *l*, young haddock (*Gadus
aeglifinus*). *m*, young mackerel (*Scomber scombrus*). *n*, young witch (*Pleuronectes
cynoglossus*). *o*, young herring (*Clupea harengus*). *p*, young long rough dab (*Hippo-
glossoides limandoides*).

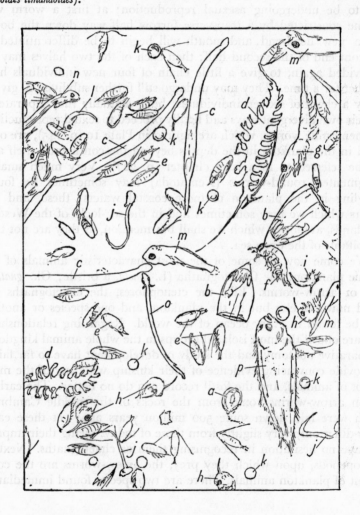

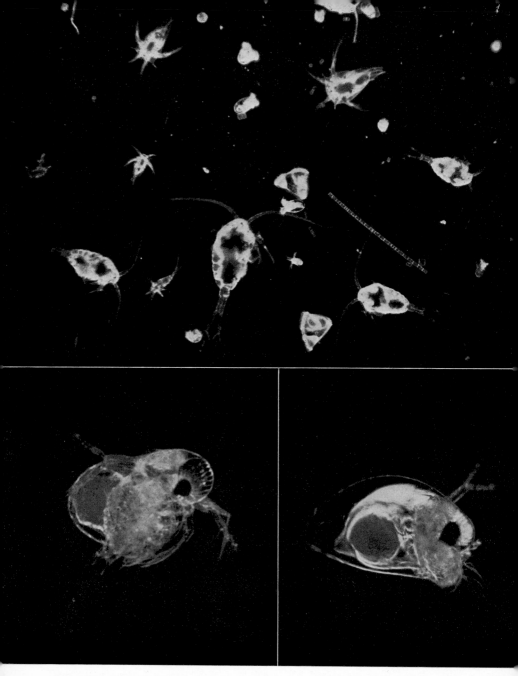

Plate XII (*top*) *a.* Living plankton including the copepod *Temora longicornis* (with forked tails), nauplii larval stages of the acorn barnacle *Balanus* (with side-pointing spines at front corners) and smaller nauplii of copepods. There are also two triangular cyphonautes larvae of a polyzoan. Note also chain of diatom cells (probably *Guinardia*), x 20. (*bottom left*) *b.* The cladoceran *Podon intermedius*, x 40. (*bottom right*) *c.* The cladoceran *Evadne nordmanni*, x 55. (*a* and *c* by electronic flash). (*Douglas Wilson*)

around our shores, each about ¾ of an inch long when fully grown: *Sagitta elegans* and *Sagitta setosa*;[1] they are shown in Plates IX and XI respectively (pp. 132 and 140). As explained in Chapter 2 (p. 29) they are characteristic respectively of oceanic and more coastal waters and, because they are so useful as indicators of these different types of water, we shall presently distinguish them. First however let us get a more general idea of what an arrow-worm is like.

Sagitta is perfectly transparent save for two very small black eyes; it is long, straight and slender like an arrow and can only be seen by slight differences in refraction between its body and the surrounding water. It is in three segments: a short head furnished on either side with stout curved bristles which act as powerful jaws, a long trunk segment, and a shorter tail-piece. There are two pairs of horizontal side-fins and the tail ends in a fin shaped like the tail of a fish, but horizontal instead of vertical.

Sagitta is another of the important carnivores of the plankton. In the general introduction to the zooplankton, p. 68, I described its characteristic way of hanging motionless in the water until some luckless copepod or other prey comes within striking distance; then it darts forward, like a torpedo, propelled by the violent up-and-down vibration of the tail. In a flash it covers a distance of some five or six times its own length; and its great jaws, opened wide, snap to upon its victim. Sometimes it will capture a young herring almost as big as itself. Dr. Marie Lebour (1923) writes as follows:

"Unfortunately it has not been possible as yet to keep Sagitta in the plunger jars for more than a day. It was, however, taken several times in the tow-nets whilst eating the larval Herrings, and preserved immediately. These it seizes with its powerful jaws at any part of the body, and usually gets the whole of the fish inside it. Sagitta is a miscellaneous feeder, its usual food most of the year being copepods, other Sagitta and young fishes. The latter seem to be specially taken in January and early February, when the young Herring are newly hatched and freely eaten by Sagitta. A large proportion of the specimens taken in any haul are empty, but so many are eating Herring that much damage must be done."

I reproduce her drawing in Fig. 41 (p. 142).

As with the ctenophores it is only their food inside them which makes the arrow-worms visible. We must not describe too much of their anatomy—the text-books can be consulted for this—but their eyes are worthy of comment. Under the simple lens they can just be made out as a pair of very small black dots on the upper surface of the head; under the higher power of the compound microscope—if the top of the head is sliced off with a fine scalpel and placed under a cover glass on a slide—they can be seen to be subdivided into five little separate retinas pointing in different directions. Much verbal description may be saved by looking at a little diagram

[1] Formerly these two species were confused with *S. bipunctata* which does not occur in British waters.

(Fig. 42, *e* and *f*) of the eye of the right side. Above we see the eye in plan as we look down on it, and below is how it would appear if we cut a vertical transverse section through it along the dotted line XY shown in the plan. Looking down on the eye, it appears to be divided into three by pigmented partitions—each subdivision has a set of retinal cells backed by black pigment: a large one (O) looks out to the side, a smaller one (P) looks upwards and forwards, and a third (Q) upwards and backwards. When, however, we examine transverse sections, as the one shown in the diagram, we find that below P and Q are similar sub-eyes, R and S, pointing downwards and forwards and downwards and backwards respectively. The animal is so transparent that its eyes can see down through its own body! They cannot be expected to form any very sharp images but they must serve to record the movements of creatures—the possibility of prey—in all directions about them. These details of the eye were first made out by Dr. S. T. Burfield who has given us such a splendid account of the animal in one of the Liverpool Biological Memoirs (1927)

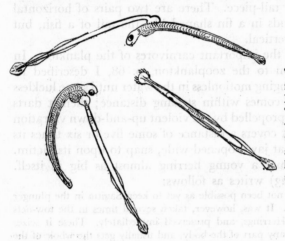

FIG. 41

The arrow-worm *Sagitta* feeding, from drawings by Dr. Marie Lebour (1923). Two are shown capturing young herring as large as themselves and a third has a partly digested young herring in its gut; note also two copepods in fore gut of one specimen.

The naturalist afloat with a tow-net and a lens may well wish to know if he is in water which has flowed in from the main ocean—and such water may extend in late summer into the North Sea from the north down to areas off the Yorkshire coast—or if he is only in the more coastal water typical of the southern North Sea or eastern Channel; all he has to do is to take a plankton haul with his net and examine the *Sagitta*. He should look at several: if they are all *S. elegans* or all *S. setosa* he is definitely in oceanic or coastal water respectively, but if he should find both then he will know that he is in one of the boundary areas of mixture.

The difference between the two species is most easily explained by reference to the drawings in Fig. 42 *a* and *b*; it partly concerns relative size, shape and position of the fins—this is easily seen—and partly the position and shape of a pair of little structures at the hind end of the body between the second pair of fins and the tail. These are part of the male reproductive apparatus—the seminal vesicles for containing ripe sperm (the animals are hermaphrodite); they are wedge-shaped in *S. setosa* and conical in *S. elegans*. If you are still not certain drop them into weak formalin; when preserved *setosa* always remains transparent whereas *elegans* very conveniently goes opaque. An important study of the growth and breeding of the two species in the western Channel has been made by Russell (1932–33) who shows that *elegans* has four, if not five, broods in the year and *setosa* at least six.

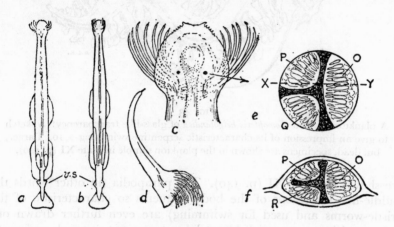

FIG. 42

The arrow-worm *Sagitta*. *a* and *b* showing the differences between *Sagitta setosa* and *S. elegans* (×2), note the position of the second pair of lateral fins in relation to the male organs *v.s* (vesiculae seminales) which are wedge-shaped in *setosa* and conical in *elegans; c,* enlarged view of the head; *d,* one of the bristle-jaws more enlarged to show attachment of muscles; *e,* the right eye highly magnified and seen from above; *f,* a section along the line XY in *e.* *e* and *f* are diagrams based on Burfield (1927).

In the more open ocean to the north of Scotland, in the North Atlantic current, several other species of *Sagitta* may be met with: particularly *S. serratodentata* and the larger *S. maxima*, which is about two inches long and has the lateral fins continuous with one another. In the great ocean depths are the beautiful orange forms *S. macrocephala* and *Eukrohnia fowleri* shown in Plate 10 (p. 145).

The great phylum of segmented worms, the Annelida (L. *annulus*; dim. *annellus*, a ring) has produced an animal most beautifully adapted to pelagic life: *Tomopteris* (Fig. 43, p. 144). It is related to the common *Nereis*, the rag-worm, one of the bristle-worms (Polychaeta); in fact it belongs to the same suborder—but how altered! Preserved specimens

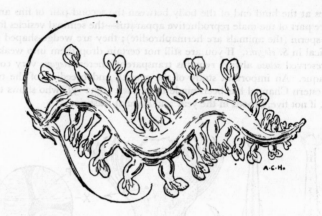

FIG. 43

A planktonic worm *Tomopteris helgolandica* of glass-like transparency: a sketch
to give an impression of its characteristic serpentine swimming × 10. Larger,
but dead, specimens are shown in the plankton sample in Plate XI (p. 140).

are shown in Plate XI (p. 140). The parapodia (in other words the
paddle-like extensions of the body segments so characteristic of the
bristle-worms and used for swimming) are even further drawn out
than usual into very much larger lobes to present as much surface to
the water as possible and to assist in the animal's support. Like
Sagitta and many other planktonic animals it is in life perfectly transparent; and likewise it is a carnivore, feeding upon other planktonic
animals including young fish. It uses its bi-lobed paddles almost like
limbs, and appears to run through the water as a centipede runs along
the ground, throwing its body into a series of rapid sideways undulations. On each side of the head are what appear to be very long
tentacles; they are in fact modified parapodia drawn out and supported
by exceedingly long bristles. In time of danger *Tomopteris* will at once
roll itself up into a ball and rapidly sink to a safer layer below. When

Plate 9. A PLANKTONIC MIXTURE: SIPHONOPHORES, 'COMB-JELLIES' AND A 'SEA-
BUTTERFLY' (PTEROPOD)

1. The ctenophore *Pleurobrachia pileus* (nat. size). Lowestoft, August 1954
2. The siphonophore *Physophora hydrostatica: a*, whole animal natural size, *b*, one
   of the swimming bells enlarged. Drawn on the R.R.S. *Discovery* in 1925
3. The siphonophore *Muggia atlantia* (× 3), from a preserved specimen from Plymouth
4. The pteropod *Clione limacina*, a young specimen (× 4), drawn at Cullercoats,
   November 1954
5. The ctenophore *Beroe cucumis* (nat. size) Lowestoft, August 1954

1

2b

2a

3

4

5

A.C.H.

Plate 9

1

2

3

4

5

6

7

Plate 10

A·C·H

caught it may often remain for a time rolled up at the bottom of the tow-net jar, and so be unrecognised; but after a time it will uncurl and run with all its elegance through the water: a thrilling sight. Between the two lobes of each parapodium is a small light-producing organ; occasionally when an animal is brought up in a tow-net at night it may be brilliantly luminous, but not often—I have only once seen it lit up. There are several species of *Tomopteris*. *T. helgolandica* is the one most commonly found in our plankton; in the depths of the ocean there is the wonderful crimson form illustrated in Plate 10, opposite.

Apart from *Tomopteris* there are some other polychaete worms to be found in the plankton, but not so specially adapted to its conditions. Several of the small Syllid worms are pelagic, being supported by enormously long bristles; some of these undergo a form of asexual reproduction by dividing into a series of smaller individuals with the production of new heads at intervals, exactly as in the flat-worm *Microstoma* already mentioned. Then, occasionally in the coastal areas, the plankton may be invaded by some of the bottom-living bristle-worms which in the breeding season develop specially large parapodia to help them swim up to scatter their eggs and sperm. The famous palolo worms of the tropics which swarm up into the plankton in their millions at certain phases of the moon—at different times of the year in different localities—provide the best examples of a lunar synchronization of breeding in marine animals; similar outbursts on a smaller scale occur among related worms in our own waters, although not in any obvious association with the moon.

Molluscs are represented in our plankton by those curiously modified sea-snails, the pteropods or sea-butterflies, whose bodies are drawn out on either side into wing-like extensions which assist in their support. It is now realized that in many species, but not all, these wing-like lobes also form an important part of the feeding mechanism; their surfaces are highly ciliated and so set up currents sending diatoms and other small organisms in their neighbourhood streaming towards the mouth. The pteropods were at one time thought to be a single natural

---

Plate 10. WORMS AND MOLLUSCS FROM THE DEEP-WATER PLANKTON

1. The arrow-worm *Eukrohnia fowleri* ($\times 1\frac{1}{2}$)
2. Another arrow-worm *Sagitta macrocephala* ($\times 1\frac{1}{2}$)
3. A nemertine worm *Bathynemertes hardyi* (nat. size)
4. Another nemertine *Pelagonemertes* ($\times 2\frac{1}{2}$)
5. The pteropod *Diacria trispinosa* ($\times 2$)
6. Another pteropod *Cio polita* ($\times 2$)
7. A remarkable crimson *Tomopteris* ($\times 4$)

FIG. 44

a-e. Outline drawings from J. E. Morton (1954), showing successive positions
of the 'wings' of the pteropod *Limacina retroversa* in swimming upwards, when
it describes a broadly spiral course, and *f* showing them held motionless and
vertical when it sinks. The figure has been slightly rearranged, and arrows
added to indicate wing movements, with Dr. Morton's kind permission and
approval. See also Plate X (p. 133).

group, some with shells and some without. It is now, however, realized
by the molluscan specialists that the two kinds have obtained their
large wing-like extensions independently along two quite separate
lines of evolution. The shelled pteropods are more commonly met
with in tropical than in temperate seas, and in some parts of the world
the ooze on the floor of the ocean contains so many of their shells,
which have sunk from the layers above, as to be called Pteropod Ooze.

Two pteropods are common in our seas, one from each group: *Limacina retroversa*
with a shell and *Clione limacina* without. *L. retroversa* is just like a little snail with
wings and has a typical snail-like shell; at its maximum size its shell is hardly 1/8 inch
in diameter and is usually much smaller. It is entirely planktonic and drifts in vast
swarms into the North Sea from the Atlantic—or off the west coast of Scotland.
Thousands may at times be taken in one tow-net haul; their shells, although very
thin, are heavy enough to make them fall at once to the bottom of tow-net jars,
to form a thick deposit when they are no longer swimming; this makes one realize
how much energy they must be expending in nature to keep themselves afloat. In
Plate X (p. 133), Dr. Wilson has caught with his camera two phases of their aquatic
'flight'. An excellent description of the general biology of *Limacina retroversa* has
recently been published by Dr. J. E. Morton (1954) and in Fig. 44 I reproduce his
sketches of the successive positions of the wing in swimming upwards.

Occasionally other more oceanic forms may come in with them into the North
Sea. On the first plankton-collecting cruise I ever made, as assistant naturalist

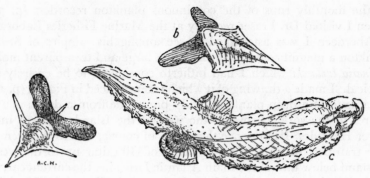

FIG. 45

a and *b* pteropods *Clio cuspidata* and *C. pyramidata* taken off the north of Scotland
× 1½; the heteropod *Carinaria lamarcki*, natural size, caught 70 miles west of
Barra, Outer Hebrides. All drawn from Dr. Fraser's specimens at the Scottish
Fishery Laboratory, Aberdeen.

in the Ministry of Fisheries, in the autumn of 1921, I came across swarms of another species of *Limacina* off the Yorkshire coast; this was identified by the late Dr. G. C. Robson of the British Museum as *L. leseuri*, which is typical of the open and more southern Atlantic and had never been recorded in the North Sea before. There were also a few young specimens of a little pteropod *Peraclis* with beautiful reticulate markings on its shell; these too had not been previously met with here. It turned out to be a year of exceptional Atlantic influx (Hardy, 1923). It was beginner's luck to come across so unusual an example of planktonic indicators of water movement; I have not heard of either of these species being recorded in the North Sea since. To the west and north of our islands two or three species of another very beautiful genus of pteropod may be taken: *Clio* with its tapering and sharply pointed shell. Two cosmopolitan oceanic species which in some years are to be found to the north of Scotland are *C. pyramidata* and *cuspidata* (Fig. 45, above). Then as part of the deeper Lusitanian fauna, already referred to (p. 35), is a most striking dark violet form, *Clio polita*, with a perfectly glass-like shell. I was lucky enough when out on the *Discovery II* to see living specimens of this and another very beautiful deep-water pteropod, the chestnut coloured *Diacria trispinosa*, which may also be taken in Scottish waters; I have included coloured drawings of them both in Plate 10 (p. 145).

Coming into the northern North Sea from the Atlantic in most years are specimens of the other type of pteropod *Clione limacina* (without a shell) shown in Plate 9, p. 144; in general appearance it is like a little slug, with pointed wing-like extensions on either side of its body. A full-grown specimen is about an inch long and pale grey, but usually much smaller ones are found, and these have their little wings and tentacles a delicate shade of salmon-pink. Like *Limacina* they are often useful indicators of the extent of the inflow of Atlantic water during the late summer and autumn as they are sampled by the monthly runs of the continuous plankton recorders (p. 311). When I visited Dr. Fraser recently at the Marine Fisheries Laboratory at Aberdeen I was astonished to see among his samples of Scottish plankton a magnificent specimen of that large and transparent mollusc *Carinaria lamarcki* which I had hitherto considered to be entirely subtropical. I made a drawing of it which is reproduced in Fig. 45 (p. 147). It came from a deep plankton haul—one made from a depth of 1,000 metres to the surface—out to the west of the Island of Barra in the Outer Hebrides. Its presence there was, of course, due to that remarkable transport of water from the Gulf of Gibraltar up to the west of Scotland below that of the main Atlantic Drift: i.e. the current carrying the Lusitanian plankton which we have already discussed. This animal is not a pteropod but one of the heteropods: planktonic snail-like animals belonging to quite a different order of gastropod molluscs. Its shell is exceedingly thin and fragile, and shaped like a little "Cap of Liberty"; in addition to being transparent, its body has become laterally flattened and provided with fins to become almost fish-like in shape in harmony with its oceanic life. It swims upside-down with the shell, such as it is, hanging below as a keel. Usually when caught they will only make sluggish movements, but on the *Discovery* I have seen one swimming quite actively round an aquarium; under natural conditions they must be capable of very rapid swimming, if only for short bursts, for Mr. B. B. Woodward in his *The Life of the Mollusca* (1913) records how he "once took from the interior of one individual six small fish, each nearly as long as their collapsed captor." He also says they attack jelly-fish.

In Chapter 7 I have already mentioned another very remarkable pelagic snail which may just be included in the British fauna. On August 26th, 1954, Dr. Wilson wrote a letter to *The Times* recording the stranding on the coast of Cornwall of a number of specimens of the beautiful oceanic *Ianthina* and asking for further records of specimens from other points on the coast. It secretes a bubble float to which it clings and so hangs upside down supported on the surface of the sea. It too has an exceedingly thin shell, but it is not transparent; it is of the purest lavender blue. And this is deep in colour on what we should expect to be the underside and pale almost to white on what would normally be the upper surface of the whorls; as, however, it hangs inverted in the water the lighter colour is underneath and perfectly counteracts the effect of shadow. Its body, mainly hidden by the shell, is deep purple. I have included this *Ianthina* in the same plate (Plate 5, p. 112) as that showing the two surface floating siphonophores *Velella* and *Physalia*, because like them it is one of the rare visitors from the subtropical seas brought here by abnormal currents and winds.

Russell and Kemp (1932) found "great numbers of fresh specimens" of *Velella* on the south-western coasts of Ireland and with them were "shells of the pelagic mollusc, *Ianthina*, with the soft parts decomposing"; they also recorded a report from Miss Delap saying that in the same period *Ianthina* was plentiful in Valentia Harbour where there were enormous numbers of *Velella*. I have recorded (p. 113) Mr. David's observations of how it feeds upon *Velella*, and as it does so, exudes a violet dye which appears to act as an anaesthetic enabling it to browse upon its victims without resistance. The bubble floats are very strong, difficult to puncture, and when preserved in formalin remain as when the animal was alive. Our species, *Ianthina janthina*, is vivparous. Another kind, *Ianthina exigua*—that more usually illustrated in natural histories of the sea—attaches a long row of egg capsules on the underside of its bubble raft; it, however, only very rarely comes into British waters, which is a pity for it has a shell of a truly vivid blue.

Leaving aside the Crustacea for the next chapter we come to the great phylum of the Chordata (having a notochord)—that to which all the vertebrates from fish to man himself belong. The floating eggs of fish and their newly-hatched fry, which cannot yet migrate against currents, must of course be classed in the category of drifting planktonic life; these however will be better dealt with when we come to consider the natural history of the fish in our subsequent volume. Plate XI shows a plankton sample containing a number of young fish and fish eggs. Also in the plankton are the tunicates, remarkable creatures which are close relatives of those curious bottom- or coastal-living animals, the ascidians or sea-squirts; these are known to be related to the vertebrates because they have little tadpole-like larvae with the most definite chordate characters.

Space cannot permit of any full account of the intricate anatomy of these very unusual animals—for this the text-books must be consulted; nevertheless the merest outline must be attempted if their true nature is to be understood. The typical simple ascidian, shaped somewhat like a sack, is enclosed within a gelatinous jacket made of a substance not unlike the cellulose of plants. At the top it has two openings: the mouth, into which water is drawn, and the "atrial opening", from which water is expelled. The mouth leads into a very large cavity—the pharynx—which has a vast number of tiny slits in its side making it almost like a basket. These slits, which open into another, outer, cavity called the atrium, are strongly ciliated so that they act as pumps; they create the water-current which passes in at the mouth, through to the atrium, and out by the atrial opening. It is a matter of great interest to zoologists that these slits are essentially of the same nature as (indeed homologous with, as they say) the gill slits of fish. Up one side of the pharynx runs a ciliated groove, the endostyle, lined by glands producing mucus; this mucus is drawn off in thin threads by cilia and passed across the basket-work on either side as a moving sticky network for the capture of minute food particles from the water passing through; ciliated lobes at the opposite side to the endostyle collect up these threads of mucus with their cargo of food into one stream, which is carried also by a band of cilia, like a continuous conveyor-belt, into the stomach. The walls of the pharynx, i.e. the 'basket-work' between the slits, are highly vascular and so serve for respiration in addition to being a feeding mechanism. With this brief sketch in mind we can now pass to consider the pelagic tunicates. One or another of the small Larvacea, such as *Oikopleura*, are so common as to be nearly always present in our tow-net samples, but they are so peculiar in form that it will be better to start with other kinds more like the type just described.

Plankton samples taken in late summer in the region of Orkney and Shetland, and down the east coast of Scotland as far south as Aberdeen, may sometimes have a number of little barrel-like objects in them: these are about half an inch long and transparent except for eight colourless or sometimes blue bands—actually muscle-strands—completely encircling the body, just like the hoops of a barrel. They may sometimes occur too at the western end of the Channel. They are specimens of *Doliolum gegenbauri* which strays into our seas from the Atlantic (Plate IX, p. 132). It moves end on through the water and feeds essentially in the same manner as an ascidian but with the shape of the body modified so that the mouth opens at the front end of the barrel and the atrium at the back; thus the greater part of the barrel is half pharynx in front and half atrium behind. The rhythmic contraction of the main hoops of muscles cause the barrel to pulsate, expelling water at the back and taking more in at the mouth as its elastic sides regain their shape; the muscle-bands at the two ends act as valves, opening and closing the mouth and atrial apertures at the right moments, so that the barrel is propelled forward as the water is pumped out in a jet behind. In the course of this process the water passes through fine gill-slits in the wall of the pharynx at its hind end and so the very small plankton organisms are sieved out for food; these slits are beautifully shown in Dr. Wilson's photograph.

Perhaps the most remarkable thing about *Doliolum* is its life-history. This would take a chapter to itself to describe in all its fascinating detail; but some indication

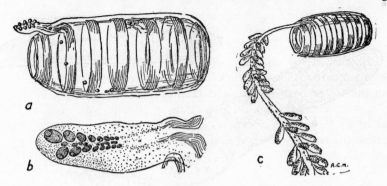

FIG. 46

The (asexual) so-called nurse form of *Doliolum: a*, a specimen brought me from Naples by Mr. Q. Bone, showing the small buds migrating from the underside to the upper appendage or cadophore, × 10; *b*, an enlarged view of a cadophore showing the buds attached in three rows; *c*, a much older specimen with a very long cadophore from which many of the developed buds have become detached. (*b* and *c* drawn from specimens in the Oxford collection). Note the absence of the gut and gill slits in the nurse form and compare it with the photographs of the sexual form in Plate IX (p. 132).

of its nature may be given in brief outline. Asexual reproduction by budding is common among the more typical ascidians. In *Doliolum* there is an alternation of generations between sexual and asexual forms. Let us start the story with the little tailed tadpole-like larva developed from the fertilized egg of the sexual form; this, when it loses its tail, becomes barrel-shaped. As it swims end on and does not rotate, we can speak of an upper and an under side; it is prevented from rotating by a little fin-like process that grows backwards from its upper hind end. From another little projection—the stolon—on its underside, it now buds off, one after the other, a number of little embryonic groups of cells which, when they have separated, creep in procession over the surface of the barrel up the right side and over the top travelling all the while towards the hind end. They are in fact marching with extraordinary precision towards the little fin-like projection at the upper hind end just referred to; it will now be convenient to call this by its technical name: the cadophore (Gk. *kados*, an urn, i.e. the body carrying the little urns or buds). The little marching buds multiply further by division and attach themselves in three rows along the cadophore: a row along the top and a row along each side. (Fig. 46, above). Meanwhile the cadophore is increasing in length. The main parent or 'nurse-form' now becomes merely a perambulator pulling the tail of buds behind it, as shown in the figure; these buds meanwhile develop into little doliolid-like creatures, but those along the top play a different rôle from those at the sides. The latter take over the business of feeding and respiration for the whole colony and develop no further, while those on the top develop into individuals which will eventually break away to live a free life: these are called phorozoids (from the Gk. *phora*. movement). Before they are liberated one or two further buds from the stolon of the old nurse settle on their stalks so that each takes away its stalk with the little buds attached

152 THE OPEN SEA

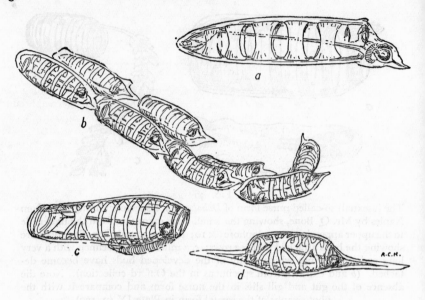

FIG. 47

Salps from British waters drawn from specimens at the Scottish Fishery
Laboratory, Aberdeen: *a* and *b*, solitary and aggregate forms of *Iasis* (*Salpa*)
zonaria, a member of the Lusitanean plankton (see p. 34); *c* and *d*, solitary and
aggregate forms of *Salpa fusiformis*. Natural size.

to it. Each phorozoid is now a sort of secondary nurse and perambulator and the
buds on the stalk undergo further secondary budding! It begins to look like one of
those endless stories that go on and on with slight variations of repetition; but no,
the end is at last in sight. The products of the last budding are eventually set free
and these become the hermaphrodite sexual individuals which will in turn produce
the eggs to be fertilized (probably by sperm from another individual) and so start
another complex round of development.

Sometimes, also in late summer, in the regions of Orkney and
Shetland one or two species of the related form *Salpa* may be found,
and they too may penetrate a little way into the North Sea; they may
also occasionally be taken at the western end of the Channel. *Salpa*,
two species of which are shown in Fig. 47 above, is larger and less
barrel-shaped than *Doliolum*; and its muscle-bands do not form continuous hoops. Mouth and atrial openings are terminal, and the water
is pumped through as in *Doliolum*, but in place of the large number
of small slits in the wall of the pharynx there is just one huge opening
on each side so that the pharynx and atrium form practically one large
chamber. Along the upper midline of the pharynx is the series of

ciliated lobes which carry the mucus network, drawn from below, in a stream towards the gullet; because the pharynx and atrium are almost one, this structure, which is the actual top of the pharynx, forms a sort of longitudinal bridge—the median bar as it is called—spanning the cavity fore and aft.

Salpa has also an alternation of generations, sexual and asexual, but nothing like so complicated a life-history as Doliolum. The sexual form is also hermaphrodite but the eggs and sperm of one individual do not ripen at the same time so that the egg must be fertilized by the sperm of another. Usually only one, or at most two or three, eggs develop at a time with the aid of a kind of placenta of tissue through which they receive nourishment from the mother. Each of these when liberated becomes the solitary or asexual form. This produces from its underside a stolon which now buds off groups of little new individuals; these remain linked together in chains which eventually break away. They are the 'aggregate' forms which in due course will separate and become the sexual salps. Salpa asymmetrica and S. fusiformis are those more frequently reaching our northern waters; of the former species both the solitary and aggregate forms are small—only about ¾ of an inch in length—whereas of the latter the solitary ones are quite large—nearly three inches long—and the aggregate ones about an inch. S. zonaria, is another large 'salp', also shown in Fig. 47, which, with Cyclosalpa, may form part of the Lusitanian fauna (p. 35) to the north of Scotland. Dr. Fraser (1949) gives an account of the occurrence of these different salps and Doliolum in Scottish waters from 1920 to 1939. Salpa fusiformis is the species which is more usually taken at the western end of the English Channel sometimes reaching the waters off Plymouth in numbers (Russell and Kemp, 1932).

Now let us come to the Larvacea, whose chief representatives, belonging to the genus Oikopleura, are, as I have already indicated, one of the commonest elements of the plankton. They are very small; the species most frequently met with has a little oval body only 1/6 of an inch in length with a little undulating tadpole-like tail about four times as long. One is seen in the photograph on Plate X (p. 133); it is shown without its gelatinous 'house' which will be described in a moment. The anatomy is relatively simple: only one pair of gill-slits lead out of the pharynx and these open directly to the outside world, for there is no atrium. The Larvacea are generally regarded as having been evolved from the little larval forms of typical ascidians by the process called neoteny. This means that by a precocious development of the reproductive organs they have become sexually mature while still in the free-living larval stage, and so have cut out the sedentary life of their past ancestors. This subject is more fully discussed in the chapter on larval forms (p. 194), for it seems to be a process that has taken place more than once in the animal kingdom, with some very surprising results. The late Professor Walter Garstang, who made a special study of this matter (1928), came to the conclusion that Oikopleura and the other Larvacea were evolved by this process, but not from the larvae of sedentary Ascidians but from those of forms like the Doliolids just discussed, which had already become pelagic.

Oikopleura, while so simple in anatomy compared with other tunicates, has developed a remarkable method of feeding. It fashions for itself a most elaborate plankton-capturing machine which rivals any of those produced by man, and catches organisms far smaller than man can catch in any form of net. From its surface it secretes an exceedingly thin, transparent and elastic envelope round itself; then, having used its tail as a tool to separate the envelope from all round the body, it sets the tail undulating to create a stream of water inside the envelope which inflates it like a large balloon around itself. This thin gelatinous 'house' (as it is usually called) is a marvel of construction; a diagrammatic sketch is shown in Fig. 48 opposite. At its hind end it has two large openings which are covered by a network of threads which act as a grid to prevent all but the smallest particles from entering; towards the front, but within the balloon-like envelope, are two conical nets of the finest threads, just like microscopic tow-nets, leading to the mouth of the animal which fashioned them. The undulations of the tail set up a strong current system within the 'balloon' drawing water in through the openings at the back, driving it round inside the envelope to pass through the meshes of the conical collecting nets and then out through an aperture at the front. As the water circulates round, the tiny specks of life, minute flagellates and protozoa, are sieved from it by the tow-nets and passed down to the mouth. At the back of the house is a little doorway through which its owner may escape in a time of crisis to swim away and then produce a new one.

This wonderful mechanism was first described by the German planktologist Dr. Lohmann; by studying what he found within that net he revealed the existence in the sea of a quite unexpected population of flagellates so small that they had never been caught or seen before. These are now often spoken of as the 'μ' flagellates: 'μ' being a symbol for a measurement of $1/1000$ of a millimetre; some of them are only 1 or 2 μ in diameter! Man has not yet succeeded in devising suitable means for the capture and estimation of these tiny members of the plankton.[1] That *Oikopleura* should have solved the problem so efficiently makes one marvel all the more at the astonishing powers of adaptation through natural selection.

[1] Since the first edition was published, Professor Yount, of the University of Florida, has kindly drawn my attention to the satisfactory use of " Millipore Filters " for collecting these small flagellates from sea water (Goldberg and Fox, *J. Mar. Res.*, *11*, 197-204, 1952).

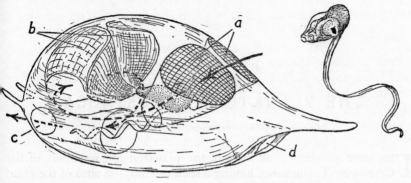

FIG. 48

Oikopleura. *Left,* in its transparent gelatinous 'house', or feeding apparatus, and *right* swimming freely. The 'house' is drawn as a perspective sketch based on the original plan and elevation drawings of Lohmann (1899): *a,* protective grids over water entrances; *b,* very fine filtering nets; *c,* water outlet; *d,* emergency exit. The arrows represent the water current on one side only.

We must not expect to see *Oikopleura* swimming in its 'house' in a plankton-sample collected by a tow-net; the 'house', which is exceedingly fragile, is always broken to pieces and usually unrecognisable, so that we find the little animal jigging about without it, quite unnaturally. It is said that under good conditions they can secrete a new one in a very short time. When you find, by using a tow-net, that there are quite a lot of *Oikopleura* in the very top layers of water, you should take samples of it by dipping in a large glass jar from your boat, or the steps of a pier, and examine it with a hand lens. It is best to let a strong light come from the top or one side and hold the jar up against a dark background. Presently, if you keep on trying, you may be rewarded by seeing the light catch the thin filmy exterior of the 'house' and the rapid undulations of the little tadpole tail within. I first saw it to perfection by chance, in such a glass jar full of water collected for another purpose, just by dipping into the sea from the steps of the pier at Millport on the Isle of Cumbrae. Several species of *Oikopleura* occur in our waters; *O. vanhoffeni, O. dioica, O. fusiformis* and *O. labradoriensis* have all been recorded from the North Sea, in addition to the less common *Appendicularia sicula* and *Fritillaria borealis,* which have broader and shorter tails than *Oikopleura.* The name Appendicularia has often been used instead of Larvacea for the whole order; this is incorrect for, as just mentioned, the name belongs to one genus of the order.

CHAPTER 9

THE PLANKTONIC CRUSTACEA

THE MOST prominent animals of the plankton are members of the Crustacea (L. *crustaceus*, having a shell or rind)—a class of the great phylum Arthropoda (animals with jointed legs). The members of the other classes, such as the insects, spiders and centipedes are essentially terrestrial animals. The crustaceans are the arthropods which swarm in the sea as the insects swarm on land. A survey of them, to be at all adequate as a guide to the naturalist meeting all this profusion of different kinds for the first time, must take a whole chapter to itself; even so, it can be no more than an outline, leaving the identification of all but the commonest for further study in the appropriate monographs.[1] Just as there are a great many orders of insects, so too are there a great many orders in the class Crustacea. To help us in distinguishing the many different kinds and allotting them to their particular groups, we must say a word or two about their classification. The older arrangement followed Latreille, who in 1806 divided the class into two subclasses: the Entomostraca (Gk. *entoma*, an insect; *ostrakon*, a shell) to include all the small 'insect-like' forms, and the Malacostraca (Gk. *malakos*, soft) for the larger more familiar types such as shrimps, prawns, crayfish, lobsters and crabs—a name which refers to the fact that in these animals the new shell is soft when it is first exposed after the old shell has been cast. While the Malacostraca undoubtedly form a true natural division, it is now realised that the Entomostraca are an artificial assembly of four distinct groups: the Branchiopoda, the Ostracoda, the Copepoda and the Cirripedia, each of which has as much right to be called a sub-class as has the Malacostraca. We will deal with them in turn and, in addition to the entirely

[1] The best systematic account for the identification of the planktonic Crustacea is the six volume work by the great Norwegian naturalist G. O. Sars: *The Crustacea of Norway* (1895 to 1928). For a full account of the classification, the morphological characters and the relationships of the different groups there is no better treatise than that by W. T. Calman (1909), who has also written an excellent more general natural history of the group: *The Life of Crustacea* (1911).

pelagic forms, will include the various larval stages of the bottom-living representatives which often form a striking feature of our plankton.

The Branchiopoda (Gk. *branchia*, gills; *pous*, *podos*, a foot) are very primitive crustaceans most of which inhabit fresh water or inland salt pools or lakes; only one of its orders has marine representatives: the Cladocera (Gk. *klados*, a branch; *keras*, a horn). The greater part of the zooplankton of lakes and ponds is made up of cladocerans—such as *Daphnia*, the common water-flea—but only a few are marine. Their name refers to the branched second antennae which stick out prominently in front of the animal and serve as the main swimming organs. All crustaceans have five pairs of typical head appendages: the first and second antennae which are usually sensory, and the mandibles, with the first and second maxillae,[1] which are modified for feeding; these are followed by a varying number of trunk and abdominal limbs. The crustaceans, of course, like all other arthropods, are enclosed within a jointed armour of the horny material, chitin (Gk. *chiton*, a tunic), secreted by the outer layer of cells; in the larger forms this may be much strengthened with salts of lime. In the Cladocera the body, except for the large antennae, may be almost entirely enclosed within a carapace, which takes the form of a bi-valve shell; it (the body) is very much abbreviated so that as a rule there are only some four or five trunk limbs. These create currents of water, and with their fringes of comb-like bristles filter out small particles of food which are carried to the maxillae and so on to the mandibles and mouth. Between the folds of the bivalve shell and the top of the body there is a large cavity serving as a brood-pouch for the developing young, which are hatched out of the large eggs at a late stage, as little miniatures of their parent.

All Cladocera possess a very characteristic single large compound eye, formed by the fusion of two such eyes, one from each side. There are only two species which are common in our plankton and they are both very small: *Evadne nordmanni* and *Podon intermedius*, which are shown on Plate XII (p. 141). Another species, *P. leuckarti*, may occasionally be taken. The two common species may be very abundant in the seas all round our coasts; but they appear to swarm in great numbers for rather short periods—generally about midsummer—and their distribution may be very patchy.

Next we come to the sub-class Ostracoda (Gk. *ostrakon*, a shell):

---

[1] The first antenna and first maxilla are often, in more technical language, spoken of as the antennules and the maxillulae respectively; the terms antennae and maxilla being used only for the second antenna and second maxilla.

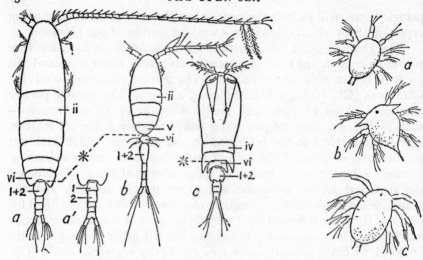

FIG. 49 (*left*). Showing how to distinguish the two main types of planktonic copepod: *a Calanus*, a typical member of the Gymnoplea, compared with *b* and *c*, *Oithona* and *Corycaeus*. members of the Podoplea. The movable articulation between the front and back divisions of the body is marked by an asterisk; the roman and arabic numerals mark some of the segments of the true thorax and abdomen. The first and second segments of the abdomen are always fused together in the females but distinct (as shown in *a'*) in the males.
FIG. 50 (*right*). Showing the similarity of the early larvae, the *Nauplius* stage, in three widely different groups of Crustacea: *a*, copepod; *b*, barnacle (*Balanus*); and *c*, euphausiid.

they are, with one exception, all very small and enclosed even more completely than the Cladocera in a bivalved shell. They are never a conspicuous feature of our plankton. They have the smallest number of appendages of all crustaceans: never more than two pairs behind the typical five of the head.[1] *Conchoecia*, which is one of the commonest of the ostracods of the sea has, in common with other members of its order (Myodocopa), the last appendage modified into a remarkable flexible organ like a miniature elephant's trunk; this can be twisted in all directions and has round its end a little group of bristles like a bottle-brush which is said to be used for cleaning the inside of the shell and the other limbs. As can be seen under the microscope, this organ is

[1] Some authorities have considered that the second maxillae are missing and that there are three pairs of thoracic limbs; this however is disproved by Professor Cannon's discovery of a *maxillary* excretory gland opening at the base of the limbs in question in *Cypris*. The excretory glands in Crustacea always open at the base of *either* the second antennae or the second maxillae.

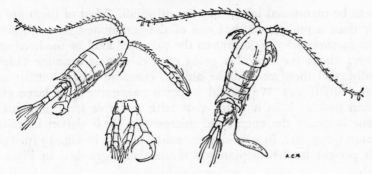

FIG. 51

Male and female of the copepod *Centropages hamatus*. Note the hinged antenna of the male for embracing the female and the last pair of his limbs (also shown enlarged) modified as a pair of forceps for attaching his sperm sac to her; the female is shown just after pairing with the sperm sac safely attached.

continually on the move and one cannot help wondering if its owner can really be using all this energy in constantly brushing out a shell which appears to be perfectly clean! Three deep-water species, includ-ing the exceptionally large *Gigantocypris*, are shown in Plate 12 (p. 161); but these will be discussed in Chapter 12 (p. 229). Incidentally, the flexible bottle-brush appendages can just be made out in the upper drawing of the younger and more transparent *Gigantocypris* in the plate.

The crustaceans of the marine plankton *par excellence* are the Cope-poda (oar-footed, Gk. *kope*, oar); not only are there a great many different species, but the number of individuals of some of the commoner kinds is beyond calculation. It has been suggested that there must be more copepods in the world than all other multicellular animals put together; I think this is probably no exaggeration. Hardly anywhere on land—not forgetting the swarming insects—can you capture as many animals in a few minutes with a small net as you can in the open sea, and more often than not some ninety per cent of these will be copepods. If we look at an average catch with a tow-net and realize that such numbers are stretching through the waters of the vast oceans for thousands and thousands of miles in every direction—then we must recognize that for sheer abundance the Copepoda have first place. Who dare estimate the probable number of these little animals in the world at any one time?

So important are the copepods in the general economy of the sea and so many are their different kinds that we must give them special attention. Except for the beautiful *Euchaeta* and some other deepwater

forms to be mentioned later, they are all small. Most of them are no
bigger than a pin's head, but one of the commonest in our waters,
Calanus finmarchicus[1]—important as the principal food of the herring—
is larger: about the size of a grain of rice. Let us examine *Calanus*
carefully, and then we shall be able to compare others with it and
note the differences. We should watch it swimming in a large glass
jar, then look at it in a watch-glass with a simple lens, and finally
examine it under the compound microscope. It is shown in colour
in Plate 11 opposite, in semi-diagrammatic outline in Fig. 49 (p. 158),
and is present in the zooplankton sample photographed in Plate II
(p. 33).

The main part of the body—the head together with what we may call a thorax
—has a long oval outline; behind this is a comparatively short and much narrower
tail-piece or abdomen with a forked ending. The first antennae are of great length
and are usually kept stretched out on either side of the body; they are provided with
feathery side-bristles which help to keep the animal poised in the water. *Calanus*
usually hangs head upwards in this fashion and would gradually sink—as it sometimes does—if it were not for the action of the second antennae, which may keep
it up, or drive it slowly upwards; when it is actively swimming the oar-like thoracic
limbs propel it forwards and if it makes a sudden spurt the first antennae fold back
to offer the least resistance. When we look at the animal's upper surface we see
five transverse divisions which mark off segments of the thorax; actually there are
six such segments, but the first is completely fused with the head. On the underside are the six pairs of thoracic limbs; the first pair are modified in relation to
feeding and the rest are typical biramous—two branched—crustacean limbs, only
flattened like paddles for swimming. In the male the abdomen has four segments
followed by the little tail-flap or telson (Gk. *telson*, extremity) with its two lobes;
these segments carry no limbs, but the tail lobes each carry five long slender feathery
bristles; in the female the first and second abdominal segments are fused together.

[1] In the waters round Britain there are two slightly different forms, *C. finmarchicus*
and *C. helgolandicus*, and it is not yet known whether they are distinct species; the
latter is more commonly found in the more southern parts of the area. Their differences and distribution are discussed by Dr. C. B. Rees (1949).

Plate 11. OUR COMMON COPEPODS, WITH A RARITY FROM LOCH FYNE (*all* ×6)
1. *Pseudocalanus elongatus*, female, carrying eggs
2. *Oithona similis*, male
3. *Acartia longiremis*, female
4. *Calanus finmarchicus*: *a*, normal male; *b*, a male parasitised by *Paradinium* which
   causes it to appear bright red (see p. 161)
5. *Euchaeta norvegica*, female, with egg sac, from the deep-water of Loch Fyne
6. *Anomalocera patersoni*, female
7. *Caligas rapax*, female, with pair of egg sacs
8. *Temora longicornis*, female; *a* and *b*, seen from the side and from above
9. *Candacia armata*, female

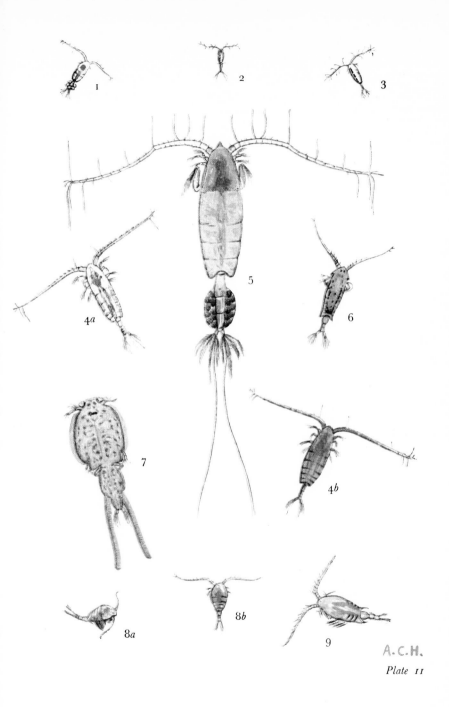

1

2

3

4a

5

6

7

4b

8a

8b

9

Plate 11

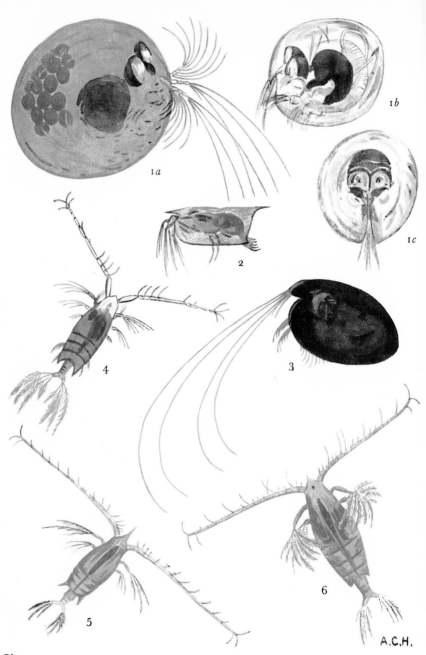

Plate 12

Calanus is beautifully transparent except for a few spots of bright scarlet pigment which may vary very much in their size and disposition, and may or may not extend into the antennae; a typical male *Calanus* is shown in Plate 11, Fig. 4*a*. Sometimes *Calanus*—in the Clyde sea-area usually about one in a hundred—may be a vivid scarlet all over, as shown in Fig. 4*b* of the same plate; this is either caused by a curious excess of pigment formation due to some disturbance caused by an internal parasite, such as small tapeworm-like creatures or protozoa, or may be due, as in the one here painted, to colour in the parasite itself—in this case the protozoan *Paradinium*.

The typical copepod eye is placed in the middle of the head and is quite different from the compound paired eyes of other Crustacea. It usually consists of three units or ocelli, each consisting of a single lens with a small group of retinal cells within a cup of dark pigment behind; and each is supplied by a separate nerve from the brain: two look forward and upwards and the third downwards. Most of the copepods feed on the very small unicellular plants of the plankton, which are carried to the mouth by currents set up by the swimming action of the second antennae and sieved out by the fine plumes on the maxillae and other feeding appendages; some however are carnivorous and may prey upon smaller copepods. While *Calanus* feeds largely on the little plants, it also takes protozoans and the young *nauplius* stages of copepods and other crustaceans.

In reproduction, internal fertilization is brought about by a remarkable device. When sexually mature the male produces from a gland a plastic-like substance with which it forms a kind of 'bottle', the spermatophore, with an exceedingly long tubular neck; it passes its sperm into this and then swims off to attach it to an opening on the first abdominal segment of a female. Some male copepods have their last pair of thoracic limbs modified into little forceps to perform this delicate operation, such as *Centropages* shown in Fig. 51 (p. 159); and quite a few, like this species and *Anomalocera* (Plate XIII) have the right first antenna hinged so as to form a prehensile clasping organ to hold the female still during the process. The sperms now pass from the spermatophore into the female and enter a little storage cavity which in turn connects with the ducts leading from the ovary;

Plate 12. OSTRACODS AND COPEPODS FROM THE DEEP-WATER PLANKTON

OSTRACODS:

1. *Gigantocypris mülleri* ($\times 3\frac{1}{2}$): *a*, adult with eggs; *b* and *c*, two views of a young and more transparent specimen
2. *Conchoecia ametra* ($\times 5$)
3. *Cypridina* (*Macrocypridina*) *castanea* ($\times 5$)

COPEPODS:

4. *Arietellus insignis*, female ($\times 5$)
5. *Gaetanus pileatus*, female ($\times 5$)
6. *Euchirella maxima*, female ($\times 5$)

(The ostracods were kindly identified for me by Dr. E. J. Isles and the copepods by Mrs. M. Fontaine) here they are stored ready to unite with the eggs as they pass down to the outside. *Calanus* extrudes its eggs one at a time into the water, but the females of most species carry their fertilized eggs attached to the underside of the abdomen, either in a simple cluster or in a pair of egg-sacs which hang one each side. The eggs are often a most beautiful translucent green or blue—like clusters of emerald or sapphire beads— or of a very deep opaque blue as in *Euchaeta* (see Plate 11, p. 160).

The copepods, like most other crustaceans, start their active life with that characteristic larva, the *nauplius* (Plate XII, c p. 141, and Fig. 50 p. 158). It is tiny, with only three pairs of limbs: the first un-branched or uniramous, the other two biramous; they are now all used for swimming but will later become the antennae and mandibles of the adult. The little body is usually kite-shaped with its point behind and characteristically has a single conspicuous eye placed towards the front. All crustaceans, because they are encased in their protective armour of chitin, cannot grow in one gradual process of increasing size: from time to time they 'cast their skins', and grow in the short interval before their new soft covering hardens; thus they grow in little spurts casting away their old armour as they do so. This is not such an extravagant process as at first sight it might appear, for a great deal of nitrogenous waste matter goes into the formation of chitin: each casting of the 'skin' is helping the body to be rid of waste-products which would otherwise have to be passed out by special organs of excretion. The little nauplius thus grows by repeatedly bursting out of its clothes. As it gets larger, rudiments of the first and second maxillae appear, followed by some of the thoracic limbs, and it is now called a metanauplius; next comes a more striking change to the first 'copepodid' stage, when it has the appearance of a miniature adult, only with fewer segments, and this is followed by four other such stages, each approaching more nearly the final adult state. In the Arctic Ocean *Calanus* is said to breed only once a year, but in our own warmer seas it has three or four generations in the same time. Typically, as at Plymouth and in the Clyde sea-area, it remains over winter in deeper water in the copepodid stage V which becomes adult and breeds in February; the next three generations become adult in May, July and September, and it is the offspring of the last which carries the race over the winter. A great deal of research has been carried out on *Calanus*, and Drs. Marshall and Orr of Millport, who have themselves done so much of it, have recently written a splendid book (1955) to bring together all the many different contributions to its natural history. The other common copepods in our waters, so far

s they have been studied, show a somewhat similar series of genera-
ions.

Now let us look at some of these other species. The true copepods[1] are divided
nto two groups: the Gymnoplea and Podoplea. The Gymnoplea are in general
imilar to *Calanus*, with an abdomen clearly without limbs, i.e. naked (*gymnos*).
The Podoplea, on the other hand, have a pair of limbs on what looks like the first
egment of the abdomen, as in *Oithona* shown in Fig. 49; this, however, is really
he sixth thoracic segment which has become rather distinct from the rest. It is
n the Podoplea that the females carry the pair of egg-sacs and not a single cluster.
Most of the Copepods that concern us in the plankton belong to the Gymnoplea;
he ones we are most likely to meet are included either in Plate 11 (p. 160), or in the
photographs in Plates XII and XIII (pp. 141, 164). Very like *Calanus* in appearance, only much smaller, are the common forms *Pseudocalanus elongatus* and *Para-
alanus parvus; Microcalanus* is also similar but still smaller. About the same size as
Pseudocalanus are the very transparent and delicate members of the genus *Acartia*
(Plate XIII) characterized by their rather spiky and feathery antennae; *A. clausi*
s a more open ocean species than *A. longiremus*, and *A. bifilosa* with a beautiful
pouch of deep blue at the end of the thorax is a brackish-water form, common at
he mouth of the Humber.

While *Calanus* is the dominant copepod of the northern North Sea and to the
north and west of our islands, it is relatively scarce in the more coastal waters of the
southern North Sea and eastern Channel; here its place is taken in the diet of the
herring by the very common *Temora longicornis*. Two views of its very characteristic
orm are shown in colour in Plate 11 (p. 160) and it is also included in the photograph
n Plate XII (p. 141); its two very long tail-prongs and its body shape—rather fat
and pear-like when seen from above, deep and arched when seen from the side—
make it easy to recognise. When alive it shows a great wealth of colour: its bright
ed eye and other red and orange pigment contrast with a distinctly blue hue of
the rest of the body—except the gut, which (as often as not) appears a bright green
from the small plants it has been eating.

Two species of *Centropages* are commonly met with: *C. typicus*, more oceanic, and
C. hamatus, more coastal; the head is somewhat square in front and the last segment
of the thorax has large hook-like spines at its two outer corners. The prehensile
right antenna of the male has already been commented on, likewise the remarkable
pincer-like last thoracic legs for handling the spermatophore. Then there is the
large and handsome bright green-blue *Anomalocera pattersoni* (Plates 11 p. 160 and
XIII, p. 164, which usually inhabits the layers of water just below the surface, and the
closely related pinky-blue *Labidocera wollastoni*: they have a pair of eyes close together
which glow red like coals of fire, a large hooked spine on each side of the head and,
n the males, a strongly prehensile right antenna. *Candacia armata* and *Metridia lucens*
are Atlantic forms coming into the northern North Sea.

The last of the Gymnoplean copepods we have space to mention is
the exceptionally large and beautiful *Euchaeta norvegica* which forms
the centre-piece of Plate 11 (p. 160). Its more usual home is in the
deep water off the coasts of Norway and in some of the fjords; it

[1] The class Copepoda is actually divided into two orders: the true copepods (the
Eucopepoda) and another very aberrant group (the Branchiura) of freshwater and
semi-parasitic forms which will not concern us.

occurs, however, in a similar habitat on the west of Scotland: particu
larly up Loch Fyne between Strachur and Inveraray, but also in th
deeps near Tarbert and off the north of Arran. Its large size, its rose-re
pigment, deep blue eggs and the rainbow colours of its iridescen
tail-plumes make it a never-to-be-forgotten animal. Loch Fyne, o
Strachur, is over 70 fathoms deep; if we let a tow-net down to some 6
fathoms we may haul up large catches of this copepod. At night w
can take it much nearer the surface, for, like so many other planktoni
animals, it then makes a marked upward migration; it is luminous an
in the dark may be seen flashing brilliantly in the tow-net bucket
These two phenomena, vertical migration and bioluminescence—stil
so much a mystery—are discussed in Chapters 11 and 13 (pp. 199; 248)
It is perhaps sacrilege to mention that I have hauled up sufficien
Euchaeta not only to serve the needs of science but also to provide
delicious addition to the supper table. Boiled in sea water for a moment
strained and then fried in butter and served on toast, *Euchaeta* is
delicacy which one day might support a small fishery to supply a luxur
market. Its deep blue clusters of eggs turn a brilliant orange when cooked
Calanus, served in the same way, also makes a pleasant shrimp paste.

Since I have introduced the topic of plankton as human food,
might here make a brief reference to the question that has been discusse
a good deal in recent years—whether it would be profitable to harves
plankton on a big scale to supplement our food-supply. In the earl
days of the last war I thought it might be possible, by using large net
swinging in the strong tidal flows of the sea lochs on the west of Scot
land, to collect sufficient plankton to be dried as a protein-rich mea
to feed to poultry and other live-stock. For two years we made
survey but failed to get plankton in large enough quantities to mak
the venture worth while. The plankton is so very uneven in its distribu
tion and such an enormous quantity of water has to be filtered to giv
sufficient yield in even a very rich area. Experimental work has als
been done in America and there too they have come to the sam
conclusion. More recently the whole question has been carefull
investigated from an engineering point of view by Mr. Philip Jackso
of the Seaweed Research Institute, who has just published an accoun
of his findings (1955). Plankton may help to save the marooned sailo
or airman on a raft from starvation but it is unlikely to support a fisher
in our own waters, except perhaps as a luxury food as I have alread
suggested. It is possible that some harvesting of plankton in the
extremely rich areas of krill—the planktonic food of whales (p. 172)—
in polar seas might prove a success; see Addendum B on p. 177.

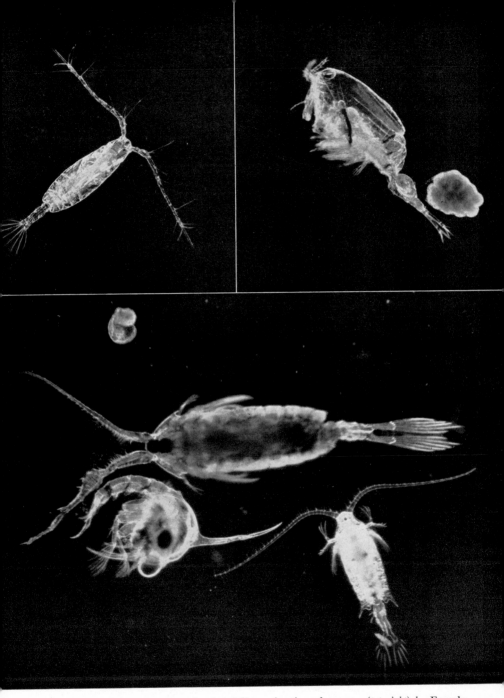

Plate XIII. Living planktonic copepods. *(top left) a. Acartia* male, x 30. *(top right) b.* Female *Corycaeus anglicus* carrying egg sac, seen from the left side; note the telescopic eyes, x 55. *(bottom) c. Anomalocera patersoni,* male (facing left) seen from the underside: note right antenna modified for clasping female; and *Centropages typicus,* female, facing upwards, showing a spermatophore which is attached to abdomen by a strand too fine to be seen; a zoea larva of a crab is also shown in the photograph, x 18. (*c* by electronic flash).   *(Douglas Wilson)*

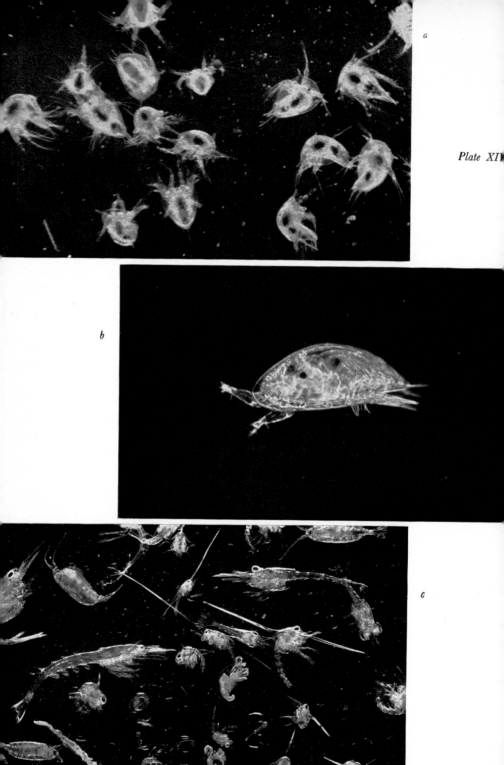

The second group of copepods—the Podoplea—are not so well represented in the plankton. The commonest is the little *Oithona similis* (Plate 11, p. 160 and Fig. 49, p. 158) which is very similar in general appearance to the familiar freshwater *Cyclops* to which it is closely related. Another which deserves special mention is *Corycaeus anglicus* shown on Plate XIII (p. 164) and also in Fig. 49; it is common in the English Channel and may get carried in considerable numbers through the Straits of Dover into the southern North Sea, where it is a good indicator of water of Channel origin. It is our only representative of a group of more tropical copepods which have very remarkable eyes. The two side elements of the typical simple eyes are widely separated on either side of the body and carried far back from the front of the head, where, however, two large lenses are developed; each of these lenses focuses on the retinal elements further back and the animal swims forwards looking like a living pair of opera-glasses! To the Podoplea also belong the large number of parasitic forms, some of which have such surprising life-histories and finish up by looking very different from a copepod or even a crustacean. They are mostly parasites on fish and spend only a limited time in the plankton, but a few, such as *Caligus rapax*, may often be found swimming freely in transit from one host to another; the females carry a pair of long egg-sacs as shown in Plate 11 (p. 160) and around the body is a delicate transparent rim which refracts light in faint rainbow colours.

The next sub-class, the Cirripedia (L. *cirrus*, a curl: i.e. with feet like curls of hair)—the barnacles—are of course only planktonic in their larval stages[1]: first a nauplius and then a little bivalve form known as the cypris stage from its superficial resemblance to the ostracod *Cypris*. We have mentioned in the introductory chapter (p. 7) how Vaughan Thompson discovered the true nature of the barnacles by finding these little larval stages. Their nauplii are always characterised by horn-like spines at the two front corners of the body; those of the little acorn-barnacles, *Balanus*, as well as its cypris stages, are characteristic of coastal plankton samples and are shown in Plate XIV, a

[1] With the interesting exception of *Lepas fasicularis* (p. 114).

Plate XIV (top) a. Nauplii larvae of the acorn barnacle *Balanus*, × 20. *(centre) n.* Later cypris stage of *Balanus*, × 80. *(bottom) c.* Mixed zooplankton containing mainly various Decapod larvae, including *Porcellana* zoea (with very long spines), crab zoea and megalopa, but also several pilchard eggs and larval worms, × 5. (All living and taken by electronic flash, but *c* narcotised.)

(Douglas Wilson)

and b, (p. 165). The nauplii and meta-nauplii of the ship's barnacle, *Lepas*, have enormously developed spines and present a most bizarre appearance.

We come now to the last and largest sub-class of the Crustacea, the Malacostraca, including all the shrimp-, prawn-, lobster- and crab-like creatures which live on the sea-bed and only send up their larvae into the plankton; in addition some groups, while mainly bottom-living, have permanently planktonic representatives and one order (Euphausiacea) is entirely pelagic. The most primitive malacostracan *Nebalia* cannot be regarded as a member of the plankton: it is an inhabitant of the rock-pools and waters of the tidal zones; it has however a planktonic relative *Nebaliopsis*, in the depths of the ocean over the edge of the continental slope; this we shall refer to in Chapter 12 (Fig. 78, p. 231). Passing over certain primitive forms confined to freshwater the first great division to concern us is the Peracarida (Gk. *pera*, a pouch; L. *caris, caridis*, a shrimp): shrimp-like crustacea that carry their eggs and young in a protective brood-pouch. There are five orders in this group, but only four concern us in the plankton: the Mysidacea (referring to *Mysis*, the principal genus), the Cumacea (Gk. *kyma*, a wave), the Isopoda (Gk. *isos*, similar, i.e. with feet of the same kind), and the Amphipoda (with two kinds of feet). Before we can compare them we must have some idea of the general malacostracan characters. We can distinguish three regions: head, thorax and abdomen. The head carries compound eyes which are sometimes but not always mounted on stalks[1]. At the risk of being a little technical, I will enumerate the typical limbs; to have them clearly set down will make subsequent descriptions easier. The head bears the usual five: first and

[1] See Addendum note on p. 177.

Plate 13. SHRIMP-LIKE CRUSTACEANS OF THE PLANKTON: MYSIDS (1, 4, 5, 6 AND 7) AND EUPHAUSIIDS (2 AND 3)

1. *Leptomysis gracilis* (G.O. Sars), adult male (×5) taken off the Lizard, August 1952
2. *Meganyctiphanes norvegica* (×2) from Loch Fyne, August 1951
3. *Thysanoessa rachii* (×2) Millport, Firth of Clyde, August 1951
4 and 5. *Anchialina agilis* (G.O. Sars), females (×5), *a*, adult carrying eggs, *b*, immature, showing remarkable range of colour characteristic of this species. Falmouth Bay, August 1952
6. *Gastrosaccus normani* (G.O. Sars), adult male (×5) taken off the Lizard, August 1952
7. *Praunus flexuosus* (Müller), female carrying eggs (×2), from Plymouth, October 1954
(The mysids were kindly identified for me by Mrs. O. S. Tattersall)

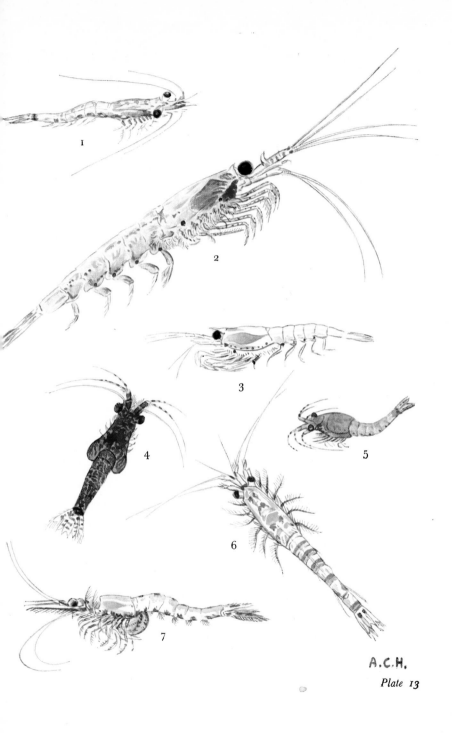

Plate 13

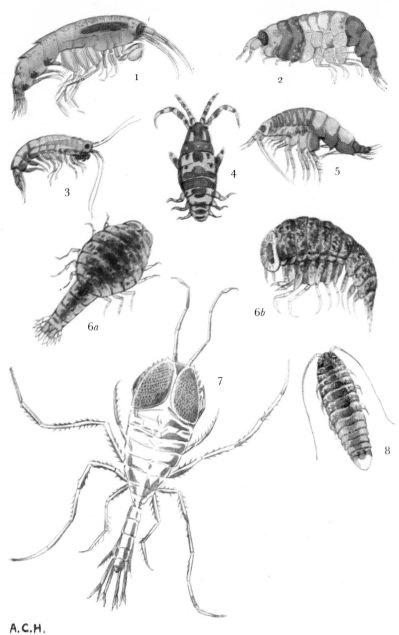

second antennae, mandibles, first and second maxillae; but unlike those of all other Crustacea, the first antennae may be biramous. The thorax is made up of eight segments, although one or more of these may be fused with the head, and it always has eight pairs of limbs; these are often uniramous walking legs, but not always, and sometimes the first pair, or the first three pairs, are modified in relation to feeding and called maxillipedes. The abdomen is usually fairly well marked off from the thorax and has six segments, each typically bearing biramous swimming limbs except the last; the last pair of limbs—the uropods as they are called—are usually flattened to form a tail-fan in combination with the terminal tail-flap or telson.

Let us begin our review of the planktonic malacostracans with those delicate little shrimp-like creatures *Mysis* and their near relatives, or "Opossum Shrimps" as they are sometimes called—because they carry their young in a pouch. While they are typically bottom-haunting creatures they are often encountered in our shallow seas swimming up to mid-water during the night and a few are entirely pelagic. Plate 13 (p. 166) shows several of these very graceful animals. Note the biramous first and second antennae, and the compound eyes standing out on quite long stalks; then see the large carapace (i.e. the jacket-like extension of the head and front part of the thorax) which reaches backwards as a cloak over the whole of the thorax but does not actually coalesce with more than three of its segments. This carapace is richly supplied with blood-channels, and is the animal's main organ of respiration—acting, in fact, like a gill.

Plate 14. MORE CRUSTACEANS FROM THE PLANKTON: AMPHIPODS (1-7) AND AN ISOPOD (8).

1. *Eusirus longipes* Boeck (× 5), this and the next three species taken off Falmouth, August 1952
2. *Lysianassa plumosa* Boeck, male(× 5)
3. *Apherusa oralipes* Norman and Scott (× 10), male
4. *Jassa marmorata* Holmes, young male, seen from above (× 7½). (This is not a strictly planktonic amphipod, but may be taken in a tow-net either accompanying floating weed or from the weed off the ship's bottom)
5. *Urothoe elegans* Bate, male (× 10) taken off the Lizard, August 1952
6. *Hyperia galba*, female (× 4), one of many taken from under a large jelly-fish (*Cyanea*) in Loch Fyne, August 1951. *a*, seen from above, and *b*, seen from side
7. *Cystosoma neptuni* (× 2) taken from 200-100 metres depth on the R.R.S. *Discovery*, September, 1954
8. *Eurydice pulcra* (× 5), off Falmouth, August 1952
   (The amphipods 1 to 5 were kindly identified for me by Mr. G. M. Spooner)

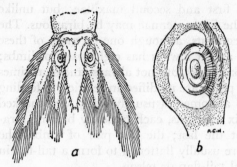

FIG. 52

a, the tail-fan of a Mysid, showing the two statocysts, × 10; *b*, the left statocyst seen from below and more highly magnified; note the nerves running from the sensory cells. The whole animal is shown in plate 13.

The thoracic limbs, except the first which functions as a maxilliped, are in a primitive biramous swimming form; from the bases of these limbs on the female are plates extending inwards to form the brood-pouch in which the young are protected until they are well-developed as miniatures of the parent. The abdominal segments of the male carry biramous swimming limbs, but those of the female are greatly reduced or sometimes absent. The tail-fan, formed, as already explained, by the uropods and telson, is well-developed and has a character peculiar to the group: a statocyst or balance-organ situated in the inner branches of the uropods. This, which is illustrated in Fig. 52 above, is in principle not unlike the organs described for various medusae. Within a little chamber is a weight, or statolith, which has an organic core surrounded by a heavy shell of calcium fluoride; this is balanced upon little sensory bristles springing from the floor of the cavity and embedded in its substance, and, as its weight presses this way or that, it keeps the animal informed as to its position in relation to gravity. There are some beautifully coloured mysids in the plankton (Plate 13, p. 166) and in the great depths beyond the edge of the continental shelf we find the bright scarlet forms which will be referred to in Chapter 12 (p. 231).

The next group, the Cumacea, we shall only briefly mention, as they are really more bottom-living than the Mysids, but, like them, tend to come up into the plankton at night; of some species only the males swim up, leaving the females burrowing in the mud below. Fig. 53 opposite shows the male of *Diastylis*; note the characteristic long, forked uropods and small telson, and also that the two sides of the carapace extend forwards so as to form a false rostrum or frontal spine. They are small animals, most of them not more than a fraction of an inch in length.

The Isopoda, looking like our common woodlice—which have so successfully invaded the land from the sea—are typically bottom- and shore-living crustaceans; but I must mention a form which from time to time may be found in a plankton sample, and would present a puzzle to the naturalist who is not familiar with it. It is *Gnathia* (=*Anceus*) a parasite of fish in its larval stages, but free-living as an adult; the male, with its enormously developed mandibles and broad head, has

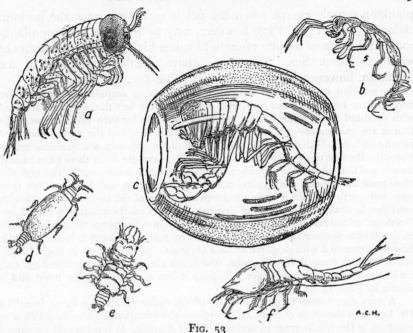

FIG. 53

A selection of Crustaceans not included in the plates: *a*, *b* and *c* are amphipods—*Themisto gracilipes* (×7½), *Phthisica acaudata* (×12), and the remarkable *Phronima sedentaria* (×2) in its "barrel of glass" (see p. 170); *d* and *e* larval and adult male of the isopod *Gnathia* (a fish parasite) (×5); *f* the cumacean *Diastylis*, male. (×7½).

a very different appearance from the female. It is shown in Fig. 53 above). The significance of this remarkable sexual dimorphism is not at all clear. They are said to feed little, if at all, in the adult condition, continuing to subsist on the food-reserves built up when they were parasitic larvae. Another isopod not infrequently found in coastal plankton is *Eurydice pulchra*, shown on Plate 14 (p. 167); in a plankton sample it may often be seen swimming at a fast speed and causing a great disturbance in the jar. There are some deepwater planktonic isopods which, instead of being red like the other crustaceans of the depths, are black like the fish.

The Amphipoda are close relatives of the isopods, but whereas the latter tend to be flattened from above downwards, these instead appear as if squeezed flat by pinching from the sides. The most typical amphipods are like our common freshwater shrimp, *Gammarus*, and belong to the sub-order Gammaridea; but few of these are to be found in a plankton-sample except when the net is used very near the bottom. *Apherusa* (Plate 14, p. 167), however, may be regarded as truly planktonic and *A. cleevei* is quite often to be taken high up in the water in the southern North Sea. The more characteristic amphipods of the plankton, however, belong to another sub-order, the Hyperidea.

In addition to the lateral flattening, the following amphipod characters are noteworthy: no distinct carapace, eyes sessile, and the first thoracic segment fused with the head (or in a few forms the second also). As for their limbs, those of the thorax are uniramous—the first pair being maxillipedes, and the second and third usually ending in little prehensile claws; those of the abdomen are biramous and in two sets—the first three pairs being multi-articulate and the other three pairs usually simple and flat (all three in fact like uropods); the telson is very small. The typically planktonic amphipods, the Hyperidea, have the characters just described, but their most noteworthy feature is their enormous eyes. One of the commonest is *Themisto gracilipes*; in some years it occurs in immense numbers off the coasts of Scotland and Northumberland and may be quite an important item in the diet of the herring caught in the Shields summer fishery. It grows to three-quarters of an inch in length and is heavily pigmented with deep red stellate chromatophores (Fig. 53, p. 169). Closely related to this form is *Hyperia galba*, which has already been mentioned (p. 128) for its association with the larger jelly-fish; it has a much rounder body and is shown in Plate 14 (p. 167).

A very rare visitor to our seas—one of the exotic Lusitanian forms found by Dr. Fraser to the north of Scotland (p. 35)—is the remarkable *Phronima* which is so common in the Mediterranean and subtropical Atlantic. It is a hyperid amphipod with huge compound eyes, but the whole animal is perfectly transparent. It attacks salps and other similar animals,[1] and from them, by eating away parts, fashions a gelatinous barrel-like house in which it shelters and rears its young. One may sometimes be seen swimming and pushing a barrel full of young ones in front of it. A specimen taken by Dr. Fraser in north Scottish waters is shown in Fig. 53 (p. 169).

The third and last main sub-order of the amphipods is the Caprellidea which contains those remarkably modified forms such as *Caprella* which feed upon hydroids. We should hardly expect to find the caprellids in the plankton; but in fact, in the shallower regions of the southern North Sea, one of these little forms, *Phtisica acaudata*, is very commonly met with. It is particularly common after rough weather when the bottom has been stirred up; but apart from such an accidental lift into the plankton, the species must at times make frequent use of the currents for dispersal.

[1] Particularly the colonial *Pyrosoma* (see p. 255).

Plate 15. PLANKTONIC LARVAL STAGES OF DECAPOD CRUSTACEANS

1. Larva of the prawn *Processa edulis*, × 10, taken with the next one at Sherringham, July 1954
2. Larva of the common shrimp *Crangon vulgaris*, × 10
3. The curious larva of *Jaxea nocturne*, with a long 'neck,' × 10, taken near Millport, Firth of Clyde, July 1954
4. Larva of *Caridion gordoni*, × 10, taken at Millport, July 1954
5. The flat and very transparent Phyllosoma larva of the crayfish *Palinurus*, × 2½

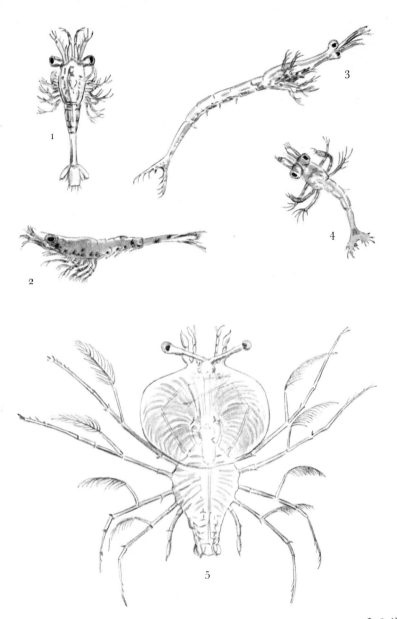

1

2

3

4

5

A.C.H.

Plate 15

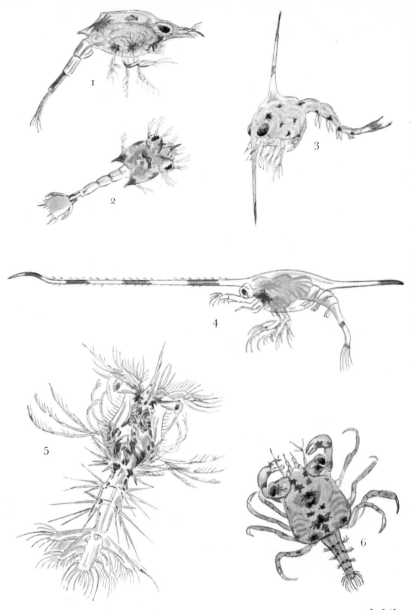

1

2

3

4

5

6

A.C.H.

Plate 16

A specimen is shown in Fig. 53 (p. 169); we see the first two thoracic segments fused with the head, next the other thoracic segments three to eight; and then—where is the abdomen? Where indeed! It looks as if it must have broken off; but no, it is vestigial, represented by a tiny little button stuck on behind. Typical amphipods, as we have seen, have abdominal limbs for swimming, but the caprellids, having taken to an 'arboreal' life—climbing little tree-like hydroids—would find the 'swimming' abdomen an encumbrance; those varieties in which it was smaller could climb and feed the better—thus can we imagine variation and natural selection bringing about its loss.

We come now to the most important malacostracan division: the Eucarida (Gk. *eu*, true, i.e. the true shrimp-like forms) containing the two closely related orders Euphausiacea (Gk. *phausis*, a shining light) and Decapoda (i.e. with ten walking legs). They have stalked eyes like a mysid but can be distinguished from this and all other forms in that the carapace coalesces with *all* of the thoracic segments. The Euphausiacea, as has already been mentioned, are entirely planktonic; there are not many species but one or another of them is to be found in all the oceans of the world. Like the arrow-worms (*Sagitta*), the various Euphausian species are very good indicators of oceanic waters of different origin. They have an appearance superficially very like that of the mysids, as will be seen in Plate 13 (p. 166) where members of the two groups have been placed together; indeed for a very long time they were classed together in one group called the Schizopoda, and will still be found referred to by this name in many books even today.

There are, however, a number of points which distinguish them, but we will only mention one or two. The carapace, already mentioned is the easiest test; when in doubt put a needle under it and try to raise it: if it comes up as a cloak showing at least five of the thoracic segments free from it, then we are looking at a mysid; if it is fused with them all we have a Euphausiacean. In the latter the carapace is not respiratory as it is with *Mysis*; instead, all its thoracic limbs bear little feathery gills on the outsides of their basal segments and these become larger and larger as one passes from the front backwards; in fact the last thoracic limb itself is vestigial—it is practically all one large plume-like gill. While the Euphausiaceans lack the statocysts so conspicuous in the uropods of a mysid, they have something else

---

Plate 16. MORE DECAPOD LARVAE FROM THE PLANKTON

1 and 2. Zoea stages in the development of a hermit crab *Anapagurus laevis* side and top views respectively, × 10
3. Zoea stage of the burrowing crab *Corystes cassivelaunus* × 10
4. Zoea of the porcellain crab *Porcellana longcornis* × 10
5. A remarkable unidentified prawn larva which is not, as it might appear at first sight, a sergestid (× 4) taken on the R.R.S. *Discovery*, September 1954 (p. 335)
6. Megalopa stages of common shore crab *Carcinus maenas* × 10
(All except No 5 were taken at Millport, Firth of Clyde)

even more striking, which the mysids lack: a wonderful set of luminous organs
These appear as prominent red spots: two thoracic pairs (at the bases of the second
and seventh thoracic limbs), four on the abdomen (one in between the limbs on each
of the first four segments) and one on the inside of each of the eye-stalks. They are
beautifully constructed with a reflecting layer behind the light-producing element
and a lens in front of it, as shown in Fig. 84 (p. 256). At times they can flash with
a brilliant light. What is their significance? We must at present admit that we are
entirely ignorant: it is one of the unsolved mysteries and will be discussed in Chapter
13 (p. 248). The thoracic limbs are biramous and none are modified as maxillipedes
the abdominal limbs are biramous little paddle-like swimmerets, except the
last—which are typically large uropods taking part with the telson to form a
tail-fan.

Our largest and most handsome Euphausiacean is undoubtedly
Meganyctiphanes norvegica, shown in Plate 13 (p. 166); it occurs in the
open ocean to the north of Scotland, but is also found in the deep
water in some of the Scottish sea-lochs such as Loch Fyne. The more
common forms in our plankton are *Nyctiphanes couchi*, which is the
one most frequently met with in the North Sea and English Channel
and *Thysanoessa rachii*, so common in the Clyde sea area, also shown
in Plate 13. *N. couchi* is a not inconsiderable item in the diet of the
North Sea herring, particularly in the late summer. It is the Euphausiaceans too which form the food of the whalebone whales—the Blue,
Fin and Humpback whales—in the polar seas; here these shrimp-
like crustacea occur in prodigious numbers—often in densely packed
swarms—and are known as 'krill' by the Norwegian whalers.

The females do not carry the eggs and young in a brood-pouch as
do the mysids. The eggs are fertilized internally by sperm which are
conveyed by the male to the female in little flask-shaped spermato-
phores similar to, but larger than, those of the Copepoda. To do this
the male uses its first abdominal limbs (not the last thoracic as in the
copepods), and these are quite extraordinary. They are provided with
a series of devices which can be unfolded like the various blades of a
complex jack-knife—devices which are prehensile, like fingers, for the
presentation and insertion of the little casket of sperms which is the
culminating copulatory act (see Fig. 54). The fertilized eggs are
scattered and hatch out as very round nauplii (Fig. 50), which develop
through[1] a metanauplius and several calyptopis and fucilia stages[1] to
become gradually adult; some of these are illustrated in Fig. 55
(p. 175), and may often be met with in our samples.

The order Decapoda contains the best known of all crustacea:
the shrimps, prawns, lobsters and crabs. They are, on the whole,

[1] The name *Cyrtopia* used to be given to the late furcilia stages, but there seems no
reasonable need for an extra name.

Plate XV. Planktonic stages in the development of the common shore crab *Carcinus maenas*: 1st and 3rd zoea (side view) and megalopa (seen from above), all living, x 40, x 35 and x 25 respectively. (*Douglas Wilson*)

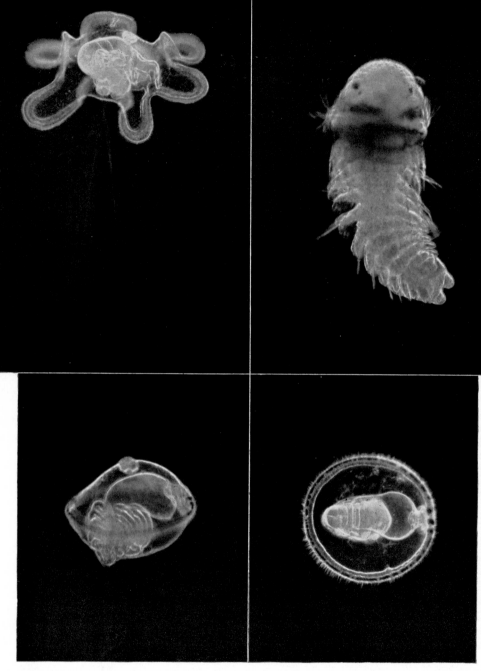

Plate XVI. Planktonic larvae of marine worms. (*top left*) *a.* The remarkable mitraria larva of *Owenia* with folded ciliated band (see account in text), x 50. (*top right*) *b.* Late larva of *Phyllodoce*, x 80. (*bottom left*) *c.* Late trochophore of *Polygordius lacteus*, side view with mouth facing to right: note ciliated prototroch (seen in plan in *d*) and invaginated and coiled up posterior part of body, x 55. (*bottom right*) *d.* The same seen from above, x 55. (All living and taken by electronic flash). *(Douglas Wilson)*

bottom-living creatures, and cannot be included as adults in the plankton of the seas immediately round our islands: they do, however, send up a host of curiously-shaped larval forms which are often most prominent objects in our tow-net samples. You will note my little qualification regarding the adults in the plankton; there are indeed a great variety of beautiful scarlet prawns in the deep water plankton out beyond the continental shelf—and these will be treated in a special chapter devoted to this bathypelagic life (p. 218). Here we will deal only with the larvae of the decapods. I am including them in this chapter, and not in that devoted to the planktonic larval forms of other bottom-living invertebrates, because it is more convenient to review them here together with the other Crustacea with which they might be confused. There is another reason too: the crustacean larvae are so different from those of all other groups of marine invertebrates; they have no cilia for one thing and they are unmistakably Crustacea from the moment they hatch. All other invertebrate larvae are quite unlike their parents and are supported in the water by various arrangements of bands of beating cilia. It is a truly remarkable fact that no crustaceans possess cilia at any time of their lives—indeed no arthropods do, except the most primitive of all: *Peripatus*; it would seem that the production of cilia, at least on the outer surface of the body, and that of a chitinous exoskeleton are achievements not compatible one with the other.

It will be a help, I think, if I tabulate a classification of the decapods for reference:

Order DECAPODA
    Suborder NATANTIA (Swimming kinds)
        Section 1 Penaeidea (prawn-like forms)
            „   2 Caridea (true prawns and shrimps)
    Suborder REPTANTIA (crawling kinds)
        Section 1 Palinura (sea-crawfish)
            „   2 Astacura (lobsters)
            „   3 Anomura (squat-lobsters, hermit crabs, etc.)
            „   4 Brachyura (true crabs)

The Natantia are the swimming shrimp- and prawn-like decapods and the Reptantia are (as adults) the walking crawling kinds. The Penaeidea are the only ones which have a complete development from a nauplius stage upwards; all the rest hatch out at a later stage—either as a *zoëa* or in an even more advanced state, the so-called *schizopod* stage. The Penaeids, however, do not concern us here: *Penaeus* is the Mediterranean prawn not found in our seas and the nearest other members

FIG. 54

An enlarged view of the remarkable hand-like appendage (shaded) found on the first abdominal limb of a male emphausiid, used for attaching a sperm sac to the female. Only part of the outer lobe of the limb is shown; the whole animal (*Meganyctiphanes norvegica*) is shown on Plate 13, p. 166.

to our coasts are the deep ocean *Sergestes* (p. 230) and the sub-tropical *Lucifer*.

The Caridea include the common shrimp *Crangon* and the edible prawns *Leander* and *Pandalus*, as well as the beautiful little *Hippolyte varians* which has such remarkable powers of chameleon-like colour change. They all hatch as a zoea. The zoea stage of *Crangon vulgaris*, the common shrimp, is shown in Plate 15 (p. 170).

The zoea always has paired stalked eyes, a small carapace and, in addition to the normal head appendages, the first three thoracic limbs; the latter are well defined as biramous swimming legs; but the remainder of the thoracic limbs are mere rudiments. Although the segments of the abdomen are defined, the abdominal limbs have not yet appeared, except the last pair—the uropods—which develop precociously to form part of the tail-fan. In the following stage—the metazoea—the rest of the thoracic limbs become bilobed, although still only in a rudimentary condition. In the following so-called schizopod stage all the thoracic limbs are well developed biramous swimming organs, which take over the main task of locomotion which has hitherto been largely performed by the second antennae; the abdomen increases in size and its appendages become functionally developed. Such a stage does at first sight look like a mysid but can only be mistaken for it in a hurried glance. The form of the antennae, the absence of the prominent mandibular palp of the mysid and the nature of the tail-fan—lacking the statocyst—all mark the difference. The zoea of the Caridea is easily distinguished from those of other decapods for it lacks the backwardly pointing spines on the hind corners of the carapace so characteristic of the anomuran zoea and likewise lacks the prominent median spine seen on the back of the brachyuran zoea; the Palinura and the Astacura both hatch out in later stages.

The larva of *Palinurus*, the sea-crawfish, and its allies, is among the most remarkable of all crustacean forms. It is illustrated in Plate 15 (p. 170). It is as flat as a piece of paper and as transparent as glass: it might almost have been fashioned from the coverslip of a microscope slide: I have dropped one on a page of fine print and been able to read every word as if the animal had not been there at all. It is called the phyllosoma larva and is really to be interpreted as a modified

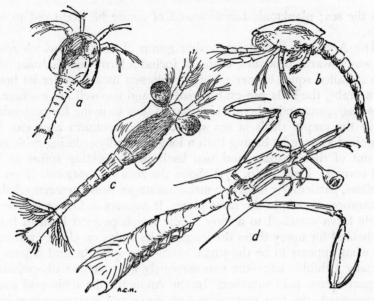

FIG. 55

Some crustacean larvae not shown in the plates: *a*, *b* and *c*, larval stages of a
euphausiid (*Thysanoessa inermis*)—i.e. first and second calyptopis and furcilia
stages, (×30, ×15, ×12 respectively); *d*, the larva of the mantis shrimp,
Squilla (×8).

schizopod stage in which the first two thoracic limbs are rudimentary
(or absent) and the next four greatly enlarged for swimming and support. The whole surface of the extremely flattened body assists in
support by offering a high frictional resistance to the water relative
to its mass. The abdomen is in a curiously rudimentary condition.
Early, newly-hatched, stages are narrower than the later, which are
very wide and round. Before taking on the adult form it passes through
another transparent but more ordinary late schizopod-like stage,
which must be of very short duration, for it is one of the rarest objects
in a plankton collection.

The Astacura, the lobsters, hatch out as a late schizopod stage and
before long become little miniature editions of the adult. The fresh-
water crayfish, it may be noted, like many secondarily fresh-water
types, has cut out a larval stage altogether and hatches from the egg
direct as a tiny crayfish; this cutting out of the free-swimming larvae
must have been a necessary preliminary to its invasion of the rivers

from the sea; planktonic larvae would of course be swept out to sea again.

The Anomura is that interesting group of decapods which seem half-way towards becoming crabs; it includes forms like *Galathea* (the little so-called squat lobster with its abdomen tucked under its body like a crab), the little pea-crab *Porcellana,* and the well known hermit crabs *Anapagurus* and others, which have taken to living for protection within the empty shells of sea snails. The anomuran zoea can be recognised at once by having both a long forwardly-pointing spine out in front of the carapace and two backwardly-pointing spines at its hind corners. Plate 16 (p. 171) shows the zoea of *Eupagurus*. That of *Porcellana*, particularly in its late metazoea stages, presents a remarkable appearance, as seen in the same plate. It appears at first as if it were a little larva attached to a long needle which projects both in front and behind for many times the length of its body; on closer examination what appears to be the single needle-like spine behind is seen to be really a double one—the two extremely long backwardly-pointing carapace spines held together. In the Anomura the schizopod stage is suppressed, the late metazoeas sink down and metamorphose into small adult-like forms.

The last decapod group, the true crabs or Brachyura, have an equally characteristic zoea, also provided with conspicuous spines: a simple rostral spine in front pointing downwards, and another pointing directly upwards from the middle of the back of the carapace; here there are no spines on the hind margin of the carapace. A typical brachyuran zoea, that of *Carcinus*, the shore crab, is shown in Plate XV (p. 172), also that of *Corystes*, the burrowing crab, on Plate 16 (p. 171). What can be the significance of the extraordinary development of these long spines: here in a vertical direction and in the *Porcellana* larva horizontal? They must assist in support by offering a greatly increased resistance to the water; but I think that a further explanation is that they have been produced by the selection of those varieties having longer and larger spines which proved to be more and more of an obstacle to would-be predators. The long spine of the *Porcellana* larva (Plate 16) is banded with red like the bar of a level-crossing gate or surveyor's pole to attract attention—no doubt a warning to keep clear. The brachyuran zoea becomes a metazoea then passes through a young megalopa stage (Plate XV, p. 172) to later megalopa stages in which the spines get shorter and shorter and the carapace, thoracic limbs and abdomen more crab-like (Plate 16 p. 171). It has now sunk in the plankton to near the bottom and finally has only to tuck in its relatively very small abdomen under its thorax and to begin crawling to become, at last, a little crab.

We end our survey of the planktonic crustacea of our seas with just a brief mention of the rather rare larval forms of *Squilla mantis*, that strange burrowing crustacean of the order Stomatopoda which is chiefly tropical in distribution. They occur in odd localities up the English Channel; and the little larvae are sometimes carried as far as the southern North Sea (Fig. 55 p. 175). Some of the tropical ones are among the most beautiful of larvae, being perfectly transparent except for the very margin of the carapace: this may be picked out in rose, purple or green—or sometimes it sparkles with a metallic gold. They have big clasp-like claws on their greatly enlarged second thoracic limbs; in the adult these are held out in front of the body, projecting outside its burrow, ready to seize any passing creature suitable for food. They are a striking parallel to the forelimbs of the praying mantis, the insect which sits on a flower head awaiting its victims with its cruel 'hands' raised as if in innocent prayer.

ADDENDUM A

I find that, inadvertently, I have done no more than mention the crusteceans' compound eyes which are such important sense organs for their life in the pelagic environment. They are not unlike those of insects, each being made up of a vast number of tiny tubular eyes massed together. These little components are not parallel to one another, but radiate so that they look out over an exceedingly wide field of view, but split it up into a mosaic of little pieces. The varying light intensity entering the little tubes will produce a pattern of light and shade on the mosaic which however crudely, will represent a picture of the outside world; its detail will, of course, depend on how many little units there are. Any movement outside will produce a change of pattern and so be detected. A crustacean can presumably, like an insect, turn towards an object by adjusting itself so that the 'image' of the object comes to occupy a corresponding but opposite position in the mosaic of each eye; it can now judge its distance from the object, for the nearer it is the closer to the mid-line will come the two 'images'.

When such compound eyes are mounted on stalks on either side of the head, as they often are in planktonic crustaceans, the two hemispheres (or sometimes more than hemispheres) of little facets will look at their surroundings in every possible direction at once. This is indeed an invaluable equipment for a plankton animal whose prey—or predators—may be above or below, as well as on any side of it.

ADDENDUM B

Since the first edition was published Dr. Isabella Gordon informs me, in relation to my remarks at the foot of page 164, that the Japanese have a flourishing fishery for two planktonic Sergestid shrimps, one of which is no larger than an average Mysid.

T.O.S.--N

CHAPTER 10

PELAGIC LARVAL FORMS

W E HAVE JUST remarked upon the great difference between the young planktonic stages of the Crustacea and those of the other invertebrates on the sea-bed. The present chapter will be devoted to a review of all these other larval forms. Bottom-living worms of many kinds, gastropod and bivalve molluscs, sea-urchins, starfish and sea-cucumbers—as well as less familiar animals such as *Phoronis* and the Polyzoa—all spend their early lives as little transparent larvae swimming in the plankton and keeping themselves aloft by the activity of powerful ciliated cells arranged in bands of varying design. They often occur in vast numbers off our coasts and may easily be collected with a fine tow-net for examination under the microscope; some are little waltzing spheres with just simple ciliated girdles, others have these girdles duplicated or drawn out into wing- or arm-like processes.

The study of these little forms is one of the most fascinating in the whole field of marine zoology. What do these early stages really represent ? We find that many groups of different animals have rather similar larvae. Until comparatively recently many zoologists believed that these various groups of animals were most likely all descended from little ciliated creatures similar to the larvae, that once existed as *adults* swimming in the plankton of some far-off pre-Cambrian sea. Surely, they have said, we can well imagine the different stocks evolving from such pelagic ancestors, as new varieties arose which were able to leave the over-crowded open waters to colonize the sea-bed or coastal regions; they would diverge along their separate paths of adaptive modification to become the highly differentiated invertebrates of today. If we see such a transformation from a planktonic to a benthic habit in the span of a single life-time, could it not also have taken place in the long and gradual evolution of the race ? Is not this indeed the story their life-history is telling us ? Is it not likely, too, that earlier still these little engirdled forms were in turn evolved from yet simpler little jellyfish-like ancestors one step further back and nearer to the distant dawn of life ? Such were the views of those who believed in the theory of recapitulation, or as some would prefer to call it, follow-

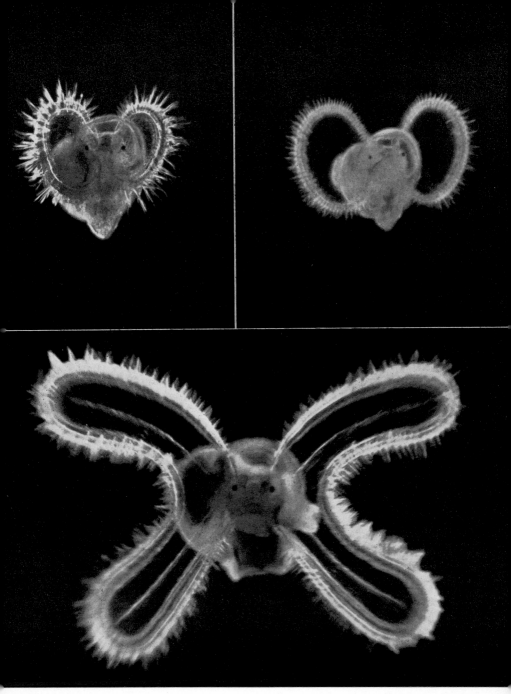

Plate XVII. Planktonic veliger larvae of some gastropod molluscs photographed alive by electronic flash. (*top left*) *a. Rissoa sarsii*, x 60. (*top right*) *b.* Another species of *Rissoa*, x 60. (*bottom*) *c. Nassarius incrassatus*, x 40. (*Douglas Wilson*)

Plate XVIII. Various planktonic larvae, all living. (*top left*) *a.* Young stage of the razor shell *Ensis,* x 190. (*top right*) *b.* Cyphonautes larva of a polyzoan, x 60. *c, d* and *e.* Three stages in the development of the actinotrocha larva of *Phoronis,* x 40. (*a, c* and *e* by electronic flash.) (*Douglas Wilson*)

ing Haeckel, the Biogenetic Law. Let us remind ourselves of the force of this argument. How reasonable it seemed that the animal during the course of its individual development from egg to maturity should be passing through, in broad outline, the stages it had passed through in the long history of its race in past aeons of time. Could not development be looked upon as a kind of speeded up continuous modification of past ancestral adult forms ? Imagine a photograph of each adult in the same position, generation after generation, arranged in sequence of time from the beginning of life to the present day, and imagine them joined one after the other like consecutive shots in a cinema film; now if such a film was projected at speed, would it not give a representation of the development of the modern adult from the egg ? Evolution was thought of primarily as a modification of *adult* forms. What else was worth considering, since it was, of course, the adults that gave rise to each new generation ? Were not such new modifications in the adult to be thought of as something added to the pre-existing form ? Did not evolution consist in always adding something new to the adult end of development, and was not development gradually compressed, or speeded up, to make room for each new addition ? Early in the century, when our knowledge of genes and their modification was not what it is today, the majority of zoologists had some such faith. The late Professor MacBride was perhaps the champion of this view; in the introductory chapter in his *Text-book of Invertebrate Embryology* (1914), which deals largely with these planktonic larvae, he wrote as follows:

"This theory is the so-called *fundamental law of biogenetics*, and is summed up in the phrase '*the individual in its development recapitulates the development of the race.*' If this 'law' can be substantiated the interest in embryology becomes immense, it binds all the innumerable phenomena of development into one coherent scheme, and opens the door to the hope that we may yet be able to sketch the main history of life on earth."

At the end of the book he summed up his conclusions thus:

"The first and most far-reaching conclusion we may draw, is that, in general, the *larval phase* of development represents a former condition of the *adults* of the stock to which it belongs. This, in substance, is of course the *recapitulatory* theory of development, the famous biogenetic law of Haeckel."

The late Professor Walter Garstang also made a special study of these larval forms; it is a great pity he never produced the book upon them that he always intended to write. His ideas were put forward in important but scattered papers in different scientific journals or, for the private amusement of his zoological friends and pupils, set out in comic verse.[1] I shall later quote an example of his wit and skill in

[1] A little volume of his verses entitled *Larval Forms* has recently been published by Blackwells of Oxford.

rhyme, for in truth he often made his points with greater force in his light-hearted verse than in his more technical scientific prose. As an undergraduate from Oxford attending the class for students at the Plymouth Laboratory in the Easter vacation of 1920 I had the good fortune to hear Garstang give a lecture on his views as to the nature of these larval forms. I vividly remember it and I am frankly basing this chapter upon it as nearly as I can; nowhere else have I heard or read such a useful bringing together of the main facts of planktonic larval life as he presented them, in a most informal manner, to a group of students sitting round the laboratory. Did these planktonic larval stages represent a primitive ancestral pelagic *adult* type, or did they represent special larval adaptations ? He left one in no doubt that the second view was correct. With rapid sketches on the board he made a brilliant survey of the larvæ of many different groups and showed how in each there is a compromise and adjustment between two rival needs—or in other words, two competing selective advantages: on the one hand to grow up as soon as possible so as to reproduce the species, and on the other to remain floating as long as possible so as to distribute the species over the largest areas. He showed clearly how the larval stages are as much adapted for dispersal as are the seeds or fruits of plants; he also showed how, with all manner of devices in different groups, these two rival needs are met. He was one of the first to realize that selection acts just as powerfully upon the young stages in development as upon the adult: that in fact the developmental stages may—nay, must—become adapted to their particular mode of life just as much as the adults are to theirs. He realized that the young, varying as much as the adults, may be modified in quite a different direction if they inhabit different surroundings. Any such great contrast between the adaptations of the young and adult forms has, of course, been accompanied by the evolution of a metamorphosis or sudden change of form in the life-history which alone makes the transition from one to the other possible.

For distribution by the moving waters certain characters are necessary. They must be floating, therefore it will be best if they are both small and light; the smaller the body, we must remember, the greater is its surface area compared with its mass, and so the greater is its frictional resistance to sinking. As it is an advantage to be small, it is the earliest stages which are likely to be selected for the purpose. As they are part of the plankton near the surface it will be an advantage for them to be as transparent as possible in order to avoid being seen by predators; as with so many permanently planktonic animals, the more transparent varieties will tend to survive better then others. From our study of other animals we see that cilia are the simplest and most efficient engines of locomotion for very small animals. It is not surprising then that the larvæ of so many different invertebrate animals are planktonic, small, transparent and ciliated. I am not suggesting that they have all arisen independently: the study of early development indeed indicates clearly that some of them have been derived from the same stock. The fact, however, that two groups of larvæ appear to be related does not mean that they were descended from the same ancestral *planktonic adult* type—but that both are descended from a group which had evolved a planktonic larval form for dispersal before the two stocks diverged.

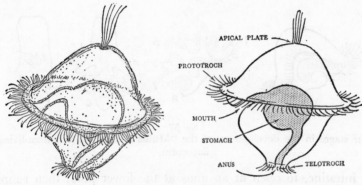

FIG. 56

Left, a sketch of a simple trochophore larva as found among the less specialized polychaete worms; *right,* a diagrammatic version of it for comparison with the series of similar diagrammatic representations of other larvae which follow. In each the gut will be shaded.

We have also seen that natural selection will always be tending, other things being equal, to preserve the races which grow up to reproductive age more quickly. It is the compromise and adjustment between these two needs—that of remaining afloat and that of becoming adult—that makes these little larvæ such interesting objects of study. Again and again in all sorts of different ways we see evolution arriving at a balance between the two; we shall now make a rapid survey to see how it has been done.

If Garstang in his studies has destroyed any faith we may have had in the concept of recapitulation, he has, I believe, given us in its place a more important evolutionary conception, the significance of which has not yet perhaps been fully grasped by more than a very few

zoologists. But this, one of his later ideas, I must reserve for the end
of the chapter; I just mention it now so that any who feel they are
losing their faith in what they may have regarded as a well established
law, may look forward to finding in its place something which I, for
one, think is much more exciting.

The larval type with which we may conveniently compare most
others is the trochophore (Gk. *trochos*, a wheel or hoop), sometimes
called the trochosphere; it is the larva of the more primitive polychaete worms and molluscs.[1] It is almost a sphere (see Fig. 56, p. 181).
At its upper pole it has a little tuft of long cilia and sensory cells—the
so-called apical sense-organ; at a point on its equator is the mouth
leading into the gut which passes via a little round stomach and a

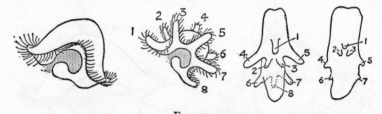

FIG. 57
Four stages in the development of the "Müller's larva" of a bottom-living
flat-worm.

short intestine, to open at an anus at the lower pole; then running
round the sphere just above the equator (and mouth) is the main
ciliated girdle, the prototroch or pre-oral band as it is variously called.
The cilia give the little sphere a spinning motion like a top; this sends
it waltzing along so that it can continually draw into the mouth, by
ciliary currents, little particles of food from fresh areas of water.

Now let us review the various larvæ group by group and see how
they compare with the simple trochophore just described. The polyclad
flatworms crawling on the bottom belong to the phylum Platyhelminthes
—animals with a mouth leading into an extensive branching gut
cavity but with no anus; they send up a larva, a primitive type
of trochophore called a Müller's larva which, like the adult, has only
a mouth opening. It starts off as a little spherical form with a wavy
ciliated band and a mouth leading into a central cavity. The band
then becomes more folded and drawn out into eight ciliated lobes
which by increasing their ciliated area are able to support a greater

[1] The two groups are clearly related; they both have essentially the same type of
'spiral cleavage' in development from the egg.

weight. Here there is actually no violent metamorphosis because the adult being flat has, like the larva, a large ciliated surface *in relation to* its mass: the creature gradually becomes more flatworm-like in shape and eventually its ciliated lobes shrink and are absorbed as it sinks to the bottom and assumes the adult form (Fig. 57, opposite).[1]

The nemertine worms also produce a little larva with a mouth but no anus; only here the mouth is central and on the underside. This is the little *pilidium* larva; it is somewhat helmet-shaped with an apical sense-organ and a ciliated band running round its rim which hangs

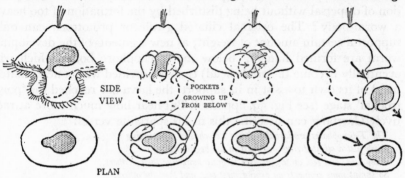

SIDE VIEW

'POCKETS' GROWING UP FROM BELOW

PLAN

FIG. 58

Stages in the metamorphosis of the Pilidium larva of a nemertine worm, seen in sideview and plan.

down as a flap on opposite sides of the mouth. Here we have a profound metamorphosis: a quick change, from a free floating life to one on the bottom, by a very remarkable device. Two little pockets, infoldings of the outer layer, grow inwards from the right and left side in front of the mouth, and two similar ones form behind it; the two in front enlarge and fuse together and so do the two behind (Fig. 58, above): then these fused cavities enlarge still further till they meet each other on either side and over the top. A point is reached when the little central mass, with its mouth and central cavity, becomes completely separated from the rest; and it now drops to the sea-floor to develop rapidly into the adult worm. The remainder of the larva, which has served as a floating perambulator to carry the precious baby as far as possible, now floats off and continues to swim for a time, but, having no means of nourishment, soon dies. Here we see a fresh start made after the distributive phase is over.

[1] My illustrations to this chapter are redrawn from those in my notes taken from Professor Garstang's blackboard sketches made at the lecture already referred to.

We pass now to the larvae of the great phylum Annelida, the segmented worms, represented in the sea largely by the bristle-worms—the Polychaeta—such as the ragworm (*Nereis*) or the lugworm (*Arenicola*) which the fishermen dig out of the sand for bait, as well as a host of others burrowing in the sea-bed, crawling over the bottom or living in tubes. It is the more primitive members of the phylum, such as *Polygordius*, of the class Archiannelida[1] and some of the less advanced polychaetes, which have a simple trochophore larva like the one I have sketched as typical in Fig. 56 (p. 181). How do *they* carry out the function of dispersal without being disturbed by the formation of too heavy a worm-body? The original ciliated girdle or prototroch can only support a certain amount of weight; as new segments of the developing worm are formed hanging below the trochophore body (which will eventually become the adult head) each is provided with a new ciliated band of its own to assist in its support: the larva has reached the 'polytrochal' stage (see Fig. 59, opposite). But hear how much more attractively Garstang can describe this in his sparkling verses:

The Trochophores are larval tops the Polychaetes set spinning
With just a ciliated ring—at least in the beginning—
They feed, and feel an urgent need to grow more like their mothers,
So sprout some segments on behind, first one, and then the others.
And since more weight demands more power, each segment has to bring
Its contribution in an extra locomotive ring:
With these the larva swims with ease, and, adding segments more,
Becomes a Polytrochula instead of Trochophore.
Then setose bundles sprout and grow, and the sequel can't be hid:
The larva fails to pull its weight, and sinks—an Annelid.

In the North Sea there is a species of *Polygordius*, *P. lacteus*, in which we see an interesting modification enabling a length of body to be prepared without upsetting the equilibrium of the spinning top: the trunk of the worm is developed but is folded up within a cavity, as is clearly shown in Wilson's photograph in Plate XVI, p. 173; when the larva settles down, this is suddenly straightened out.

On the same Plate is shown the Mitraria larva of the worm *Owenia fusiformis* which performs the same kind of trick but in a much more remarkable fashion. The intricate development of this larva has been made the subject of a beautiful study by Dr. Wilson in the *Philosophical Transactions of the Royal Society* (1932). The prototroch, which begins as a simple girdle, becomes folded into a very sinuous band to give greater ciliary power in the limited space—just as we shall see happens

[1] Strictly speaking we know now that not all the simpler worms classed in this group are really archaic; some are almost certainly degenerate or neotenous (see p. 195) versions of true polychaetes.

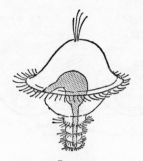

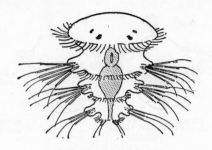

FIG. 59
A polytrocular larva, side view.

FIG. 60
A nectochaeta larva, front view.

in echinoderm larvae—to support a greater weight. From the lower end very long bristles are produced which may serve, as do the long spines of some zoea crustacean larvae, to make their owner too awkward an object to be swallowed by many would-be predators; they are just visible in the photograph, pointing down below the larvae. The developing hind segments are tucked up within a cavity formed by the curious folding of those in front, which are, to use Wilson's words, "turned inside out and drawn back over the succeeding ones much as the top of a stocking can be turned inside out and drawn back over the foot"; the head is almost separated from the rest of the body. When the time for settlement comes there is a sudden and violent metamorphosis in which the worm straightens out in a matter of seconds; the head is drawn down to take its normal place at the front of the body, the long bristles are thrown away, and the prototroch and other larval structures, which rapidly break down, are eaten by the worm as its first meal on the sea-bed. In this and other larvae Wilson (1948–54) has demonstrated remarkable powers of postponing metamorphosis if the bottom over which they are drifting should be unsuitable for settlement; this is a very unexpected and important discovery, of special interest in considering the distribution of life on the sea-bed.

Actually only a limited number of the true polychaetes, such as *Eupornatus* and *Phyllodoce*, have a simple trochophore; let Garstang, however, continue the story in his next verse.

In this way fares Phyllodoce, *but* Nereis *can beat her:*
She gives each egg some extra yolk to hatch as Nectochaeta.
The simple stage with prototroch is by-passed in the eggs,
And each when hatched has three good pairs of parapodial legs.
With these it paddles on in jerks, and could it change its skin
Would almost be a Nauplius, *its very near of kin.*

Fig. 60, above, shows the nectochaeta larva of *Nereis limbata*; the

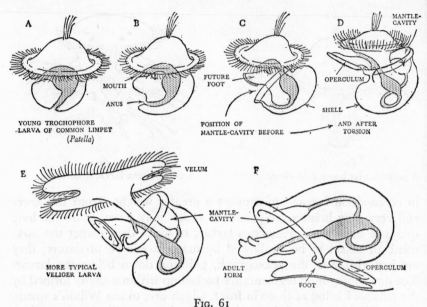

FIG. 61

A-D, stages in the development of the larva of a common limpet, showing the
remarkable twisting (torsion) of the body. E, a more typical veliger larva,
also showing torsion and F, a simplified sketch of the adult snail for comparison.

prototroch (p. 182) is now less important. The larva is developing
into an adult-like form as well as keeping up in the water; the little
limb-like parapodia are now worked by muscles and provided with
specially long bristles to assist in swimming. Here—for carrying a
heavier weight—we see muscular limbs taking over from the original
ciliated girdle designed for the support of a smaller and lighter body.
There is a size limit to the efficiency of ciliary locomotion.[1]

Garstang, in the verse just quoted, suggests how the nauplius larva
of the Crustacea may have arisen from such a little annelid larva
with muscular limb-like parapodia. It is, of course, speculation, but
not a very far-fetched idea; all zoologists believe that the arthropods
are descended from annelid ancestors. It is not difficult to imagine
a race of larvae in which the parapodia became larger and even more
limb-like—perhaps even becoming little biramous limbs; their activity
would now be sufficient to support the larva without the aid of cilia
which would be lost and so allow a chitinous cuticle to be evolved.

[1] In the Ctenophores which are comparatively large we saw (in Chapter 8, p. 134)
that the cilia had become fused into special and larger paddle-like comb-plates.

The nauplius clearly represents the *larval* form of some early anthropod ancestor, and *not* an ancestral adult type.

The primitive molluscan larva, such as that of the common limpet, *Patella*, is also a simple trochophore. In most gastropods or sea-snails, however, the prototroch grows out into a pair of large lobes, one on either side; these extensions not only support a greater weight by increasing the ciliated area, but stop the larva rotating and enable it to direct its movements on a perfectly even keel. It now becomes known as a veliger. In some veligers the lobes are drawn out into great arm-like processes, giving the animal a most remarkable appearance, as seen in Plate XVII, p. 178.

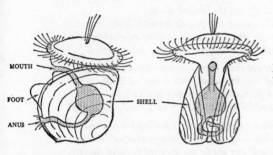

MOUTH

FOOT

ANUS

SHELL

FIG. 62
Larva of oyster with little bi-valve shell, side and front view

In the development of the veliger a very extraordinary event occurs. Perhaps the most characteristic feature of the anatomy of the gastropod molluscs is the curious fact that the main part of their viscera is twisted through 180° in relation to the rest of the body; this is quite independent of the spiral twisting of the shell and has nothing at all to do with it. This torsion actually occurs in the larval veliger stage and occurs quite quickly—in some species the whole process is completed in a matter of only a few minutes. The significance of this strange happening had for long been a puzzle to naturalists until Garstang put forward the following suggestion. Below the prototroch and between the mouth and the anus is the rudiment of the future foot of the snail. On this develops the operculum—a little lid that can be closed down upon the opening to the shell which will be developing on the other side of the larva; this is the arrangement before this twisting takes place, as we see in Fig. 61 opposite. The prototroche and the top of the larva cannot in this, their original, position be drawn completely into the developing shell for protection; behind, however, where the anus opens, there *is* a large space—the mantle cavity. If the head and foot could change places, then the head could be drawn into this space and covered by the operculum in time of danger.

Garstang supposed that a variation occurred which caused the muscles of one side to develop much more than on the other, so that the body would be twisted round with the head and foot facing back to front.[1] This would at once have a great advantage in that the vulnerable head-region could be tucked safely into the space behind. He believed that those larvae having this new property would escape destruction very much more than the others and so became the dominant type— one with its anatomy modified throughout life because of this great advantage in its early phase of development. But again let Garstang put forward his theory in his own inimitable way in "The Ballad of the Veliger or how the Gastropod got its twist."

> The Veliger's a lively tar, the liveliest afloat,
> A whirling wheel on either side propels his little boat;
> But when the danger signal warns his bustling submarine,
> He stops the engine, shuts the port, and drops below unseen.
>
> He's witnessed several changes in pelagic motor-craft;
> The first he sailed was just a tub, with a tiny cabin aft.
> An Archi-mollusk fashioned it, according to his kind—
> He'd always stowed his gills and things in a mantle-sac behind.
>
> Young Archi-mollusks went to sea with nothing but a velum—
> A sort of autocycling hoop, instead of pram—to wheel 'em;
> And, spinning round, they one by one acquired parental features,
> A shell above, a foot below—the queerest little creatures.
>
> But when by chance they brushed against their neighbours in the briny,
> Coelenterates with stinging threads and Arthropods so spiny,
> By one weak spot betrayed, alas, they fell an easy prey—
> Their soft preoral lobes in front could not be tucked away!
>
> Their feet, you see, amidships, next the cuddy-hole abaft,
> Drew in at once, and left their heads exposed to every shaft.
> So Archi-mollusks dwindled, and the race was sinking fast,
> When by the merest accident salvation came at last.
>
> A fleet of fry turned out one day, eventful in the sequel,
> Whose head-and-foot retractors on the two sides were unequal:
> Their starboard halliards fixed astern ran only to the head,
> While those aport were set abeam and served the foot instead.
>
> Predaceous foes, still drifting by in numbers unabated,
> Were baffled now by tactics which their dining plans frustrated.
> Their prey upon alarm collapsed, but promptly turned about,
> With tender morsel safe within and the horny foot without!

The bivalve molluscs have a trochophore-like larva, but one which has formed a bivalve shell; an oyster larva is shown in Fig. 62 (p. 187),

[1] The recent embryological researches of Dr. Doris Crofts (1955) have given much support to this hypothesis.

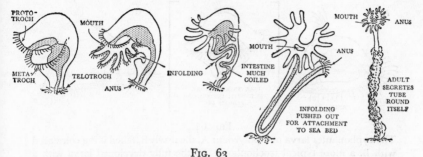

FIG. 63

The development and metomorphosis of the actinotrocha larva of *Phoronis*.
The tentacles of the adult actually grow out afresh and are not those of the larva.

and in Plate XVIII, p. 179, is a photograph of the larva of the razor-shell *Ensis*.

The *actinotrocha* is one of the most beautiful larvae, and not at all uncommon at various points in the North Sea or Channel where the coasts are rocky. It is the young planktonic stage of that very curious animal *Phoronis* which, with its horseshoe-shaped crest of ciliated tentacles for bringing food to the mouth, and its long body secreting a gelatinous casing round itself, looks for all the world like one of the tube-forming worms. A closer examination, however, shows that it is very different: the gut, after running to the far end of the body, doubles back on itself and comes to open at an anus in the 'back of its neck' as it were! But the puzzles of this animal's anatomy and relationships must be studied elsewhere; here we are only concerned with its little larva which we may meet in the plankton. It starts like a trochophore, but with a difference; it has only a partial prototroch above the mouth, the main girdle being a metatroch, i.e. one passing *below* the mouth in front; there is also a small telotroch or girdle round the anus which so far is in the usual position at the bottom of the larva. Soon the prototroch extends forwards as a hood over the mouth and the metatroch becomes folded to produce a ring of long finger-like processes or tentacles to give extra support (Fig. 63, above).

Now a marked infolding of the lower side of the body occurs; at the same time, a rapid growth in the length of the gut causes it (the gut) to become much coiled. This gives another example of a compromise between extended floating powers and an early arrival at the mature form. The animal remains afloat as long as possible; finally, just as it comes to settle down, the infolding on its underside turns inside out, and the long coiled gut shoots out with it, bent into

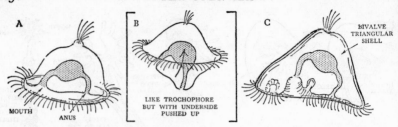

FIG. 64

The cyphonautes larva of a polyzoan: A, the newly hatched larva compared
with B, a more typical trochophore; C, the fully developed larva with a
triangular bi-valve shell.

a 'U'—like the tube of a trombone—and there we have a miniature
adult. The metatrochal processes are replaced by tentacles to form
its horseshoe crest. Three stages in this development are beautifully
photographed by Dr. Wilson in Plate XVIII (p. 179).

Among the larvae most frequently met with along our coasts are
the little triangular-shaped cyphonautes larvae which are enclosed
within a small bivalve shell: they are the planktonic young of the
Ectoproctous Polyzoa—those little animals which, by budding, form
encrusting growths almost like a moss (and are sometimes called the
Bryozoa or moss-animals). The larva, also shown in Plate XVIII,
starts life like a little trochophore but with the underside tucked up
inside like a little bell (Fig. 64, above); then it becomes flattened
sideways and develops little three-cornered shells on either side, and
so takes on its characteristic appearance.

While we have no permanently pelagic members[1] of the great
phylum of the Echinodermata—the sea-urchins, starfish, sea-cucumbers and their kin—our plankton samples are frequently enriched by
their beautiful and very interesting larvae. They supply us with the
supreme example of a larval 'double life'. Let us first consider them in
their rôle as floating perambulators for distributing their babies. We have
only space here to deal with the common forms and will omit reference
to the early planktonic larva of the crinoid *Antedon*: the feather star;
the larvae of all the other classes can be derived from one simple type.

To minimise verbal description we will compare the different types
in diagrammatic form in Fig. 65 (p. 193). In A we see—in side view—
a floating egg which has developed into a little larva something like a

---

[1] In the very deep water plankton of the south Atlantic Ocean there is a curious
floating sea-cucumber-like animal *Pelagothuria*, but we cannot claim it for our home
seas.

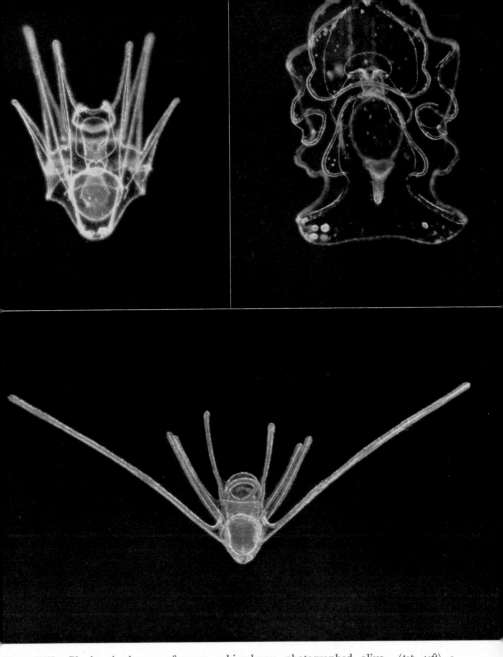

Plate XIX. Planktonic larvae of some echinoderms photographed alive. (*top left*) *a.* Echinopluteus larva of the sea urchin *Psammechinus miliaris*, back view, x 55. (*top right*) *b.* Auricularia larva of the holothurian (sea cucumber) *Labidoplax digitata*, x 50. (*bottom*) *c.* Ophiopluteus larva of the brittle star *Ophiothrix fragilis*, x 50. (*a* and *b* by electronic flash.)

(*Douglas Wilson*)

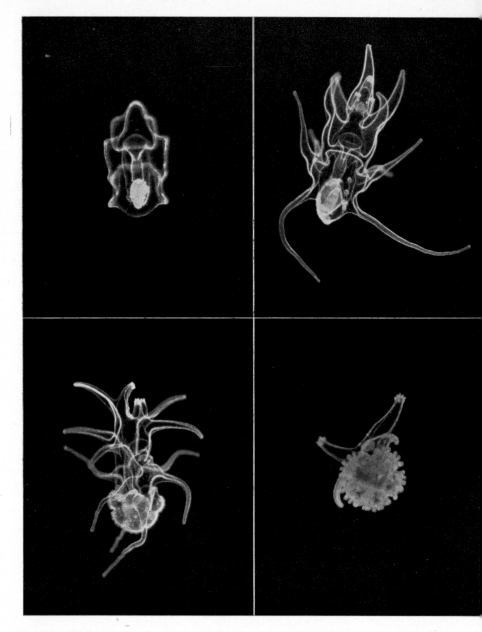

Plate XX. Planktonic larval stages in the development of the common starfish *Asterias*, taken aliv by electronic flash. *(top left)* a. Bipinnaria larva, front view, x 35. *(top right)* b. Brachiolari larva, front view, x 15. *(bottom left)* c. Brachiolaria larva, later stage, side view; starfish bod beginning to form, x 15. *(bottom right)* d. Late stage in the metamorphosis of the brachiolari larva into the young starfish, x 20. *(Douglas Wilsor*

rochophore, but instead of having a prototroch in the form of an equatorial girdle, it has a continuous wavy band of cilia; this runs down each side and loops round the front above and below where the mouth is beginning to form. It differs from the prototroch of the trochophore in that it passes at one point *between* the mouth and anus. In the succeeding diagrams the ciliated bands are shown as heavy black lines. Before the mouth opens the gut has already developed as a little stomach and intestine opening by an anus at the lower pole of the larva. In B the mouth has joined up with the stomach and a little 'adoral' ciliated band has developed round the mouth and passes in a fold inwards along the gullet; this is a feeding mechanism drawing in the little particles of the micro-plankton. In C the longitudinal band has grown longer by becoming more wavy; the loops on each side towards the top now come very close to one another. In D and E we see side and front views of the later development of the typical *auricularia* larva of the sea-cucumbers or Holothurians; the folding of the longitudinal ciliated band becomes more and more complex as it increases in length to support the increasing weight of the larva. In F we see a side view of the developing *bipinnaria* larva of the starfish; here the upper loops of the longitudinal bands on either side have pushed up so high as to meet one another over the top; they have now fused and separated again in a new way to form a smaller continuous preoral band above the mouth and a much longer postoral band behind. In G we see a front view of a much later and more familiar *bipinnaria* larva showing the bands increased in length still more and thrown out into projecting ciliated processes to give greater support. These processes become even longer in a still later *brachiolaria* stage (not shown in the diagram). In H we see a front view of the *echinopluteus* larva of the sea-urchin; the longitudinal band is continuous but extended into very much longer outgrowths than those seen in the starfish larva, processes supported by slender skeletal spicules of calcium carbonate. In a later stage than H (not shown), the arms are longer and in addition ciliated 'epaulettes' have developed on the upper surface to assist in giving support to the larva which is rapidly gaining in weight. The name *pluteus*—meaning an easel—was given to this form because when seen upside down (and the early naturalists did not know which was the right way up) it looks rather like an artist's easel. In I we see the *ophiopluteus*, the larva of the brittle-star, similar to the last but with different parts of the band drawn out into the arms; these arms are much more opened out, with one pair of them making a very wide span in advanced stages.

Having looked at the various larvae in diagram, we are now in a
better position to interpret Dr. Wilson's wonderful photographs of
them. In Plate XIX (p. 190) we see beautiful examples of the *echinopluteus, auricularia* and *ophiopluteus* larvae, in which the ciliated bands
are shown to perfection. Plate XX (p. 191) shows the development
and metamorphosis of the common starfish, *Asterias:* first the *bipinnaria*
larva and then the later *brachiolaria* stage with its ciliated bands drawn
out into long processes. In the side view of this stage (XX, c) we see
a very remarkable event beginning to take place on its lower left side.
The development of the starfish or sea-urchin is one of considerable
complexity. The true body-cavity, or *coelom* as it is called—i.e. that
space (filled with fluid) which is formed between the gut wall and the
outer muscular body wall—becomes in part, in these animals, a
remarkable ciliary pumping system; this maintains a water-pressure
to be used in the operation of those hydraulic locomotory organs—
the tube feet—which are one of the peculiar and most characteristic
features of the adults. It is not within the scope of this book to follow
the details of development which will be found in standard works on
embryology; but this point I have mentioned because the formation
of the coelom, which develops asymmetrically on the left side in these
larvae, has a most remarkable sequel. I wrote of the double life of
these larvae. We have so far dealt only with one part—that which I
likened to a perambulator; now let us deal with the other—the baby
which is to grow into the adult. Instead of trying to describe it myself
in a few words, I am going to quote part of an account of it by the
late Professor F. W. Gamble in his delightful little book *The Animal
World*. I know of no more graphic description. He writes of these
larvae thus:

"They are almost dual animals, for there grows out of their left side a 'coelom'
which is almost as foreign to the rest of the larva as a parasite. This sac has within
its sphere of influence a portion, and a portion only, of the larva. Around it the
tissues are moulded into the form of the future star or echinus, whilst beyond that
modifying influence the larva still pursues its own devices. Presently the star
within it acquires a mouth, a nervous system, and locomotor organs, whilst the
larva on which it hangs has still its own mouth, its own nervous system, and its
own ciliated bands. This organized growth, however, soon exhausts the larva
that bore it. "My need is greater than thine" is its motto, and the larva is presently
depleted of all its material in order to feed the growth that is, as it were, imposed
upon itself. The birth of Eve is no stranger a story than is the development of a
starfish or sea-urchin out of the left side of a larva."

It is in Plate XX, c and d, that we see remarkable photographs
of this very process in the developing starfish; in d the rapidly growing
star has absorbed the greater part of the larva which bore it.

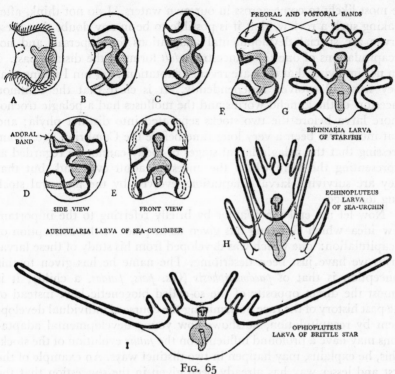

PREORAL AND POSTORAL BANDS

BIPINNARIA LARVA
OF STARFISH

ADORAL
BAND

ECHINOPLUTEUS
LARVA
OF SEA-URCHIN

SIDE VIEW   FRONT VIEW

AURICULARIA LARVA OF SEA-CUCUMBER

OPHIOPLUTEUS
LARVA OF BRITTLE STAR

FIG. 65

A comparison of the different main types of echinoderm (starfish, sea-urchin etc.) larvae; for fuller explanation see text.

There is another kind of larva in the plankton which in its young stages is extraordinarily like an echinoderm larva: that is the tornaria larva of *Balanoglossus* which, in spite of its worm-like appearance, is held by the majority of zoologists to be the most primitive known member of the stock of chordate animals: the stock to which the vertebrates belong. Fig. 66 (p. 196) shows young and later stages of the tornaria; note how like the young stage is to that of the young starfish larva: the longitudinal band is divided into preoral and postoral bands just as in the young bipinnaria; the only difference is that the tornaria larva has an additional lower band—a telotroch. The other primitive chordates of our waters, the Ascidians, send up, as we have already noted in Chapter 8, little tadpole-like larvae.

It is impossible in a book like this to deal with all the different larval forms, but we have described those which the naturalist will

TOS—O

be most likely to come across in our own waters. I do not think, after
making such a review, that it is possible to be in any doubt that these
larvae are special developmental adaptations for dispersal, and not
recapitulations of one time ancestral *adult* forms of the distant past. I
am not implying that they are recent adaptations, nor am I saying that
they were all evolved independently. It is clear that the common
ancestors of the annelid worms and the molluscs had a pelagic trochophore larva before the two stocks separated into distinct phyla; and
that must have been a very long time before the Cambrian age. I am
stressing that the pelagic larval stages of today cannot be regarded as
representing the form which the ancestral adult ever had, but that
they are surviving larval adaptations evolved by the ancestral stock
long long ago.

Now let me end the chapter by briefly referring to the important
new idea which Garstang has given us in place of the conception of
recapitulation: one which was developed from his study of these larval
forms we have just been describing. The name he has given to this
conception is that of *paedomorphosis* (Gk. *pais, paidos,* a child); it is
almost the direct opposite of the so-called biogenetic law: instead of
the past history of the race determining the course of individual development by recapitulation, he shows how young developmental adaptations may have a profound influence on the *future* evolution of the stock.
This, he explains, may happen in two distinct ways. An example of the
first and lesser way has already been given in the suggestion that the
torsion, so characteristic of the gastropod molluscs, became incorpor-
ated in the adult anatomy of the race because of its great adaptive
significance in the larval life of the animals. It is the second way,
however, which seems to me to be of such profound importance in the
consideration of evolution as a process.

Until Garstang put forward the views I am about to express, it was
generally considered that the Ascidians or sea-squirts were curious
degenerate chordate animals that had taken to a sedentary life; that
they had lost their original freedom of movement was thought to be
indicated by their free-swimming tadpole-like larvae—supposed
recapitulations of an active ancestral stock. Garstang has shown how
much more likely it is that the Ascidians are not degenerate at all—
but are representatives of an ancient stock of sedentary animals from
which the free-swimming chordates arose, by a process to be described
in a moment. There is no evidence that the Ascidians were derived
from more motile ancestors. Garstang believed that the little tadpole-
like Ascidian larva is in no way a recapitulation of the past, but a larval

laptation. Just as the crustacean larvae have superseded ciliary
:tion to keep themselves aloft, by developing little limbs which paddle
y muscular action, so the Ascidian larva developed a muscular
ndulating tail as a powerful organ of propulsion. He put forward the
ovel and striking idea that the free-swimming chordates—the fish-like
•rms—and all their higher descendants are in fact evolved from such
ttle tadpole-like larvae which were once sent up by bottom-living
nimals—not actually Ascidians but distantly related forms. Here he
oupled the idea of paedomorphosis with that of *neoteny*. We have been
ressing his early realization of the rival selective forces acting upon
elagic larvae: (a) that favouring long floating or swimming powers
or maximum distribution and (b) that favouring the more rapid onset
f maturity to lead to more prolific reproduction. The latter effect
1ay be brought about by the acceleration of the development of the
eproductive organs in relation to that of the rest of the body; this *is*
a fact what is technically known as neoteny.

A more familiar example of neoteny may make Garstang's ideas
learer. The Mexican *Amblystoma* in its fully developed condition is a
alamander, which has lost its gills, breathes by lungs and spends most
f its time on land; but more frequently this same species becomes
:xually mature and breeds in a condition which is essentially larval:
his is the so-called Axolotl (an Aztec word), which keeps its external
ills and gillslits and remains in the water. Actually in the Axolotl the
evelopment of the reproductive organs has not been accelerated in
elation to the rest of the body but the reverse: the metamorphosis to
he adult condition has been relatively retarded or inhibited to produce
he same effect. Here we see a species which is only partially neotenous,
or it may at times be converted, either in nature or experiment,
nto the fully adult salamander type. Some other newt-like forms such
s *Necturus* and *Proteus* have become permanently neotenous: an
volutionary change brought about by the fixing of this change in the
elative rate of development of body and gonads. It has been known
or some time, that mendelian genes can affect the *rates* of different
levelopmental processes: accelerating or retarding some in relation
o others; this would explain the mechanism of neoteny. Animals
nay become adult at a *younger stage* than hitherto.

Let us now return to Garstang's view of the origin of the chordates
ind vertebrate stock. He imagined a bottom-living invertebrate
inimal like an Ascidian, sending up a planktonic larva. He then
magined the dual forces of selection acting upon this larval form to
uch an extent that (a) its powers of remaining afloat were prolonged

by the production of more and more mobile forms and (b) the develop
ment of the gonads was greatly accelerated. Finally he imagined thi
prolonged larval stage actually becoming sexually mature while sti}
swimming in the plankton. Now the more dramatic step can easily
be taken; the former bottom-living adult stage will be eliminated—
as in the amphibia the once lung-breathing former adult *Necturus* wa
surely eliminated from the race. Evolution takes the remarkabl
step of leaping from an apparently lowly sedentary form to an activ
free-swimming type—the forerunner of a new race to lead on to th
vertebrates—to fish—to Man! Instead of the conception of past adul
forms determining the course of development of later members of th
race, we have from Garstang the new idea of developmental noveltie
changing a major line of evolution.

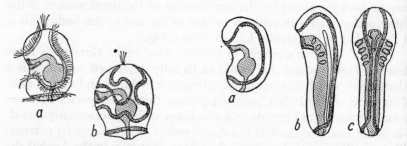

FIG. 66 (*left*). Young and older tornaria larva of the primitive chordate animal
Balanoglossus. FIG. 67 (*right*). A reproduction of Garstang's original figure to
compare a generalised protochordate (*b* and *c*) with an echinoderm larva such as
auricularia.

We have seen that the Ascidians actually do send up little tadpole
like larvae, and, furthermore, that one group, the Larvacea (see p. 153
have every appearance of having done just what Garstang supposed th
ancestors of the vertebrates to have done: they are permanently littl
tadpole-like, swimming creatures produced by neoteny from larva
forms. As I have already hinted, Garstang imagined that the mair
chordate stock was not actually derived from an ancestor like th
present-day Ascidians, but from some sedentary Ascidian-like inverte
brate of the distant past which had had a common origin with th
echinoderms. We have already noted the close resemblance betweer
the larvae of the latter group and the primitive chordate tornari;
larva. Garstang went further; he drew a remarkable comparison be
tween the very early stages in the life of a more typical chordate anc
the early echinoderm larvae. Fig. 67, above, is a reproduction o

Garstang's comparison. He showed how the folds which rise up on the upper surface of an embryo, to form the typical vertebrate tubular nerve-cord, has in the early stage of development exactly the same relation to the rest of the embryo as has the long ciliated band of the echinoderm larva. The adoral ciliated feeding band leading into the mouth of the latter can be compared with the endostyle and the band round the mouth of an early chordate. But I must not enter into the technical refinements of his theory: I must state it briefly. In short it is this: if we imagine (a) the echinoderm-like larva to be somewhat lengthened, (b) gill-slits to be formed inside the mouth to allow a better flow of water for respiration and feeding, (c) muscles and a stiffening rod (the notochord) to develop to allow the little larva to swim like a tadpole, and (d) the ciliated bands to rise up and fuse to form the basis for a nervous system—then we have a larva resembling a small free-swimming chordate; now if by neoteny we get such a larva becoming sexually mature and cutting out the old adult stage, then indeed a little chordate animal has been produced from an original larval form.

Sir Gavin de Beer, in his brilliant book *Embryos and Ancestors* (1940) was one of the first to appreciate the importance of Garstang's conception of paedomorphosis; he himself coined another term to stand beside it: *gerontomorphosis*, to signify evolutionary changes brought about by modifications of adult structure. He goes on to explain how the larval form of a millipede, soon after hatching, bears a great resemblance to an insect: how walking legs are first developed upon the three segments immediately following the head, just as on the insect thorax, while the remaining segments bear only rudimentary ones, as do the abdominal segments of the most primitive living insects of today. To see a larval millipede is to be at once reminded of an insect: and since the evolution of the insects from millipede-like ancestors is supported by so many other considerations, it is not so wild a speculation to suppose that the insects, like the vertebrates, have arisen by a process of paedomorphosis involving neoteny. We have already seen in our account of the Siphonophores (p. 110) how Garstang had showed that they were most likely derived from floating actinula larvae which became mature before settling down; this indeed is another example of his principle of paedomorphosis. The late Dr. Robert Gurney followed him in producing good evidence to suggest that the copepods, one of the most successful of all crustacean groups, were also derived by neoteny from the planktonic protozoea larvae of some bottom-living decapod-like forms; the similarity of the two is most striking (Gurney 1942).

What can be the value of such exercises in evolutionary speculation? They cannot be proved, as can an hypothesis in physiology. Many of them may be entirely wide of the mark. The importance of Garstang's idea, which is no more speculation than the old idea of recapitulation it replaces, seems to me to be independent of any particular one of them being true or false. The point I want to make is that his concept of paedomorphosis—derived from his study of these little larval forms—has altered our outlook on one important aspect of the process of evolution. In a wide imaginary view of the evolutionary streams of life we see the various lines branching again and again as they advance in time; branch after branch tends to become so well adapted to some particular way of existence that it becomes eventually highly specialized. Such specialized lines have been thought to be doomed to eventual extinction as conditions must in the long run change; the lines of new development have been thought always to spring from less specialized forms. This indeed must often be so. But is the rule as inevitable as it was thought to have been only a little while ago? No, Garstang's conception has changed this; he has enabled us to see that specialization need not always be the end of evolutionary progress. However specialized the *adults* of a stock may have become, it is still possible for their younger stages to be modified in new ways, and then, by neoteny, to produce a new paedomorphic line to blossom forth to become—who knows?—perhaps a whole new order—class—or even phylum. The siphonophores, ctenophores, cladocerans, copepods, insects and the very vertebrates have been shown possibly, if never certainly, to have had a paedomorphic origin from something very different.[1]

[1] I have discussed the evolutionary significance of paedomorphosis more fully in recent essay (Hardy 1954).

CHAPTER II

THE PUZZLE OF VERTICAL MIGRATION

THERE ARE many unsolved puzzles of pelagic natural history, but one seems perhaps more baffling than any other: that of vertical migration. This is the name given to the curious habit possessed by so many plankton animals of rising towards the surface at night and sinking—indeed often swimming—away from it in the daytime. From the very early days of tow-netting, naturalists have noticed that they tended to get much larger catches of plankton near the surface at night than in the daytime.

At first it was suggested that there was no actual up-and-down movement, but that the little animals saw the net approaching in the daylight and were able to dart out of the way of it, whereas at night it came upon them unawares. We still hear these views expressed by some today, but only by those who are ignorant of the vast amount of work that has now been done upon this problem. It is, of course, likely, indeed certain that some of the larger and more powerful swimmers, and those with good eyes, are able to avoid a net more easily in the daytime near the surface, but it is quite certain that this does not explain the phenomenon in general. By taking hauls with tow-nets at different depths at the same time it is quite easy to show not only that there are more animals in the upper layers at night but also that there are *fewer* in the *lower* layers than in the daytime. Moreover vertical migration has now been demonstrated experimentally.

What an extraordinary thing it is that so many different animals will expend so much energy in climbing up towards the surface at night only to sink or swim down again in the daytime; some species seem regularly to climb a height of two or three hundred feet or more, every twenty-four hours. It is true, of course, that the great source of food for these animals, the phytoplankton, is in the upper layers, growing only in the sunlit zone; but then why do the animals not stay up in this layer with their food-supply, instead of sinking to climb again next day ? It is a habit that has been evolved quite independently

in almost every major group of animals in the plankton, some making
shorter, others longer migrations: protozoöns, coelenterate medusae
siphonophores, ctenophores, arrow-worms, the polychaete worm *Tomop*
teris, the pteropods, nearly all the many different groups of crustaceans
salps, doliolids and *Oikopleura,* and most young fish.

This vertical climbing uses up so much energy and has been
developed so frequently in the animal kingdom, that it must clearly
be of some very profound significance in the lives of these animals
What can be the meaning of it ? We do not yet know for certain; we
can only make some guesses. It is surely the planktonic puzzle No. 1
Let us consider the main facts and some of the ideas put forward to
explain them; then we shall be in a better position to discuss the
possible value of the habit to the animals concerned.

Apart from a few very early observations, the great *Challenge*
Expedition of 1872-75 gave us the first extensive evidence of this move
ment, as of so many other aspects of marine natural history. In working
on the collection of ostracods and copepods Professor Brady recorded
how they were only to be found in the surface samples which were
taken at night, and Sir John Murray in his narrative account of the
voyage gave us the first intimation of the surprising range of this migra
tion. "The great majority of the plankton organisms", he wrote, "live
at various depths down to and even deeper than 100 fathoms during
the day . . . and only come to the surface at night." Dr. Herbert Fowle
was one of the first to make a special study of vertical migration by
undertaking an expedition for the purpose in H.M.S. *Research* to the
Bay of Biscay in 1900. Using closing nets of his own design down to
100 metres he showed that euphausiaceans and ostracods had a range
of migration of at least 100 metres; then G. P. Farran, working on
Fowler's collection of copepods, found that nine species made extensive
migrations, while three which were caught at the surface at night were
never taken in the nets above 100 metres in the daytime. Sir John
Murray and Professor Johan Hjort in their famous Atlantic voyage
in the *Michael Sars* in 1910 also showed that some pelagic decapod
crustacea migrate from depths of 800 metres towards the surface. An
increasing number of observations were accumulated to show how
widespread the habit is; and naturalists then tried to find out what
were the factors in the animal's environment which could explain it

The changing light intensity naturally seemed to be one of the most
likely agents governing this difference in day and night behaviour. It
had long been known that if a sample of living plankton is put out in
a glass dish in the laboratory the animals will tend to sort themselve

out into those which swim towards the light and accumulate at one side of the dish, and those which swim away from it and accumulate at the opposite side. The experimentalists, following Professor Jacques Loeb, began to talk about tropisms: animals were said to be positively or negatively phototropic according to whether they moved towards or away from the light. Groom and Loeb working at Naples in 1890 were the first to experiment on these lines, when they worked with the little nauplius larvae of the barnacle *Balanus*. They found that the larvae would move towards a moderate light, if they had first been kept in the dark; but that if the light were increased they would swim away from it. Vertical migration was at first thought to be due to just such a simple change in reaction to light; but soon it was realized that there was more to it than that, and Loeb introduced 'positive and negative geotropism'—reaction to gravity. It was thought that most plankton animals reacted negatively to both gravity and light and that the latter reaction took precedence over the former: in the daytime, they said, the bright light forced the animal away from the surface but when its influence faded at evening 'negative geotropism' made the animal swim up against gravity.

Professor Ostwald in 1902 tried to account for the migrations as due to changes in the viscosity of the water[1]: in the daytime when the surface layers are warmer than at night the viscosity will be less, i.e. the water is more fluid, and the animals will sink more quickly, he thought; at night when the water cooled and the viscosity increased he thought the animals would be able to swim upwards more easily because they could, so to speak, get a greater grip on the less fluid water. Actually, however, the changes in temperature and viscosity, between day and night are far too small to have any such effect, and even if they were sufficient we should expect the animals to arrive at the surface well after midnight when the water is coldest and not just after dark as they usually do.

E. L. Michael in 1911 was the first to suggest that vertical migration is controlled by a shifting up and down in the water of a particular light intensity most suitable for the life of the particular species considered. He was working with the arrow-worm *Sagitta bipunctata* off the Californian coast. He believed it found the best light conditions during the day at depths between 15 and 20 fathoms and then migrated towards the surface at sundown because light conditions in the topmost layers were then similar to those it experienced lower down during the day. As the sun goes down, the zone of a particular light intensity

[1] Viscosity is discussed on p. 47.

will move upwards in the water and as the sun rises it will sink; he believed that the *Sagitta* followed this rise and fall of its special 'optimum' light intensity. In addition he considered the picture was complicated by the animals also reacting to the most suitable conditions of temperature and salinity; this further idea however was disproved a year later by C. O. Esterly. Nevertheless, Esterly, working with copepods, obtained more evidence for Michael's idea of an optimal light intensity. "The variations in the intensity of light", he wrote "are both constant and periodical, and it is my belief that light is the primary cause for movements in this species (*Calanus*) and also the main factor in determining its vertical distribution."

It is of course impossible, in this brief general account, to refer to all the different observations that have been made or all the ideas that have been suggested; very good reviews of all this work, with full bibliographies, have been published by F. S. Russell (1927) and D. H. Cushing (1951). I want merely to give an outline of some of the various factors that have been considered in trying to find a solution to this puzzle. I must include in my sketch the suggestion made by Miss Eyden in 1923. Vertical migration had been shown to be just as striking a phenomenon in the plankton of freshwater lakes as in the sea and Miss Eyden, working with the little 'water-flea' *Daphnia* (p. 157) found that there was a change in the specific gravity of the animals, correlated with feeding. She drugged her specimens so that they would sink quite passively and then measured their rates of sinking; those which had recently fed sank more quickly than those which had been starved. She suggested that vertical migration might be brought about simply by feeding reactions: the animals come up to feed upon the phytoplankton near the surface, have a good meal, become heavier, and sink; then, when the food is digested and the unwanted material expelled, they will be able to swim to the surface again because they are lighter! The actual difference in density and rate of sinking between a fed and a starved water-flea appears to be far too small to be of significance as a direct causative factor, but we must certainly not rule out the conditions of hunger and repletion from consideration as possible factors governing their swimming behaviour. Dr. George Clarke (1932) later showed that *Daphnia* which were well fed tended to be negatively phototropic and those which were starved to be positively phototropic.

For a number of years F. S. Russell (1925–34), who is now Director of the Plymouth Laboratory, carried out an immense amount of work on the vertical distribution of a great many different plankton animals and young fish in the Plymouth area. He took many series of tow-nettings with a standard net towed in exactly the same way at a number of different levels below the surface; at the end of each tow the net was closed by a special mechanism before being hauled up, so that he knew that all the animals in the sample had actually come from the depths at which the net fished and were not caught on the way up. In addition he had an automatically recording depth-gauge attached to the rope against the net so that at the end of each tow he had a tracing on a clockwork drum of the exact depth at which the net was fishing all the time. He repeated such series of hauls at different depths at several times during the twenty-four hours, and so he was able to compare the numbers of a particular animal present at the various levels as daylight gave way to darkness and back to daylight again.

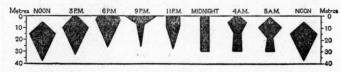

FIG. 68

Diagrams to show the typical vertical distribution of the copepod *Calanus finmarchicus* at different times in the twenty-four hours in the Plymouth area when the sun sets about 8.0 p.m. The varying width of each diagram represents the percentage proportion of the population at the different depths at each particular time. From Russell (1927).

He estimated the percentage of the population at each depth, and so could express his results in a quantitative graphic fashion—as shown for the copepod *Calanus* in Fig. 68 above; the varying widths of each black geometrical figure represent the relative proportions of the population found at different depths on each occasion through the 24 hours. In Fig. 69 (p. 204) are seen the results of similar experiments made by Dr. A. G. Nicholls (1933) in the deeper waters of the Clyde sea area where *Calanus* makes a much more extensive vertical migration; here, to give a more realistic picture, I show the percentage distribution at different depths by a series of dots, each representing one *Calanus* out of a total of 100 in the whole column of water sampled.

Russell developed Michael's original view that vertical migration is governed by a shifting light intensity which is most suitable (optimal) for the animal concerned, and drew attention to the phenomenon of midnight sinking which is shown in the diagram. He explains vertical migration in the following way. Members of a particular species tend to aggregate in the sea during the daytime at the level of a particular

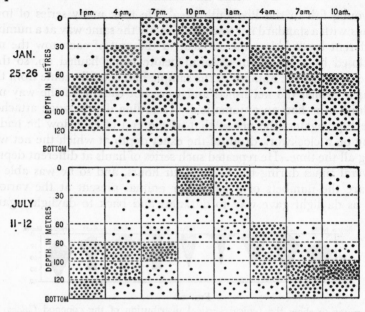

FIG. 69

The changing vertical distribution of the copepod *Calanus finmarchicus* (adult
females) throughout two 24 hour periods (January and July) in the Clyde sea
area as found by Dr. A. G. Nicholls (1933). Every three hours he sampled
the plankton with a tow-net hauled vertically from the bottom to the surface in
six stages: bottom (135m) to 120m, 120-100m, 100-80m, 80-60m, 60-30m
and 30m-surface; the net being automatically closed at the top of each stage.
In the diagram there are 100 dots in each column; these represent the percentage number of Calanus at each depth; a marked vertical migration is shown
and they are seen to go deeper in summer, when the light is stronger, than in
winter.

light-intensity which is most suitable to them; then as the sun sinks
lower and finally sets, this favourable light-intensity moves upward to
the very surface and so is followed by the animals which appear at
the top at dusk. When complete darkness comes they sink and scatter
at random in the upper layers because they are no longer drawn upwards. Then when the dawn comes they will move up once more
towards the best light intensity which is now at the surface again and
then sink down with it to their daytime level once more. Further in
support of this view, he was at times able to demonstrate that the
plankton in the daytime was lower in the water on bright sunny
days than it was on overcast and cloudy days. The laboratory experi-

ments of M. Rose also supported the view that the movements of plankton animals may be governed by an optimum light intensity; he showed that if copepods are placed in a long horizontal glass tube, which for half of its length is screened by black paper, they all collect around a point midway along the shaded portion, i.e. around a particular light intensity.

It would be nice and tidy if one could just pin down the operating factor to some such relatively simple thing as the movement up and down of a particular light-intensity—although it would in no way tell us what was the actual reason, in terms of advantage to the animal, for such a reaction being evolved. Optimum light intensity may certainly be an important part of the story but it is clear from subsequent work that there must be a number of other factors involved. Sometimes animals which usually stay down in the daytime may be seen in numbers at the very surface in bright sunlight. Some animals such as *Calanus* appear to behave differently at different seasons of the year, different broods being higher or lower in the daytime. Dr. George Clarke of Harvard University, who correlated the vertical movements of the copepods *Calanus* and *Metridia* with measurements of the varying light-intensity over several twenty-four hour periods on the American side of the Atlantic, writes as follows (1934): "The migration of the copepods was found to be more closely correlated with changes in the submarine irradiation than with changes in the hydrographic conditions (i.e. salinity, temperature, etc.) or the phytoplankton. However, great variability in behaviour was observed. This appears to be due to differences in the physiological conditions of the animals..." It is quite certain from the results of tow-netting at different levels that all individuals in a population do not react in the same way to one particular set of conditions, light or otherwise; their behaviour varies: the majority may migrate upwards, but usually a proportion remain below.

Far from being simple, the more we investigate the more involved does the plot of the story seem to become. I must not hope that the reader will have the patience to follow me much further in the tangle of different findings that have come from the many attempts to solve the problem; however, before proceeding to consider the possible value of such vertical migrations in the lives of the animals concerned— and that is the more vital issue—it may be of interest to include some results from a new experimental approach that is now being made.

I should like to record that this new work grew out of other experiments which I began just before the war with the assistance of Neil

Paton who, as a fighter pilot, was killed over Malta after many heroic exploits which won him the D.S.C. and a mention in dispatches; his death was indeed a sad loss to marine science. At that time we did our experiments with long glass cylinders at different levels in the sea itself; by opening and closing trapdoors in them—by messenger weights sent down a wire from above to hit spring triggers—we were able to record the number of animals which had moved up or down in a given time under different conditions of light and depth. By darkening the tops of the cylinders and playing with mirrors we sent daylight

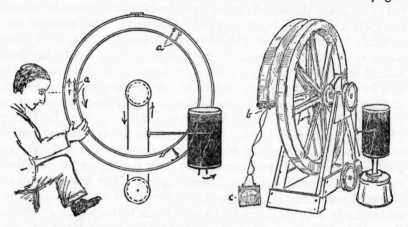

FIG. 70

The 'plankton wheel'—an apparatus for the experimental study of the vertical migration of plankton animals, *a*, weighted and buoyant trapdoors which are always open at one side of the wheel and closed at the other; *b*, position of photo-electric cell for light measurement which is recorded on galvanometer at *c*. For further explanation see text.

up from below and so fooled the animals into behaving just as if they had gone with Alice through the looking glass. The story of these and other tricks (Hardy and Paton, 1947) is however too technical and involved to find a place in such a general account as this. From them arose the idea of the device about to be described: one which gave in effect a column of water of infinite length.

Hitherto it had not been possible to see animals making extensive vertical migrations in the laboratory, or to experiment with them while they were doing so; but my colleague Dr. Richard Bainbridge and I have recently constructed an apparatus which allows this to be done (Hardy and Bainbridge, 1954). We can now study the extent and speed of the climb of different animals under various conditions. A curved transparent tube of perspex with a rectangular cross-section of 2 by 1½ inches, is made to form a complete circle of 4 foot diameter and is then mounted as a wheel; a small opening, which may be sealed up, allows it to be filled with water, and the experimental animal added. Fig. 70, opposite, shows a diagrammatic sketch of it. The wheel is now turned so that the animal is half way up one side; it can now swim either up or down if it wishes to, but as it does so the wheel is turned by hand so that it is actually kept stationary in relation to the observer. It will be seen that the animal is swimming in an endless tube; it can swim up or down for hundreds of feet. Inside the tube are little doors, some with weights and some with floats, which automatically open and close as the wheel goes round; they are so arranged that they are always wide open at the side at which the animal is swimming and so do not interfere with it, but closed on the other side to ensure that the water turns exactly with the wheel and does not lag behind it. The animal thus swims up a column of water which is actually moved downwards at the same rate, or *vice versa*. Fixed to the axle of the wheel so as to turn with it, is a pulley having a circumference exactly one-tenth that of the wheel itself; another similar pulley is mounted on a frame below it, and a cord, kept at tension by a spring, passes round the two. On this cord is mounted a needle which makes a line on a smoked drum turned by an electric clock. Thus a record on one-tenth the vertical scale is traced on the drum, recording every upward and downward movement of the animal against a time base-line. When the needle reaches the top of the drum it can be flicked to the bottom again, by pulling the cord, so as to continue its upward graph and *vice versa*. The experiments are performed with the wheel in a small glass greenhouse—with shaded sides, so that when it is covered with a white or grey sheet it will give conditions very similar to those of diffuse daylight under the sea. The intensity of light may be varied from bright daylight to complete darkness by a series of such white, grey and black sheets which can be drawn over the top. An exactly similar curved tube, filled with water and standing beside the wheel, is fitted with a photo-electric cell which records the light intensity at a point just beside where the animal is swimming.

This work is still continuing and giving new and interesting results; for the first time it is providing us with actual measurements of the swimming speeds of different plankton animals over considerable vertical distances and showing us what they can achieve in periods of an hour or more. It is astonishing what these small animals can accomplish. *Calanus*, the size of a grain of rice, can climb nearly 50 feet

in the hour, and so can the very much smaller nauplius of *Balanus*
Centropages (a little smaller than Calanus) can climb 95 feet in an hour
while the larger euphausiacean *Meganyctiphanes norvegica* (about an
inch long) may achieve as much as 305 feet in the same time. Timed
over periods of two minutes when going at their fastest, the same
animals can respectively swim vertically upwards at the following
speeds: 217, 73, 179 and 566 feet per hour. Swimming downward
for an hour *Calanus* and *Meganyctiphanes* can cover 154 and 420 feet
respectively or in short two-minute bursts can dive at speeds of 33
and 705 feet per hour. It has been suggested that plankton animals
usually swim in all sorts of directions at random, and that vertical
migration is brought about by their general speed of swimming being
increased or diminished as they move by chance into higher or lower
light intensities. The plankton animals in our wheel are not behaving
in that way when they are making an extensive migration: they are
actively swimming upwards or downwards. A great deal of the time
when they are not migrating, they are making little movements
usually up and down rather than sideways. I mentioned in an earlier
chapter, p. 97, how I had seen a small medusa feeding on *Balanus*
nauplii by capturing them individually with its manubrium; actually
this was seen in the wheel. It forms a wonderful observation chamber
in it I have seen *Oikopleura* in its 'house' to perfection, and all manner
of little larval forms, put in by chance with the water to be used
in the study of other larger creatures. The beauty of it is that, since
the water moves exactly with the wheel, you can always turn it back
to a point you may have marked, and thus observe once more some
very small animal which may not have moved far in the time.

Another new experimental line of work has recently been started
by Professor J. E. Harris (1953) who has had the ingenious idea of
producing in a jar, only some 18 inches tall, a gradient of light intensity
which is equivalent to that met with in the sea when going down to a
depth of 20 metres. He has done this by simply adding to the water a
suspension of indian ink and in this his animals, the waterflea *Daphnia*
swim with apparently no ill effects. By altering the overhead illumination from that of full daylight to almost darkness he induces a vertical
migration similar to that found in many animals over a range of 20
metres—but reproduced in extreme miniature. He has so far only
published a very preliminary account, but his suggestion—based on
the freshwater *Daphnia*—that vertical migration involves a negative
reaction to gravity in very low light intensities is not supported by
many of our experiments with marine animals in our long tubes or

our 'wheel'.[1] From some of our earlier experiments (Hardy and Paton 1947) I had thought it possible that *Calanus* in its behaviour in vertical migration might be sensitive to changes of pressure and so have a direct indication of change of depth. Later Bainbridge and I (Hardy and Bainbridge 1953) failed to obtain any change of behaviour in *Calanus* subjected to pressures equal to those at 20 metres depth; but we got a marked response in some crab larvae. Dr. E. W. Knight-Jones (1955) has very recently followed up this line of work, confirmed our decapod larvae results and demonstrated a similar pressure response in a variety of plankton animals including hydromedusae, ctenophores, larval worms and the isopod *Eurydice*; he failed as we did to get any effect with *Calanus*, but he tells me that since publishing his first results he has now got a marked response for another very common copepod *Temora*. It will certainly be interesting to discover how these animals react in this way. Have they some small pressure gauge—perhaps a tiny gas-filled vesicle expanding and contracting against a nerve ending? Or is it just the varying effect of pressure upon some vital chemical reaction within the body? Here are more puzzles to solve.

We must now leave the question of the actual factors which stimulate the animals to begin or end their migrations every day—leave it still awaiting an answer—and turn to consider what can be the meaning of such migrations in terms of advantage to the animal. We have seen that the ultimate source of food for all the animals in the plankton is in the upper layers, where alone the small plants can get enough light to allow them to grow and multiply. Can it be that the animals are climbing up into this layer to feed only at night, when the cover of darkness will hide them from their enemies? Perhaps, too, the carnivorous animals are just following the herbivores up and down, for they both migrate in much the same way. This may be so, yet it seems extraordinary that it should have become so universal a habit. It is curious, too, that some of the herbivorous animals, when making an extensive upward migration, may advertise their presence by lighting up little luminous organs, instead of concealing themselves in the darkness. In the earlier editions I stated that the euphausiaceans may be luminous when at the surface at night. Since then I have edited my original Antarctic journal (*Great Waters*, 1967) and was surprised to find no mention of seeing such lit-up patches of krill as I had thought were there; evidently my memory was faulty and I was thinking of other examples of luminosity. On coming up in the tow-net they may

[1] Since this was printed Professor Harris's second vertical migration paper has appeared; in this he gives a remarkable confirmation of the effect of a moving optimum light intensity on the behaviour of his animals, but he also demonstrates a curious depth position effect (Harris and Wolfe 1955).

indeed shine billiantly but I doubt now if they do when swimming naturally. Of course such luminosity cannot be a very serious drawback in the way just suggested, or it would soon have been abolished from the race by natural selection; but it seems strange that the animals should carry these lights at all, if the real reason for migrating up in the darkness is to avoid being seen!

Is it possible that the plants in the upper layers are giving off some substances which are injurious to the animals—having some antibiotic effect, as we should say today; and that the animals can only come into these noxious layers for a limited time each day? If so, then the animals might find it more advantageous to come up in darkness to escape from enemies: or it might be that the antibiotic effect is stronger in the daytime, when the plants are carrying out photosynthesis, than at night. At one time I thought this possible. There are small differences, for instance, in the oxygen-content and alkalinity of the surface layers between day and night, due to the plants' photosynthesis— but these are very small indeed, quite insignificant compared to the large variations due to different quantities of the little plants. I now think it unlikely that there can actually be a sufficient *diurnal* difference in the chemical nature of the upper layers to have given rise to the development of the habit of vertical migration. I still think it possible that some planktonic plants may have some poisonous or antibiotic effect, and that this may be the cause of certain patterns in the distribution of planktonic animals and plants. But it seems unlikely that any noxious effects of the plants could be the direct, or only, cause for the evolution of vertical migration. A point is that vertical migration has been shown to be performed by many planktonic animals which only inhabit the mid-water depths of the great oceans, and move between levels of 1000 and 200 metres, far below the influence of the phytoplankton at the surface. In the next chapter we shall take a look at the strange life in these depths over the edge of our continental slope. It seems clear that vertical migration must have some peculiar significance in the lives of all planktonic animals, whether near the surface or not.

I believe that vertical migration has been evolved because it gives the animal concerned a continual change of environment which would otherwise be impossible for a passively drifting creature. We have defined planktonic animals as those which are passively carried along by the moving waters; they have some power of locomotion, but it is only very small and not sufficient to carry them more, than at most, a few hundred feet in the day. Now we know that the water masses are hardly ever moving at the same speed at different depths; the surface layers are nearly always travelling faster than the lower layers, and as we have seen in Chapter 2 (p. 23) often travelling in different directions. An animal which can swim to right or left for only a few hundred feet will not get much change of environment in the sea, for all that effort; and in a diffuse light from above, it will also be difficult for it to travel far in any one direction; it will most likely curve round towards its starting point again. But if the animal sinks downwards a hundred feet, and climbs up again the next evening, it may then find itself a mile or more away from where it was the night before, for there may well be a difference of a mile or two a day between the speed of the current at the surface and that at only 30 metres depth. A drop in the morning, and a climb of a hundred feet at evening, may well give it a horizontal movement (in relation to the surface) of some ten thousand feet! Vertical migration is thus a means of providing a relatively weak and drifting organism with an extensive power of movement. It can 'hop' from one environment to another by sinking and rising again; perhaps we should say that, rather than travelling, it is being left behind by the surface waters out of which it drops for an interval each day. But *relative* to the surface it is moving along.

Man is most like a plankton organism when he is up in an old-fashioned balloon, drifting freely in whatever wind there is. But he can control his movement to some extent; he can navigate within certain narrow limits. The wind is never at the same speed or rarely in exactly the same direction at different heights. The experienced balloonist in a long-distance race will go up and down till he finds the wind that is likely to carry him farthest. He goes up by letting out his sand ballast; he comes down by valving out some gas from the balloon. Any independence he has over the environment is achieved by vertical migration. I am not of course supposing for a moment that the plankton animals make any conscious navigation; even if they were conscious of their surroundings, and were intelligent beings, they could not judge how fast or in what direction the different water-layers are moving—any more than the balloonist can when he is travelling in the clouds, cut off from the sight of land. I am suggesting that the development of a habit of regular periodic vertical migration has been evolved to enable the animal continually to sample new environments—new feeding grounds—and so to have some power of choice, of varying its behaviour within limits. For example it may not continue to ascend if it finds itself climbing into some uncongenial environment.

There is, as a matter of fact, a very remarkable case of vertical migration serving as an unconscious navigation. In the Antarctic, as also mentioned in Chapter 2, there is the cold surface-current flowing away from the melting ice of the polar ice-cap. This cold current extends down almost to 100 metres, and below it is a warmer current of water flowing in the opposite direction—towards the pole—to take its place. Some of the antarctic euphausiaceans, such as *Euphausia frigida* and *E. triacantha,* maintain their geographical position by making a vertical migration of some 200 metres every day; they spend the night at the surface travelling north and come back during the day by travelling south in the reverse current down below. Some other animals, as Dr. Mackintosh (1937) has shown, travel north in the upper current rich in plant-food during the summer, but drop down into the warmer return layer all the winter: they make a seasonal instead of a daily vertical migration.

Yes, I believe it likely that vertical migration has been evolved as a means of giving planktonic organisms a degree of independence of their environment; if so it is surely another remarkable adaptation to drifting life, brought about by natural selection acting on variations in the animals' inherited behaviour-patterns. Professor Harris in the paper I have already referred to (1953) suggests that these migrations are nothing more than "an inescapable consequence" of the use of light by plankton animals as a means of keeping within the upper illuminated layers. I find it difficult to believe that nature can be so extravagant in the use of energy as to evolve a system of swimming upwards and downwards for hundreds of feet every day merely in order to try and keep at one level; particularly does it seem unlikely when we now know from Knight-Jones's observations that a number of plankton animals are directly sensitive to pressure changes and so to changes in depth.

I would like now to return for a moment to the question of the uneven or patchy distribution of the plankton, which we briefly discussed earlier in the book (p. 76). I think it possible that this day-and-night oscillation between the lower and upper layers may have a more marked effect upon the distribution of plankton animals than is generally recognised. In the shallower regions of the sea, over the continental shelf, where the bottom may be hollowed into troughs or raised up into banks, the pattern of water movement near the bottom may be very different from that at the surface. And where two currents meet, or where one is deflected by a mass of land, we may sometimes get a surface swirl produced: an upper layer of water rotating like a disc above another layer which is being guided along

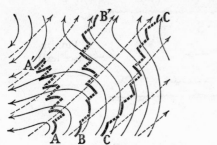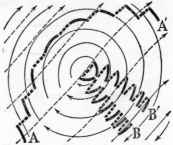

FIG. 71a and b.

Examples, in plan, of the paths taken by plankton animals A-A', B-B', etc., migrating daily between surface waters (fine continuous arrows) and lower waters (fine broken arrows) which are moving in different directions. The night and day paths of the animals are shown in continuous and broken heavy line respectively.

in some particular direction by the contours of the sea-bed. Plankton animals migrating every day upwards and downwards between the upper and lower systems are going to be carried in all sorts of directions which we should never expect if we did not have an exact knowledge of the relative movement of the two water-layers. If we work out their paths in detail, as in the examples in Fig. 71 above, we see how in some regions they may be held up, and in others hurried on.

Now let us consider what might happen if the animals were to vary their migrational behaviour: for example, under some conditions going down deeper than under others, or staying down for longer, or not rising so high in the water at night, and so on. If the water at different levels is travelling at different speeds—and we have seen that there is nearly always a gradual reduction in speed from the surface downwards—then such changes in behaviour as those just suggested might have very considerable effects upon the animals' distribution. At present we know very little about the extent of such changes in behaviour or of the factors that govern them; we do know, however, that the range of vertical migration in a species—or for an individual— is by no means constant.

Let us also look at this in diagrammatic form, as in Figs. 72 and 73 (p. 214), which represent a section of the sea. In relation to a point outside the water-system, e.g. a point P on the sea-bed, the water might be moving to the left; let the speeds of the water at three levels, X, Y, and Z, be represented by the relative lengths of the arrows drawn in continuous line. Now, for reasons that will become apparent, the

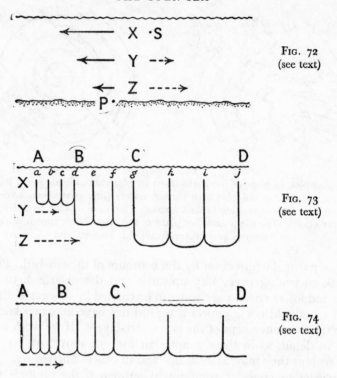

FIG. 72
(see text)

FIG. 73
(see text)

FIG. 74
(see text)

effects of variations in the animals' behaviour upon their distribution will be better understood if we consider the movements of the water, and so of the animals contained in it, in relation to the *surface* layer rather than to the fixed sea-bed. Now, in relation to a point S in the surface layer X, the lower layers Y and Z will be moving to the right at speeds in proportion to the lengths of the arrows in broken line. Now let us follow the path of a vertically migrating animal, observing it from the surface layer. In Fig. 73 let us start by observing it at night when it is up in the surface layer X at a point *a*; at dawn it will begin to sink down to remain during the daytime, let us say, at a level approaching that of layer Y, and on the next night it will rise again into the surface layer at a point *b*. The distance of *b* from *a* depends of course on the difference in speed of flow between the two water-layers. To the observer in the surface layer it remains stationary during the night at *b* until it sinks again at dawn. It will reappear in the surface layer at *c* and *d* on the next two nights; that is assuming that it went down in the daytime to the same depth each day. Now let us suppose that conditions in the area BC differed in some way from those in the former area AB and that this difference caused the animals to sink more deeply during the day; in relation to the surface layer they would be carried along a greater distance each day than before and on successive nights would occupy positions *e, f* and *g*. Suppose further that in area CD they met with conditions which made them migrate still more deeply: then on the following nights they might occupy positions *h, i* and *j*.

In the figure we have so far followed the path of a single animal; now let us consider what would happen to three separate animals behaving in just the same way as the one just followed, only let us imagine them separated from one another on one night at the points *a, b* and *c*. In three nights' time they would respectively be at the points *d, e* and *f*, and in another three nights at *g, h* and *i*. We see that if the majority of the members of a species behave in the same way under one set of conditions, and then together have their behaviour modified under other conditions, we may expect them to be *closer together* in some areas than in others. Exactly the same effect would be produced if the animals, instead of altering the depth to which they sank during the day, altered the length of time that they stayed below, as in Fig. 74. We do not yet know enough about the behaviour of such animals in the sea to know whether they are varying their behaviour in exactly this sort of way; we know, however, as we have already seen, that they *do* vary their behaviour in respect to these migrations and consequently such changes in movement must have some effect upon their relative concentration in various areas if the water layers are travelling at different speeds, as they usually are.

Marine naturalists have for a long time noticed that when they come across a dense concentration of the little planktonic plants, they often only make a very poor catch of animals in their tow-nets from the same area. Some have supposed that this is really a false impression, due to the little plants, in their great abundance, clogging up the fine meshes of the net and so reducing the amount of water passing through and consequently the number of animals being caught. This does, indeed, occur to some extent. But there can be no doubt that very often there is a real inverse relationship between the distribution of the animal and the plant plankton. Apart from certain exceptional cases (which are likely to be due to rather special causes, and need not concern us here) there are two different explanations which may be put forward to account for the phenomenon. When the herbivorous animals are abundant, they graze down the crop of planktonic plants, so that where there are many animals there are few plants; on the other hand, when the animals are less numerous than usual, they will allow the plants to multiply, to produce an exceptionally dense concentration. Much work has been done in the last two decades to show the reality and importance of this grazing relationship, particularly by Dr. Harvey and his co-workers (1935) at Plymouth, and in America by R. H. Fleming (1939), G. A. Riley and D. F. Bumpus (1946).

An alternative or at least additional explanation was first put forward by F. G. Pearcy in 1885 when he made a voyage on a herring lugger to the Shetland fishing grounds. It was he who first suggested that the herring tend to avoid dense patches of plant plankton. His observations and some more recent ones like them are discussed in Chapter 15 (p. 293) dealing with the relation of the plankton to the fisheries. Pearcy also observed that there was very little animal plankton in the dense phytoplankton zones, and spoke about the "exclusive effect" of the phytoplankton. When working on the plankton of the whaling grounds around South Georgia in the Antarctic we came across ample evidence of this inverse distribution of planktonic plants and animals. I noticed that while most of the animals were distributed away from the regions of dense phytoplankton, there were three species which had no relationship at all to the phytoplankton and a few other species which showed a positive relationship, being more abundant in the regions of dense phytoplankton. I then found that these three classes of animals had different types of vertical migrational behaviour. Those that showed the inverse relationship had a normal vertical migration; those that showed no relationship at all showed no trace of a vertical migration; and those that were more abundant in the phytoplankton had a vertical migration of a rather unusual kind: they came up earlier and stayed up much longer than the other species. On account of this (and a good deal of other evidence) I put forward what I called the hypothesis of *animal exclusion* (in Hardy and Gunther, 1935)—using these words because they were first used by Pearcy. The essence of the hypothesis was that the distributional relationship between the animals and the plants was due to a modification in the vertical migrational behaviour of the former in relation to dense concentrations of the latter. Most animals, it was supposed, would come up to feed for only a short time in the dense phytoplankton zones, possibly because of some antibiotic effect, and (as explained in a previous paragraph) would become distributed in larger numbers in other areas. On the other hand, the animals which stay up longest in the phytoplankton would tend to remain more concentrated in these regions.

This second explanation must still be regarded as hypothetical; my attempts to demonstrate it experimentally have so far failed, as also have experiments recently made by my colleague Dr. Bainbridge, except in the case of a few flagellates. Some of these are known to have a poisonous effect. Bainbridge, in a recent paper (1953), has stressed the importance of this modification in vertical migration in relation to distribution, but has rather reversed my emphasis; he believes that the animals become concentrated by modified vertical migration in regions of rich phytoplankton, then graze them down until they become poor and then move on by vertical migration to other areas. Space will not allow of a further discussion of this complex and difficult problem; I still think that perhaps some species of phytoplankton may at times control distribution by producing some antibiotic influence, but this is not incompatible with the grazing effect, which must certainly play a big part in curbing plant production.[1] The question of secretions into the water by both animals and plants, and their effects upon other organisms, has already been referred to (p. 63) and is clearly a very important one.

There can be no doubt that the patchy distribution of the plankton must be due to a great variety of causes. Nutritive salts, carried into the sunlit upper layers by columns of water upwelling against submerged banks, will cause patches of more intense phytoplankton growth. These, in turn, by providing more food, will give rise to uneven zooplankton production. Shoaling fish, such as herring and pilchards may graze down the zooplankton unevenly. Vast swarms of larvae may be sent up at intervals, from the various concentrations of benthic animals on different parts of the sea floor. The flow and interaction of waters of opposing current systems will swirl and stir these odd patterns of planktonic distribution into still further irregular patchiness. But in addition to all these factors the remarkable habit of vertical migration must, I believe, add to the complexity of the distribution picture, and not infrequently give rise in part to some of the particular patterns seen.

[1] Since this was written Dr. J. H. Ryther, of Harvard University, has published (*Ecology*, 35, 522-533, 1954) a most interesting study of the effects of phytoplankton on the feeding and survival of the freshwater cladoceran *Daphnia magna*. He shows that some product of *ageing* phytoplankton causes the animals to stop feeding and eventually to die; this effect, as he suggests, might well be a factor concerned in a possible 'animal exclusion'.

CHAPTER 12

LIFE IN THE DEPTHS

L ET US for this chapter imagine that we have been fortunate enough to be invited to accompany a research ship which is off on a voyage to capture some of the unfamiliar creatures which swim in the depths of the ocean, out beyond the continental shelf. Our vessel must be provided with large nets, a mile or two of steel cable and a powerful steam or electric winch. Usually only a special research ship can boast of such; but if this book should fall into the hands of some enterprising millionaire, it is my hope that I may be able to tempt him to add this equipment to his steam or motor-yacht, and so go hunting in the depths as did the Prince of Monaco in his famous *Princess Alice*. There must still be many more strange animals yet to be discovered in this hidden realm, and it is really not so very far away. Half a day's steaming to the west of Ireland or Scotland, and the ocean floor will be more than three thousand feet below us; in but a few hours it will be over a mile deep.

In this chapter we shall not be examining the actual floor of the ocean; we are going to look at what is sometimes called the bathy-pelagic life, i.e. the fish and plankton scattered through the deeper waters, perhaps many thousands of feet below the surface. It is a pitch dark or only dimly lit region, inhabited by a great variety of animals which mostly never touch the bottom or reach the surface; they spend all their days in a fluid world which must seem unending in all directions right and left and up and down. We are apt to think of these creatures as something very unusual, almost freakish; indeed many of them have bodies whose shapes appear, to our eye, fantastically grotesque. But if we do a little sum or two we shall begin to see their strangeness in a rather different light. More than two-thirds of the earth's surface—70.8 per cent to be exact—is sea and as much as 86 per cent of this is over a mile deep; expressed in another way 60.8 per cent of the whole world's surface is covered by this more-than-a-mile-deep water—and actually over half of this is more than $2\frac{1}{2}$ miles deep! Fig. 75 gives us a representation of the diagramatic

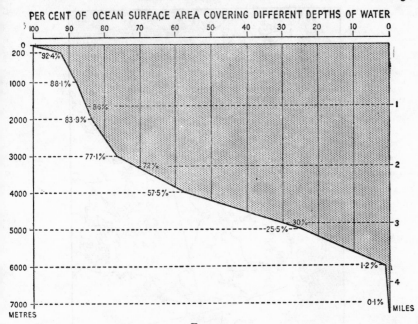

PER CENT OF OCEAN SURFACE AREA COVERING DIFFERENT DEPTHS OF WATER

FIG. 75

Diagram to show the proportions of the world's total ocean surface (including all small seas) covering water of different depths. For example 83.9 per cent. of the sea's surface area is covering depths of more than 2000 metres, and 72 per cent. is covering water over 2 miles deep. Scales of depth are given in both metres and miles.

relative proportions of the parts of the oceans lying between different ranges of depth, expressed in both thousands of metres and in miles. We know that all this water is inhabited, if only sparsely in the greater depths. How does life on land—on that less than a third of the earth's surface—compare with this in quantity? Quite an appreciable portion of the land is either ice-cap, desert or high mountain with very little life; indeed the whole antarctic continent is a barren frozen waste. The atmosphere has no permanent fauna of its own; at best the dry land has a zone of life from treetop height to but a few feet below the surface of the soil. We think again of the oceans—thousands of feet deep, inhabited at all levels and stretching over the greater part of the earth's surface,—and as we do so we begin to grasp the truth: we are setting out in our ship to get a glimpse of the most characteristic life of our globe—a realm of life almost entirely unknown to man a hundred years ago.

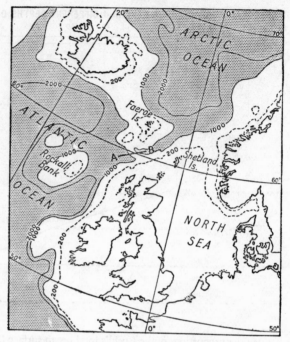

FIG. 76

Map of the Wyville-Thomson Ridge separating the deep water of the north Atlantic basin from that of the Arctic Ocean. The depth contours are shown in metres and the water over a 1,000 metres deep is shaded. The line AB, across the ridge, is that of the section shown in Fig. 77.

The deep sea areas of the world—those over a mile deep—are, with one exception, continuous with one another over all the world; the deep basins of the Atlantic, Pacific and Indian Oceans all join up with the great Southern Ocean deeps which completely surround the Antarctic continent.[1] All this is one great zone of life. The one exceptional part is the deep water of the Arctic Ocean; it alone is cut off from the deep water of all the rest of the world—cut off by the Wyville Thomson Ridge. This is a narrow ridge, like a submerged mountain range, linking the continental shelf on which stand Greenland, Iceland and the Faeroe Islands to that of the North Sea and Europe; it runs from south of the Faeroes towards Shetland and Orkney—a bar across the Faeroe-Shetland Channel—as seen in the

[1] Within the deep ocean basins there are pits and troughs of very great depth which are, of course, as isolated from one another as are the tops of different mountains on the land.

map in Fig. 76 opposite. The deep-water faunas on the two sides of the ridge, both bathypelagic and benthic, are different; we in these islands are privileged in having the two deep-water communities of the world on our door-step, in two separate basins: one to the north and one to the west. As we might expect, the basin to the west, which is that of the world at large, is very much richer in its variety of bathy-pelagic life than that of the colder and much more restricted Arctic Ocean.

The Wyville Thomson Ridge was discovered in a very unusual way: not in the first place by soundings but by thermometer readings. It was quite a dramatic event in the history of oceanography, and one worthy of recall. In Chapter 1 (p. 9) we referred to the early deep-sea dredging by Carpenter and Wyville Thomson in the *Lightning* and *Porcupine* expeditions to the north of Scotland; they took a series of temperatures from the surface down to near the bottom at different parts in the Faeroe-Shetland Channel. Down to 200 fathoms the temperatures were much the same at the north-east and south-west ends of the Channel, yet below this there was a striking difference. To the north-east at depths of 250 fathoms and near the bottom at 640 fathoms, they recorded temperatures of 34° and 30° F. respectively; whereas to the south-west—not very far away—the temperatures at corresponding depths were 47° and 42°F. That was in 1868, when there were few other sea-temperatures recorded to compare with them, so that it was not at once realized just how extraordinary they were. It was only in 1880 after considering the many series of thermometer readings taken in different parts of the world by the great *Challenger* Expedition, that Sir Wyville Thomson, as he was then, predicted that a definite barrier across the Faeroe-Shetland Channel would be found.

The Hydrographer of the Admiralty sent the surveying vessel H.M.S. *Knight Errant* with T. H. Tizard and John Murray to investigate under Wyville Thomson's direction. Thomson, who was now rapidly failing in health and had had an attack of paralysis the year before, was unable to accompany the ship but went to direct operations from Stornoway in the Outer Hebrides. He lived to see his predicted discovery made, but died before the second expedition, H.M.S. *Triton*, came back in 1882 with valuable zoological collections to illustrate the difference in the benthic fauna on either side of the ridge. In that year Murray published a paper showing that "216 species and varieties were recorded from the warm area, and 217 species and varieties from the cold area, while only 48 species and varieties were found to be common to the two areas." It was nearly thirty years later that Murray and Hjort, on the *Michael Sars* expedition of 1910, reinvestigated the region and gave us our knowledge of the difference in the bathypelagic life already referred to. Fig. 77, below, is taken from their book *Depths of the Ocean* and represents a section through the Wyville Thomson Ridge from points A to B in the map of Fig. 76 (p. 220); it shows the remarkable difference in the temperature and salinity on either side of it.

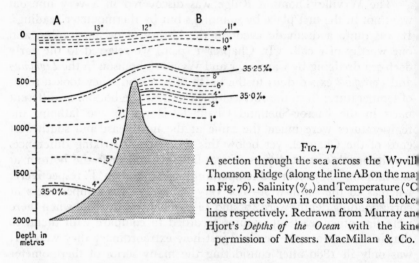

FIG. 77

A section through the sea across the Wyvill Thomson Ridge (along the line AB on the map in Fig. 76). Salinity (‰) and Temperature (°C contours are shown in continuous and broke lines respectively. Redrawn from Murray an Hjort's *Depths of the Ocean* with the kin permission of Messrs. MacMillan & Co.

    Just as the *Challenger* expedition made known the ocean bed and its benthic life, so the German Deep-Sea Expedition, in the *Valdivia*, 1898-99, revealed the bathypelagic life of the open deeps. By using large tow-nets at various levels between the surface and the bottom of the ocean in many parts of the world, they brought back a large collection of animals new to science. Then came the *Michael Sars* expedition just referred to. In the autumn of 1909 Sir John Murray wrote to Professor Johan Hjort of Oslo to say that if the Norwegian Government would lend their fishery research vessel *Michael Sars* and her scientific staff for a four months' summer cruise in the North Atlantic, he would pay all expenses. How fortunate it was that they accepted his proposal. It was only a short cruise, but it yielded great additions to marine natural history. Murray and Hjort, by using tow-nets that could be closed at the end of their tow before being hauled up (as explained on p. 72) and using them simultaneously at several different levels, gave us a much better knowledge of the depth-distribution of many different forms. They were specially interested in the various adaptations of these animals to conditions in the different zones below the surface. Their book, *The Depths of the Ocean* (1912), will always remain a classic of marine biology, a great experience to read, and a mine of facts for reference. The Danish expeditions, between 1908 and 1930, under the leadership of Johannes Schmidt in the *Dana* also gave us much information, particularly regarding deep-sea fish; their work has already been referred to (p. 21) in relation to Schmidt's famous discovery of the breeding grounds of our freshwater eels—in the far-away depths of the Atlantic.

There is something peculiarly exciting about lowering away a large net into the depths and hauling it up again from the darkness of a mile or so beneath us. What may it not bring back? To see the bulging net, the metal ring keeping it open, and the rope bridles disappearing from view into the depths, always reminds me in a curious and inverted way of the ascent of an old-fashioned Victorian balloon. On our way out to the Antarctic on the 1925–27 *Discovery* expedition we took many deep-water hauls in the South Atlantic, to compare with those taken by the *Michael Sars* in the north. On that voyage we used, on many occasions, a large net with a ring of 4½ metres diameter; it was the largest tow-net ever to be fished, and some idea of its size may be gained from the view of it in Plate XXIV (p. 289). A net with so great an opening was difficult to get in and out board in a ship so heavily rigged as was the old *Discovery*; but it was worth it because it filtered much more water than a smaller net and gave a better chance of picking up the rare animals which are so sparsely scattered through the darkness of these great depths.

Let us suppose we have now reached deep water and a 2-metre-diameter net has been lowered away to a depth of one or two thousand metres, towed for two hours, and is on its way up. I am sure no one can help being intensely curious to see what the bucket at the end of the net may contain. There is always a chance that it may bring to light—how appropriate is the phrase!—an animal which has never been seen before. Not only have such new animals never been seen by man; it is likely that their shape and colour have never previously been visible to any living creature and some of them may be of great beauty. The idea of their form as a part of objective reality would, at such moments, almost appear to be being created before our very eyes. The words from Gray's Elegy come vividly to mind:

> *Full many a gem of purest ray serene*
> *The dark unfathom'd caves of ocean bear:*
> *Full many a flower is born to blush unseen*
> *And waste its sweetness on the desert air.*

I have been lucky enough to be present on several occasions when animals new to science have made their first appearance. We can rarely be certain, of course, that our find is really new until some considerable research has been done on its anatomy, and into the literature of the group to which it belongs. That makes it really more exciting. We can pick out, from a large haul, perhaps half-a-dozen animals which might turn out to be new; then, if we have a good library on board, as we had on the *Discovery*, we may hunt through the monographs of former expeditions and learn that our treasures were taken three times by the *Valdivia* expedition, once by the *Siboga*, twice by the *Challenger*, and so on. Five out of the half-dozen we may thus identify, but the sixth looks really new; nothing like it appears to have been pulled up by anyone before. We make a colour sketch, take measurements and notes, and preserve it carefully for the specialist in that particular group of animals who will examine it when we get home; not until we have his verdict are we absolutely sure that we have played a part in the discovery of something new. I cannot resist putting into one of my coloured plates (Plate 10, p. 145) a drawing I made of one such animal on the old R.R.S. *Discovery* in 1926—a very unusual bathypelagic nemertine worm, *Bathynemertes hardyi*; it was kindly named thus by my old friend and colleague of the expedition Dr. J. F. G. Wheeler who wrote the monograph on the *Discovery* nemertines and collected so many of them himself. Actually this specimen came from the Atlantic south of the equator, but I have no doubt that it will be found in the North Atlantic as well.[1]

While we have been considering the possibility of our making a discovery, the tail of the long net has been lifted over the stern; it is

[1] Indeed, since this was written, I have recently (1954) returned from a cruise on the *Discovery II* towards the Azores on which a much smaller specimen of probably the same species was taken.

Plate 17. THE HALF-RED CRUSTACEANS OF THE DEEPER OCEAN PLANKTON, ALL × 1½, EXCEPT No. 6 (×2)
1. The euphausiacean *Nematoscelis megalopj*
2. The penaeid prawn *Sergestes vigilax*
3. Another *Sergestes: S. arcticus*
4. The amphipod *Scina*
5. The prawn *Parapandalus richardi*
6. A larval prawn, probably a hoplophorid
7. An immature specimen of *Systellaspis debilis* which is all scarlet when adult as shown in Plate 18

(All drawn on board the R.R.S. *Discovery II*, August 1952)

1

2

3

4

5

6

7

Plate 17

1

2

5

3

4

6

A.C.H.

Plate 18

now sluiced with water to wash any animals still sticking to the inside of the netting down into the bucket at the end. The cords securing the bucket to the net are next untied and its contents poured into a large flat bath for our inspection. What a sight it is! There are hundreds of specimens. Black fish and scarlet crustacea predominate—but dotted amongst them, like bright silver coins, are the curious flat hatchet-fish (Sternoptychiidae), ranging in size from a sixpence to a half-crown (Plate 22, p. 245); here and there will be some of the remarkable deep-sea medusae: red, cream and purple *Atolla*, like some exotic water-lily flowers, or deep plum-coloured *Periphylla* with long graceful tentacles. The black fish are not all of one kind—there are many different species, varying much in shape. Likewise the crustacea; there are scarlet prawns of several different families, scarlet mysids, and amphipods, copepods and ostracods of the same brilliant hue (Plates 12 and 18, pp. 161 and opposite). In addition there are species of all these groups which are partly red and partly transparent (Plate 17, p. 224); they look as if they were in the process of developing such a livery, but as Dennell has recently pointed out (1955), at any rate for the sergestid prawns, there is a fundamental difference between the colouration of the 'all-red' and the 'half-red' kinds. In the 'all-red' species the pigment is carried in the cuticle, whereas in the 'half-reds' the colour is mainly due to large red pigmented cells (chromatophores) underneath the skin.

The animals I refer to, or illustrate in the plates and figures, may all be taken in the deep water just beyond the continental shelf of the British Isles. In the late summer of 1952 I had the privilege of joining the R.R.S. *Discovery II* on one of her cruises in the North Atlantic; it was then that I made most of the colour drawings of the bathypelagic life shown in the plates. They were all drawn from freshly-caught specimens, with the exception of the Angler fish on Plate 24, and most of them were still alive—kept alive till the moment of drawing by being put at once into the darkness and cold of a refrigerator. Now, after completing the rest of the book I have been lucky enough to join the *Discovery II* again for another three weeks' cruise over the deep

---

Plate 18.   THE ALL-RED CRUSTACEANS FROM THE GREAT OCEAN DEPTHS

1. The scarlet mysid *Eucopia unguiculata*, female, × 2½
2. The penaeid prawn *Sergestes robustus*, × 1
3. The carid prawn *Acanthephyra haeckeli* (= *multipina*) × 1
4. Another carid prawn *Systellaspis debilis*, × 1
5 and 6. Deep water amphipods *Scypholanceola* and *Lanceola*, respectively, × 2½
(All drawn on board the R.R.S. *Discovery II*, August 1952)

TOS—Q

water between Britain and the Azores in September 1954; this enabled me to add considerably to my collection of drawings. Finally, just before going to press, I am glad to be able to note the publication of an important new book *Aspects of Deep Sea Biology* by Mr. N. B. Marshall of the British Museum of Natural History; it is a mine of information on bathypelagic adaptations, particularly concerning the fish upon which he is a specialist. Had it appeared earlier I should have made more extensive reference to it; I have however here and there, at the last minute, been able to call attention to some of the interesting new points he makes.

The great development of pigment, particularly black, red, orange or dark brown, is the most striking feature of the bathypelagic life in contrast to that nearer the surface. Most of the crustacea of the upper layers are protectively transparent, as are the arrow-worms, the medusae, the ctenophores and many other planktonic animals. The fish which come near the surface, such as herring, sprat and mackerel, tend to have dark greeny-blue backs, which make them merge into the colour of the depths when looked at from above, and silvery sides and bellies, which reflect the colour of the surrounding water when viewed at their own level or from below. In short, in the well-illu-minated layers most animals are invisible either because we look right through them to see the water beyond or because we appear to see it in their looking-glass sides. It is amusing to realize that, if such animals are too thick to be transparent, nature makes them adopt the well-known conjurer's disappearance trick: "all done by mirrors."

In these surface regions natural selection will always be tending to eliminate animals which are made conspicuous by the presence of yellow, red or black pigments, for these will stand out the more clearly against the blue green background of the sunlit waters. The parallel with man's warfare is obvious: the expectancy of life of a soldier who wears a bright red coat in broad daylight is much less than that of one who has a uniform matching his surroundings. At night a dark-coloured uniform will be more advantageous than a white or a pale-coloured one, and a spy may more readily creep through the enemy lines if his face is blackened; it is the same for the animals in the perpetual night of the great depths.

Light measurements at various levels below the surface of the sea have shown that the red and orange rays of the spectrum are absorbed by the water much more quickly than the blue or violet rays. Murray and Hjort made observations in the Sargasso Sea on the *Michael Sars* expedition. Writing in *The Depths of the Ocean* (p. 663) Hjort says:

"On a sunny day when the water was perfectly clear and transparent, light rays of all colours, but very few red rays, were observed at a depth of 100 metres. At 500 metres the light acted strongly on the photographic plates, especially the blue rays, but the green rays were absent; even at 1000 metres the influence of the sunlight could be traced on the plates, but at 1,700 metres no influence was noticeable." More recently the American workers R. H. Oster and G. L. Clarke (1935) have confirmed these results and given us much more exact readings. Crustacea tend to produce red (carotenoid) pigments much more easily than black (melanin), but with fish it is the reverse; in the depths below the penetration of the red rays, however, the scarlet prawns will *appear* just as black as the melanin-producing fish. In the bath up on deck we look at scarlet prawns which have never looked like that before; they, and the black fish, have come from depths of below 500 metres. The silver-sided fish are those which have come from above that level—mainly between 500 and 300 metres—caught while the net has been coming up from the greater depths; some of these silver fish, such as *Myctophum*, may migrate to the very surface at night.

Before we note some of the other more general characteristics of the deep pelagic life, let us look a little more closely at the principal types of animals which make up this unfamiliar community. We shall not, of course, find all of those which I am about to describe in one single haul, but we may get a good many of them. We might meet all of them in a week or two's fishing in the waters to the west of Ireland or Scotland. The characteristic deep-water medusae *Atolla* and *Periphylla* I have already referred to; they are illustrated in colour in Plate 8, p. 129. As a rule *Periphylla hyacinthinia* is drawn with its tentacles hanging limply down; in life they are extended in all directions feeling outwards for possible prey. Hjort records that several different colour-variations of *Atolla bairdi* were recognised on their expedition; the least pigmented was more often taken at 500 metres, and those with more pigment at 750 and 1000 metres. Another less common deep-sea form is the beautiful dark madder-brown *Nausithoe rubra* Vanhöffen which I was lucky enough to be able to watch still actively swimming after it had been brought up from a depth of 800 metres; this is also shown on Plate 8.

Among the strangest members of this deep pelagic fauna are the floating nemertines—primitive unsegmented worms, which possess a long protrusible proboscis which is said to be prehensile and used in the capture of prey. The majority of nemertines live along the coasts and are truly worm-like in shape, but *Pelagonemertes* of the depths has a body more flat and leaf-like, and is transparent except for the branching gut, which develops a brilliant orange (or sometimes brown) pigment. It is shown on Plate 10 (p. 145) together with the scarlet *Bathynemertes hardyi* to which I have already referred (p. 224); the latter in contrast is remarkably solid and heavy-looking for a pelagic worm. It was taken in a net fishing at a depth of not more than 1000 metres at a point where the Atlantic is some 5000 metres deep. The deep-water arrow-worms likewise develop bright orange pigment, as seen in *Eukrohnia fowleri* and *Sagitta macrocephala*, also in Plate 10; if we are only familiar with their perfectly transparent cousins of the surface layers, we may be quite surprised when first we see these golden shafts brought up from some 750 metres depth. The polychaete worm *Tomopteris* may also be a vivid crimson in this underworld; and the deep-water ptero-pods are dark in colour, as the beautiful purple *Clio polita* which, in its perfectly glass-like shell, is figured in the same plate, as also is the *Tomopteris*.

Now we must come to the deep water crustacea. There are a large number of bright red copepods, some of which carry brilliant yellow or orange plumes, and at least one very handsome form is rich claret colour; three examples are shown on Plate 12 (p. 161). It will be noticed that the average size of the copepods of deep water is a good deal larger than that of those near the surface. This is true of nearly all the planktonic invertebrates. The reason is not far to seek. In any natural community of animals there is what the ecologists call a 'pyramid of numbers'; we have a large number of herbivorous animals feeding on the available supply of plant food; next is a smaller number of carnivorous animals feeding upon the herbivores, a still smaller number of predators feeding in turn on the others and perhaps yet fewer still of a higher category of animals preying upon these. Each step contains a smaller number of individuals than the step below. But from the base of this pyramid towards the summit the sizes of the animals will usually tend to increase: B feeding on A will usually be larger than A; C feeding in turn on B, will usually be larger still, and so on. (Parasites are a notable exception to such a general rule). In the sea we tend to get the small copepods feeding directly on the minute plants, and rather larger carnivorous copepods feeding upon these, and perhaps others, larger still, feeding upon them.

The pyramid, of course, is essentially a metaphorical simile; in the ocean depths, however, it has some reality in space—but upside down. The plants upon which the whole pyramid is based are only found in the sunlit surface layers; and here also will be found the multitudes of small animals feeding directly upon them. All the animals in the depths below are dependent for their food either upon dead or dying forms sinking from the upper layers, or on animals which vertically migrate downwards at regular intervals. So the food-chains are directed downwards. The smaller particles of dead and dying material will tend to be eaten up in the upper layers or decay away more quickly than the larger pieces which, having less surface per volume, will sink through the water faster. The available food-particles tend to get fewer and larger as we pass downwards, and so the animals feeding upon them are also fewer and (within limits) larger. The deep water pelagic animals can never be very big, because there is not enough food to support a really large body; but they must be capable of taking large mouthfuls at relatively long intervals. We shall see many examples of animals adapted to do this, when we come to the bathypelagic fish.

Like the deep copepods, the deep ostracods are larger than those nearer the surface. The largest ostracod known comes from the lower layers of the ocean; it is the remarkable globular form *Gigantocypris* which, when adult, assumes a deep orange colour. In Plate 12 (p. 161), I have figured both young and adult specimens; in the more transparent immature forms one can see much more of the internal structure. The paired eyes have huge metallic-looking reflectors behind them, making them appear like the headlamps of a large car; they look out through clear glass-like windows in the otherwise orange carapace and no doubt these concave mirrors behind serve instead of a lens in front. How these animals capture their prey has seemed a mystery, for they feed on very active animals. Professor H. Graham Cannon (1940), who has made a beautiful study of the intricate internal anatomy of this animal, quotes a letter from Dr. Kemp writing of his experiences on the *Discovery*: ". . . we have had them in bowls of sea-water on various occasions, and I am sure they are not capable of any rapid movements—as indeed one would guess from their shape. They rock and roll a lot, finding it difficult to keep on an even keel, and swim feebly." An examination of their gut-contents, however, shows that they are capable of capturing the large copepods like *Pleuromamma robusta*, young fish and large *Sagitta*, all most active swimmers. Cannon believes they must just pounce, and grab with their mandibular palps, any unsuspecting prey which chances to pass close enough to them; but it seems questionable whether they could get sufficient food in the depths by such a means. I have thought that perhaps they secrete and spread out some kind of a sticky snare like a spider's web, of which all trace is lost as the animal is dragged up in the net; on a recent (1954) cruise of the *Discovery II*, however, we caught a *Gigantocypris* (the adult shown in the Plate) which was remarkably active and swam backwards and forwards across a large bowl as fast as any prawn. Two other smaller deep water ostracods, one bright red and the other deep brown with beautiful golden antennae, are also figured in the same plate.

Undoubtedly the most characteristic crustaceans of the deep ocean layers are the large scarlet prawns. The suborder Natantia of the order Decapoda, to which they belong, is divided into three distinct 'Tribes' as they are usually called, two of which concern us: the Penaeidea and the Caridea. My reason for introducing this refinement of zoological classification is to draw attention to the striking convergent evolution which has taken place in these two groups. On Plate 18, p. 225, we see *Acanthephyra multispina* and *Systellaspis debilis* which are carids, and *Sergestes robustus* which is a penaeid. They are remarkably similar. But if we look at their abdominal segments, we see something which at once distinguishes them. In the carids the side plate of the second segment overlaps the plate of the segment in front as well as the one behind, whereas in the penaeids this is never so; this little feature shows us that these two kinds of prawn have come by quite distinct evolutionary lines to appear at last so much alike.

The long whip-like antennae of Sergestes are worth noticing; there is a more rigid part up to the kink and then a much more flexible part like a lash. We see examples in both Plates 17 (Figs. 2 and 3) and 18 (Fig. 2). Can they be used as fishing rods? I think it possible. Along the flexible lash are a vast number of curved hooks which may perhaps be used to catch prey and draw them towards the limbs which are often provided with their cruel claw-like spines. The small hooks are really too fine to be shown in the plates; I have however indicated them in Plate 17, Fig. 2, p. 224.

Many of the deep water prawns are provided with luminous organs, such as those seen as the black markings on the sides and legs of *Systellaspis debilis* in Plate 18 (p. 225); a discussion of luminous organs in general will be reserved till the next chapter (p. 256) when we will have seen so many striking examples among the fish. Apart from having luminous organs, some of the deep water prawns produce a luminous cloud in the water when attacked; this evidently serves as a smoke-screen to hide their escape and will be referred to again later (p. 257). On Plate 18 we also see a vivid scarlet mysid, *Eucopia unguicu-*

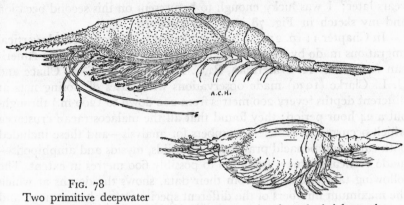

FIG. 78

Two primitive deepwater crustaceans of special zoological interest (see text): *above*, the bright scarlet mysid *Gnathophausia gigas* (natural size) and *below*, *Nebaliopsis typica* ($\times$ 1$\frac{1}{2}$).

lata; and two red amphipods, one of which is more crimson (*Lanceola*) and the other more orange (*Scypholanceola*).

There is a much more remarkable deep water mysid *Gnathophausia* which I was not able to draw in colour from a living specimen on my recent cruises; I saw several brought up on the 1925–27 *Discovery* expedition and I have made a drawing in Fig. 78 above, from one of the largest specimens we then obtained. In life it is just as red as a prawn and has a long serrated rostrum exactly like a carid. You would, at first sight, think it was a true prawn—but it is yet another striking example of evolutionary convergence. It is, however, of special interest to zoologists for another reason. If you count the number of segments in its abdomen, you will find it has seven, whereas all typical mysids have six,[1] like all other malacostracan crustaceans except the most primitive forms like *Nebalia* and its close relatives. *Gnathophausia* shows a link between the higher crustacea and a more primitive stock. One of the ancient *Nebalia*-like crustaceans, *Nebaliopsis typica*, lingers on in the great depths with *Gnathophausia*, but is a distinct rarity; and it is usually brought up to the surface dead and in a very damaged condition. Only twice has it ever been seen alive—on the Swedish *Antarctica* Expedition in 1904 and on the *Discovery II* fifty

[1] Dr. Sidnie Manton, in her beautiful study of the development of *Hemimysis* shows indeed that the typical mysids have seven abdominal segments as embryos, but the last two fuse together as they grow up. (*Phil. Trans. Roy. Soc.*, B. 216, 363-463, 1928); then she gives a detailed account of the incomplete fusion of these segments in *Gnathophausia* and its allies (*Trans. Roy. Soc. Edin.*, 56, 103-119, 1928).

years later; I was lucky enough to be present on this second occasion and my sketch in Fig. 78 is of this specimen.

In Chapter 11 (p. 210), I briefly referred to the considerable vertical migrations made by many of the bathypelagic organisms. The American workers T. H. Waterman, R. F. Nunnemacher, F. A. Chace and G. L. Clarke (1939) made observations with series of closing nets at different depths (every 200 metres from 200 down to 1200 m.) throughout a 24 hour period; they found that all the malacostracan crustacea which occurred in sufficient numbers for analysis—and these included both carid and penaeid prawns, euphausiids, mysids and amphipods—made vertical journeys from 200 to possibly 600 metres in extent. The following table, compiled from their data, shows the depths at which the maximum numbers of the different species were taken at noon and at midnight, and the approximate range of their probable migration. They must have had speeds of vertical movement ranging from 24 to 125 metres per hour.

			DEPTH AT NOON	DEPTH AT MIDNIGHT	RANGE OF MIGRATION
Carids:	*Gennadas elegans*	..	800	400	400
	Hymenodora glacialis	..	1000	800	200
	Parapasiphae sulcatifrons	..	800	400	400
	Acanthephyra purpurea	..	800	200	600
Penaeid:	*Sergestes arcticus*	..	600	200	400
Euphausiids:	*Nematoscelis megalops*	..	400	200	200
	Thysanopoda acutifrons	..	800	200	600
Mysids:	*Boreomysis microps*	..	800	400	400
	Eucopia unguiculata	..	800	400	400
Amphipods:	*Cyphocaris anonyx*	..	600	200	400
	Vibilia propinqua	..	600	200	400

Turning now to the fish, we will begin with those beautiful blue-and-silver members of the Myctophidae—or Scopelidae as they used to be called—which are not strictly deep-living forms at all; some often come to the very surface at night. If you stop your ship and hang powerful electric lights over the side, you may attract numbers of these little myctophids, and see them as streaks of silver; they also appear to flash a bright red light at you as they dart across the patch of illuminated water. Actually what you think at first must be a powerful red luminous organ, turns out to be the little fish's eye catching the light—as do cats' eyes in the light of a car—but reflecting back only the red rays. They have indeed quite prominent luminous organs studding their sides, as will be seen in a typical myctophid as figured in Plate 22 (p. 245); sometimes they are called 'lantern-fish', but they do not always shine. The specimen drawn is one of many caught by hand-nets from the deck of the *Discovery* as they came like moths to the light. Along its dark back are scales of a deep iridescent blue; and its sides are a brilliant silver, which here and there is touched with a sheen of pin, or green, like mother-of-pearl. There are a great many different species of myctophids, but most are very similar in appearance. On the 1925–27 *Discovery* expedition we collected sixteen species of *Myctophum* and fifteen, nine and three species respectively of the closely allied genera *Lampanyctus, Diaphus* and *Lampadena*, all from the North and South Atlantic; they are mostly warm water forms, but a few, such as *Myctophum glaciale* and *M. punctatum* extend as far north as the Norwegian Sea. Some are sure to be met with over the deep water to the west of the British Isles. Marshall in his recent book (1954) records that one of the weather ships, when steaming towards Glasgow from her station in the Atlantic, passed for five hours through an immense field of small fish which, on putting over a net, were found to belong to the last mentioned species.

Descending deeper, we come to representatives of one of the most remarkable groups of fish, the Sternoptychiidae, sometimes called the hatchet-fish, which mainly inhabit the layers between depths of 250 and 500 metres. Two examples, *Argyropelecus hemigymmus* and *A. aculeatus* are shown in Plate 22 (p. 245). Earlier in this chapter, when describing the appearance of the contents of a deep-water net, I said that these little fish looked like silver coins scattered among the black and red of the fish and prawns from greater depths: not only are they bright silver, generally somewhat rounded in appearance and almost the size of coins, but they are as *flat* as coins. I have never seen them brought up to the surface alive; they are always lying on their sides dead, as we see them in the tow-net bucket. It must be a remarkable sight to see them swimming in their true position, mirroring the dim twilight of the depths in their sides, and perhaps switching on their luminous organs, which form a prominent double row on the underside. It is curious that all these 'lamps' should point downwards while the large tubular eyes are for ever looking upwards; this arrangement, the odd shape of their bodies, and their relatively huge mouths with their corners drawn so sharply down, all combine to give these fish a most grotesque appearance. If they are figured in a so-called 'popular science' article in one of the less reputable magazines, their true size is almost always kept from the reader—who is left to imagine them at least the size of a codfish and possibly bigger. The specimen of *Argyropelecus aculeatus* shown lifesize in the plate is actually a large specimen for this species; the very largest member of the genus, *A. gigas*, a new species which we discovered on our 1925–27 *Discovery* expedition, did not reach a size more than half as much again as this.

Dr. William Beebe, looking out from the window of his bathysphere, was the first person to see these strange fish swimming in their natural environment. He gives a most graphic account in his book *Half Mile Down* (1935) of how at a depth of 1,050 feet he "saw a series of luminous coloured dots moving along slowly, or jerking unsteadily past, similar and yet independent." Then when he turned on his searchlight he saw that he was looking at a school of *Argyropelecus*, and "frisking in and out among them" were numbers of pteropods "like a pack of dogs round the horses."

Marshall (1954) discusses the nature of the curious tubular eyes that these hatchet-fish have in common with a number of other deep water species. He shows that they have no well-defined mechanism for altering the focus of the eye, but that, in addition to the main retina at the back, there is an accessory one placed up the side of the tube, nearer the lens. Tubular-eyed fish all have their two optical axes parallel, or nearly so, and thus have the requirements of binocular vision. "But as well as being binocular," writes Marshall "the visual system is also bifocal. The accessory retina, which is closer to the lens, will be in focus for distant objects, while the main retina will come into focus for nearby objects."

In this region of silver-sided fish, from 300 to 500 metres deep, is to be found the beautiful long and slender *Stomias boa*, perhaps the commonest stomiatid of the Atlantic. A small specimen is shown in Plate 22 (p. 245), with its iridescent silver sides, and the bright red tip to the barbel which springs from its lower jaw; this looks very like a small red copepod and presumably serves to lure other smaller fishes to destruction. Along the greater part of its length, on its underside, are four rows of small luminous organs becoming a double row towards the tail. A larger specimen would be some six or seven inches in length. Many of the fish of these regions make very extensive vertical migrations. In the daytime *Stomias boa* is most abundant at a depth of about 500 metres, and occasionally may be taken deeper still; but at night it is up in the top 50 metres, and Hjort records it once in a surface-net. Still more extensive migrations are made by another very common deep water fish *Chauliodus sloanei*, which is seen on Plate 23 (p. 252); I was just going to write 'common deep water fish of the Atlantic', but it is world-wide in its distribution, and just as common in the Indian and Pacific Oceans. In the daytime the largest number are taken below a depth of 1000 metres, whereas at night the majority of the younger ones are to be found in the top 200 metres and the older ones between 400 and 600 metres (Haffner, 1952). It is a fish which lacks scales, but the pigment forms hexagonal scale-like areas in which are deposits of opalescent crystals; those along the back give the appearance of iridescent scales, but the sides of the body, at any rate of the specimen I drew, were lustreless and grey. Its most characteristic feature is the first dorsal fin-ray which is greatly enlarged and hinged like the angler-fish's fishing rod; it no doubt serves the same function but has not moved forward, as has the angler's, to the front of its head. Marshall (*loc. cit.*) says that the end of its line is luminous, also like the lure of many angler-fish. It, too, has luminous organs like those of *Stomias*, on the underside. Two other characteristic fish of these regions are also shown in Plate 23: *Vinciguerria attennata*, with very prominent luminous organs and a patch of iridescent silver scales on the fore part of its body, which otherwise is grey; and *Valenciennellus tripunctulatus*, also with only a confined area of brilliant scales but with luminous organs arranged in groups reminiscent of *Argyropelecus*—it is like a more normal fish on the way to becoming a hatchet-fish. Here, too, we may mention the abundant all-grey *Cyclothone braueri* (Plate 23) which is found in largest numbers round about 500 metres, extending down to 1000 metres; below 1000 metres its place is taken by its near relative *C. microdon* which is black.

Before passing to consider the still more unusual fish which inhabit the greater depths, some mention should be made of an at present unsolved puzzle concerning the layers we have just been discussing; it is a problem posed by war-time research. The modern method of echo-sounding will now be familiar to most readers: sound-waves from detonations are given out from the bottom of the ship, travel through the water to the sea-bed, and are echoed back again to the ship, where they are picked up on a receiver linked to an electric recording device. The distance travelled by sound-waves in water in a given time is known and constant; thus half the time taken for the sound to go to the sea-bed and back will at once give us an accurate measure of the depth of the sea at that point. The electric recording apparatus, usually on the ship's bridge, gives a continuous tracing of the depth of the bottom of the sea over which the ship is steaming; in fact it draws a section of the sea on a slowly unrolling band of paper: the surface is represented by a straight line and the bottom by a curve which follows its rise and fall as at every second a new point on the line is automatically added to it by the echoes from below. A scale provided enables us to read off the depth at any moment. Occasionally echoes come back from obstructions in the water between the surface and the bottom; a submarine would give such an echo, hence the great value of the invention in war, as well as in peacetime navigation. It has become a useful instrument in commercial fishing as well as in fishery research; shoals of herring, sprat and cod may not only be located—the actual area they cover may be charted. Intensive research on its development was naturally undertaken in the war both by the United States and our own country as part of the anti-submarine campaign. Very soon it became evident that there was a curious 'something' in the midwaters of the open ocean—a layer of something that was scattering the sound waves and sending back echoes far above the bottom. On the recorder chart was shown the surface and the bottom, but between them was also a definite but smudgy layer stretching for hundreds and hundreds of miles; it was found in all the deep oceans of the world, except the arctic or antarctic seas, and—most important—it rose towards the surface at night and sank to a depth of some 800 metres by day. It has been given the name of the Deep Scattering Layer.

The echo-trace on the chart coming from this scattering layer is not so hard and distinct as that from a shoal of herring or larger fish— it appears as if reflected from very large numbers of something much smaller. It was not long before biologists put two and two together; the layer came up at night and went down in the daytime, just as the vertically migrating plankton did, and it appeared to be made up of many little objects that were smaller than fish. Would small plankton animals give such an echo? Opinion is still divided on the subject. At first it was thought, and by some still is, that the animals causing the echoes might be numbers of fairly large crustacea like the deep-water prawns and euphausiids, which we have seen can make extensive migrations over just such a range as that of the scattering layer; and the euphausiids certainly occur in vast numbers. There are two facts, however, that seem to tell against this hypothesis. First, the distribution of these prawns and euphausiids is often very patchy, whereas the tracing on the record shows a remarkable uniformity. Secondly, the mysterious layer has hardly ever been recorded in the antarctic seas, although many attempts have been made to try and find it; if it had been due to a crowding of pelagic crustacea, surely one would have expected it to be more pronounced in the antarctic than anywhere else, for it is here that we find the greatest concentrations of euphausiids, such as nourish the vast numbers of the huge plankton-feeding whales, which migrate thither to

feed from all the seas of the southern hemisphere. A new hypothesis
has been put forward by Marshall (1951), as a result of his recent
investigations into the development of the air-bladder among the
bathypelagic fish. A body containing an air-space acts as a resonator
and gives a much stronger echo in water than anything else—that is
why a submarine or a whale (with air in its lungs) gives such a strong
echo. This is also why fish are recorded so well; most species have
'air-bladders', hydrostatic organs, which, by means of a controlled
gas-producing gland and gas-releasing valves, enable them to adjust
their buoyancy to a particular depth or enable them to rise and
sink more easily. Now it had previously been thought that these
bathypelagic fish, which we have seen also make extensive vertical
migrations, had no air-bladders or only vestigial remains of them.
Marshall has now shown that the majority of the Myctophidae,
Gonostomatidae and Sternoptychidae have, in fact, very well developed
air-bladders. He now believes that it is these fish which cause the
scattering layer; not only do they migrate through the required
distances, but they are not found in the cold polar seas. If he is right
it means that these little fish, which rush up towards the surface at night,
are not only in very much larger numbers than we had ever imagined,
but they must also be distributed remarkably evenly through the oceans.
Beebe, from his observations through the window of his bathysphere,
had already said that he thought these fish were indeed much more
numerous than our net collections made us believe. The big tow-
nets are towed only very slowly through the water; it would not be
surprising if the great majority of these very actively swimming fish
were to make their escape through the large mouth of the net and leave
only a small sample behind. Many research ships will be investigating
this problem before long, and it will not surprise me if the answer should
be found before this book finally reaches publication.

I must here insert some news which has indeed come in just before
going to press. It does not give the solution to the problem we have
been discussing; it rather poses another which may—or may not—
have a bearing on it. I refer to the wonderful dives which have so
recently been accomplished by the two French Naval Officers, Lt.
Comdr. Georges Houot and Lt. Pierre Willm in their bathyscaphe—a
true navigating submarine of the great depths. On 17th February
1954 they reached the ocean floor 2½ miles down. On this vertical
voyage and several earlier ones to depths of over 1000 metres, they
looked out of their port-hole at the bathypelagic life lit up by their

powerful floodlights, just as Beebe had done, but going deeper. They passed through what should have been the deep scattering layer, but they did not see any mass of animals—fish or plankton—concentrated about a relatively narrow zone such as they had expected from the echo charts; they have, however, seen something instead which at first sight looks like being an even bigger puzzle. A short but vivid account of this has recently been given by Captain Jacques-Yves Cousteau (1954), the great pioneer of the aqualung, who accompanied Comdr. Houot on a dive of 673 fathoms (1230 metres) to the bottom of the Mediterranean in December 1953.

"So far as we can see," he writes, "there is, biologically speaking, no deep scattering layer, but rather a great bowl of living soup extending on down and growing thicker the deeper into the 'tureen' we go." He reminds us that Beebe, and Barton who went with him, had both described the life getting more and more abundant the deeper they went, and then he adds "but so far little attention has been paid to their statement." The generally accepted idea of an increasing scarcity of life in the great depths—the one which I have adopted in this chapter in common with all standard works on oceanography—appears, to put it mildly, to have received something of a shock. We must take these observations very seriously; they are those of five men, Beebe and Barton in the bathysphere, and Houot, Willm and Cousteau in the bathyscaphe who have been there to see for themselves.

Of course I do not doubt their veracity, but for two reasons I think we must suspend judgment upon their interpretation of what they saw, until more detailed observations have been made. There have been many series of vertical tow-net hauls taken from great depths— so operated that each net was closed after covering a certain upward distance—say 2,500 to 2000 metres, 2000 to 1,500 metres and so on; as far as I am aware these collections do not support the view that there is an increase of life with depth. Again, as Cousteau realizes and remarks, his observations are not at all what one would expect if the only supply of energy and food comes from the sunshine and plant-life of the uppermost layers. "There must", he says, "be somewhere an unsuspected link in the cycle of marine life yet to be discovered." We must of course keep our minds open for some startling new discoveries; at the same time, however, we must admit that the conception of organic food being built up in the depths without photosynthesis seems most improbable. Perhaps there is another explanation.

I am interested particularly in one remark of Cousteau's which I think may possibly give the key to this new puzzle he has set. Looking

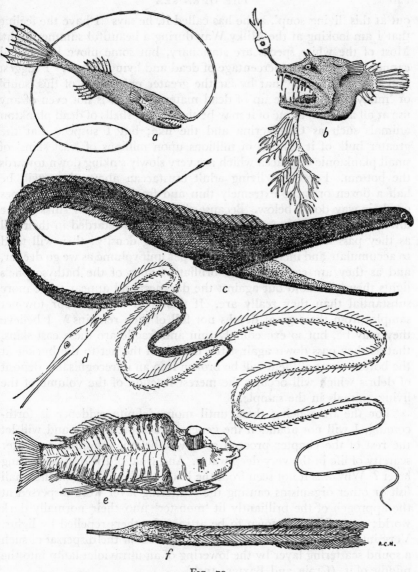

Fig. 79

A selection of deepwater fish: *a* and *b*, two remarkable angler fish *Gigantactis macronema* and *Linophryne arborifer* (×½); *c*, *Eurypharynx pelicanoides* (×1½); *d*, *Nemichthys scolopaceus* (nat. size); *e*, *Opisthoproctus grimaldii* (×½); *f*, *Cyema atrum* (×⅔). *a* and *b* are drawn from scale models in the fish gallery of the British Museum and the remainder from specimens caught by the R.R.S. *Discovery* and now in the British Museum.

out at this 'living soup', as he has called it, he says "I have the feeling that I am looking at the Milky Way during a beautiful summer night. Most of the white specks are stationary, but some move in jerks. I cannot determine the percentage of dead and living matter." I suggest that it is most likely that by far the greater proportion of this 'soup' or 'milky way' is made up of dead material which is not even of any use at all as food. Some of it may be the sinking shells of dead plankton animals such as Globigerina and the like; but I suspect that the greater bulk of it consists of millions upon millions of 'cast skins' of small planktonic crustacea which are very slowly sinking down towards the bottom. For every living adult crustacean above there will be half a dozen or more extremely thin and fragile chitinous envelopes on their way down below. Because the heavier seawater sinks to the lower layers, these tenuous trifles will tend to be retarded in their fall as they pass into water of slightly increasing density; they will tend to accumulate and increase in number per unit volume as we go deeper, and as they are caught in the brilliant beams of the bathyscaphe's lights they will stand out against the darkness and appear to be more substantial than they really are. If this is so why are our tow-net samples from these great depths not full of such remains? I believe they may be, but so exceedingly thin and fragile are these cast skins, that as they pass down against the meshes of the net to the bucket at the bottom most of them will be broken into an unrecognisable deposit of debris which will occupy the merest fraction of the volume of the living animals in the sample.

For the time being then, until more definite evidence is forthcoming, I will not abandon the generally accepted views and will let the rest of the chapter proceed as it was before, assuming a greater scarcity of life in the very deep water. And what of the deep scattering layer? Why was it not seen from the bathyscaphe? Perhaps the small fish or other organisms causing the scattering layer were dispersed at the approach of the brilliantly lit 'monster' into their normally dark world; some fish are known to be attracted, others repelled by lights. Very interesting observations have been made on the dispersal of such a sound scattering layer by the lowering of an ultraviolet lamp into the middle of it. (Craig and Baxter, 1952).

Plate 19. Sketches of Colour Change in the Oceanic Squid *Stenoteuthis* (*Ommastrephis*) *pteropus*
While drawn from two specimens in tanks on the deck of the *Discovery II*, one animal can change from the colour of the light individual to that of the dark one in a flash.

A.C.H.

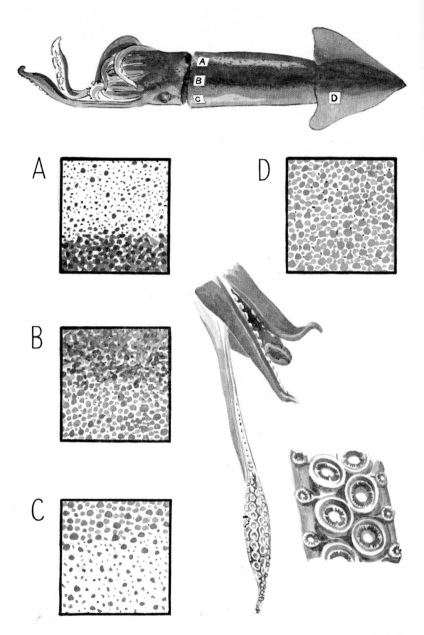

A

B

C

D

A.C.H.

Plate 20

Let us now return to our consideration of the fish in the deeper layers. It is among those living in the depths below 1000 metres that we meet with creatures having some of the very oddest of shapes; a few are shown in Fig. 79 (p. 239). Here we may find small angler-fish of many different genera, all with relatively huge mouths and well-developed 'fishing-rods', which are sometimes provided with a luminous lure. Then there are the fantastic members of the Lyomeri or Gulpers such as *Eurypharynx*, which seem to be all mouth and thin tapering whiplike body, or the Nemichthydae with their extremely long slender bird-like beaks, which set one wondering how, and on what, they can feed. Even the fish of more normal shape in these depths, such as the black Gonostomatidae, have mouths which can be unfolded and opened to a prodigious extent, enabling them to swallow prey actually bigger than themselves. Murray and Hjort figure a specimen of *Chiasmodon niger*, 5.7 cm. long, which has swallowed a larger specimen of its kind and has it curled up within its enormously distended stomach which bulges the underside of the body out like a large bag hanging below the body. I match their illustration with a similar drawing in Fig. 80 of a deep-water angler-fish *Melanocetus johnsoni* (also shown in Plate 24) which had swallowed a lantern-fish nearly three times its own length!

The late Dr. V. V. Tchernavin (1947, '48) made a special study of the mouth-mechanisms of some of these deep-water fish, and showed how the vertebral column is bent up in a sharp fold when the mouth makes its full gape, and how the various bones involved in the suspension of these jaws form a system of hinged levers which unfold upon one another to make the opening of the mouth as wide as it can possibly be. Figs. 81 a, b and c (p. 243) are taken from two of his papers, one showing the arrangement in the common *Chauliodus sloanei* which we have illustrated in Plate 23 (p. 252), and the other of *Saccopharynx ampullaceus*, one of the Gulpers (Lyomeri), of which a small specimen of a closely allied form is shown in Plate 24 (p. 253). These latter fish have, as we have already noted, an exceedingly long thin tail; this in many forms, including the one in the plate, has a

---

Plate 20. DETAILS OF THE SQUID'S COLOUR CHANGE MECHANISM
A, B, C and D are accurate enlarged drawings of the state of the expansion and contraction of the different coloured pigment cells (chromatophores)—red, blue and orange—seen in the correspondingly marked squares of skin of *Sthenoteuthis pteropus* immediately after death. Sketches of the arms and details of sucker arrangement are also shown drawn from another specimen.

TOS—R

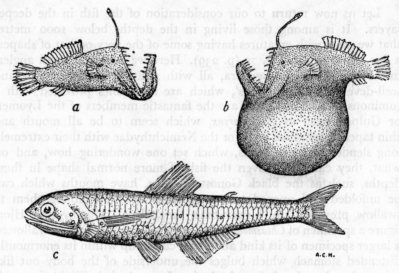

FIG. 80

a and *b*, the deepwater angler fish *Melanocetus johnsoni* before and after swallowing *c*, a myctophid fish *Lampanyctus crocodilus*, nearly three times its own length. Drawn from scale models in the fish gallery of the British Museum ($\times \frac{1}{2}$).

luminous ending. This specimen came up in the net with its tail tightly coiled round some long broken filaments which might have been part of a siphonophore. While it had most likely made contact with this animal on the way up in the tow-net bucket, it does suggest how these fish may use their tails; *i.e.* to whip round prey attracted by the moving light and so hold them tight till brought within reach of the enormous mouth. All along either side of the body, in the specimen drawn in the plate, is a row of little stalked papillae; evidently these are remarkable modifications of the lateral line organs such as are referred to by Marshall (1954, p. 242) and are probably used to guide the 'lasso' to its goal by the disturbance in the water produced by the swimming of its prey. At some parts of the tail we see these papillae and at others the dorsal and ventral fin rays, according to the way it is twisted; I have attempted to show this in the plate.

This great increase in size of mouth is one of the most striking adaptations to life at these depths. We are now nearing the apex of the inverted 'pyramid of numbers' we referred to on page 228; the deeper we go the scarcer is the food, and the fewer are the smaller forms of life. The predators can only expect to meet with a meal very occasionally, and it is likely to be a prey as big as itself; these

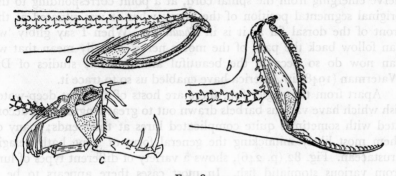

FIG. 81

Showing the skeletal jaw mechanisms which give some deep-water fish such an enormous gape; note the sharp bending upwards of the backbone. *a* and *b*, *Saccopharynx ampullaceus* with mouth closed and open (see Plate 24); *c*, *Chauliodus sloanei* (see Plate 23). Redrawn from Tchernavin (1947, 1948).

opportunities of large meals at long intervals have led in the course of time to the selection of varieties having larger and larger mouths. Relatively to size, the swallowing capacity of the boa-constrictor is rivalled by these monstrous little fish.

Another characteristic feature of the fish in the depths is the development of various kinds of angling devices. We have in passing already noted the 'fishing-rod' of the true angler-fish; it is well worth a little closer attention, for it presents us with an interesting example of the migration of an organ from one part of the body to another. It is well known that the little supporting bony rays of fins are movable, each one swivelling upon a little ball-and-socket joint as it is pulled this way or that by small strands of muscle. Sometimes these fin-rays are enlarged as offensive spines, or for purposes of adornment in courtship display; or they may be greatly lengthened as a fishing-rod to dangle a little lure—a model of some tempting planktonic morsel—which may be jerked in front of an unsuspecting prey, so drawing it nearer to the huge mouth that is waiting to engulf it. *Chauliodus*, as we have seen, has the first dorsal fin-ray developed enormously and provided with a light but still situated in its normal position. The angler-fish shows us the remarkable phenomenon of a fin-ray which has become so modified in the course of evolution as actually to move from its original position to take up one on the very front of the head. We can be sure that it has so migrated, for we can follow back the path of the motor nerve controlling the contractions of the little muscles which work the rod—follow it back till we see it join the segmental nerve emerging from the spinal cord, at a point corresponding to the original segmental position of the finray when it was situated at the front of the dorsal fin as it is in *Chauliodus*. When I say glibly 'we can follow back the path of the motor nerve' I really mean that we can now do so because the beautiful and intricate studies of Dr. Waterman (1948) in America have enabled us so to trace it.

Apart from the angler-fish, there are hosts of different deep-water fish which have various barbels drawn out to great lengths, and decorated with sometimes quite complicated lures at their ends; many of them most likely mimicking the general form of some bathypelagic crustacean. Fig. 82 (p. 246), shows a variety of different types of lure from various stomiatid fish. In most cases there appears to be a luminous bulb at the end and, close against it, a number of branched processes which may be dimly illuminated and suggest the appendages of a luminous copepod or prawn. Such barbels usually spring from the lower jaw, and as here are frequently shown curving backwards; no doubt when fishing they may be projected forwards in front of the mouth, as was that of the fresh specimen of *Stomias boa* figured in Plate 22 (p. 245).

The long interval which may elapse before one fish finds another in this dark world of the ocean depths, has had a curious consequence in the sex-life of many of the angler-fish. The young of any species of fish will be more numerous than the adults and so the young males will have a better chance of meeting a female than would an older male. Natural selection has favoured the early meeting of the sexes and the linking of the male to the female in preparation for the later act of reproduction. The young male grips the skin of the female with its jaws and remains attached—for life. Over long periods of time selection has favoured the earlier and earlier attachment of the male, and his gradual conversion into a small parasitic organism sharing the female's food supply; the smaller he is, of course, the better, considering how scarce food is. The tiny males may be attached at all sorts of places on the body of the female: to her face, her flanks or her

---

Plate 21. LITTLE DEEP-WATER SQUIDS

1. *Calliteuthis reversa* (×2) studded with opalescent photophores (light organs).
   a, from the left side, *b*, from the right side, but when lying on its back. Note the remarkable difference in the two eyes and their surrounding photophores. Brought up alive from 1,000 metres in the R.R.S. *Discovery II*, September, 1954
2. *Thaumatolampus* (*Lycoteuthis*) *diadema* (×1½)
   Drawn on the old R.R.S. *Discovery*, in 1925
3. A young specimen of *Mastigoteuthis* (× 2)

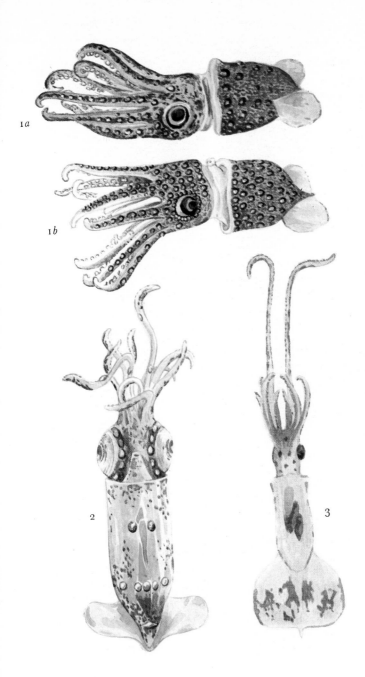

1a

1b

2

3

A.C.H.

Plate 21

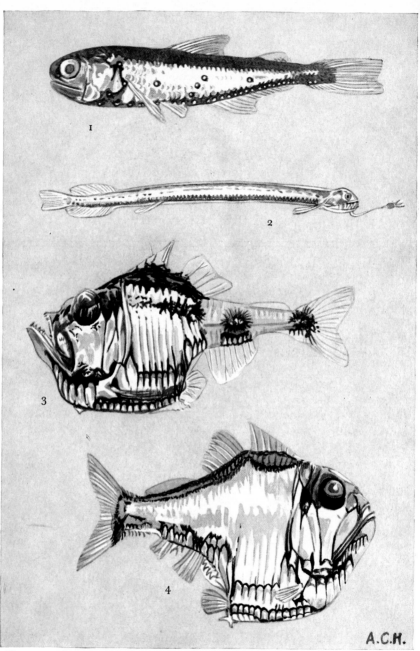

1

2

3

4

A.C.H.

Plate 22

belly; their bodies fuse on to hers like a graft, and their blood systems become continuous one with another. In return for the nourishment supplied by the blood of his spouse, and perhaps stimulated by hormones from the same source—or could they pass perhaps in the reverse direction? —the little male eventually supplies the sperm to fertilize her eggs. This, one of the most remarkable facts of natural history, was first revealed by the late Mr. Tate Regan, Director of the British Museum (Natural History), when he examined a specimen of a quite large and new species of angler-fish which had been landed by a trawler at Hull after fishing very deep down the continental slope off Iceland. A sketch of this specimen is shown in Fig. 83 (p. 247). It was probably a fish from much deeper water which had just by chance come up to within range of the trawl. The skipper realized he had found something unusual and kept it carefully for examination by an expert; his care was rewarded by this wonderful discovery. This same species has recently been found by Mr. Robert Clarke (1950) in the stomach of a sperm whale, and it has now been shown to have a much more remarkable fishing rod than was originally supposed. The rod can be pulled backwards in a most unexpected way to bring the lure, and the 'hoped-for' prey following it, up to the mouth; it slides backwards in a tube until its base, or hind end, covered by a pouch of skin, actually sticks out behind from the back of the fish and points towards the tail! The rod in the original Icelandic specimen was broken; in my illustration I have put it in the drawn-back position, sketching it from Mr. Clarke's specimen which is also in the British Museum. The unexpected discovery of the dwarf male in this species led Tate Regan to hunt through the collections of deep-water angler-fish brought back by some of the great expeditions, in the hope of finding more examples of this strange partnership; he was amply rewarded, particularly from the rich material obtained by Johannes Schmidt's *Dana* Expeditions. Dwarf males in various stages of adaptation as parasites on the females are now known to occur in four families of angler-fish, all from deep water (Regan, 1925, 1930).

Another character of bathypelagic fish, and one to which Hjort drew attention, is the progressive reduction in the relative size of the

Plate 22. OCEANIC SILVERY FISH FROM THE UPPER 500 METRES
1. A myctophid Lantern-fish *Myctophum punctatum* × 1½
2. *Stomias boa* (× 1½)
3 and 4. Two species of hatchet fish: *Argyropelecus hemigymnus* (× 2) and *A. aculeatus* (× 1½) respectively
Drawn on board the R.R.S. *Discovery II*

eye with depth. The following is a quotation from *The Depths of the Ocean* (p. 681):

"In the fish taken between 150 and 500 metres the diameter of the eye compared to the length of the head is, according to Brauer, as follows:—

Stomias	about	1:4	*Argyropelecus*	about	1:2
Chauliodus	„	1:4	*Sternoptyx*	„	1:2
Ichthyoccocus	„	1:2.6	*Opisthroproctus*	„	1:4
Vinciguerria	„	1:3			

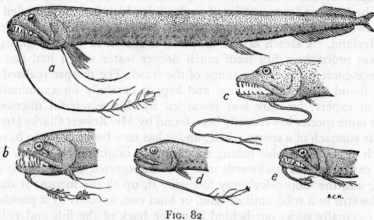

FIG. 82

Barbels of various Stomiated fish sketched from specimens in the British Museum, all natural size: *a, Flagellostomias boureei; b, Echiostoma tanneri; c, Eustomias obscurus; d, Leptostomias ramosus; e, Chirostomias pliopterus.*

If we consider *Cyclothone* and other fish which live deeper than 500 metres we find the following relations:—

Cyclothone	*signata*	1:12
„	*microdon*	1:12
„	*obscura*	1:15 or 20,

and if we inspect the figures representing *Gastrostomus bairdii, Cyema atrum,* and *Gonostoma,* we obtain a still stronger impression of the small size of the eyes. Finally our deepest pelagic hauls contained blind forms which have never been taken in the upper layers."

He partly links this reduction in eye size with the loss of light. "Although in fact," he says, "many cases as yet seem inexplicable, there seems reason for supposing that the efficiency of the eyes decreases with the decreasing intensity of light as we descend into deep water." If there was an advantage in the animals seeing at all then I should have thought, provided there was sufficient light, larger eyes would have been evolved in the dimmer regions just as nocturnal animals tend to have large eyes. When there is not enough light to see by, then, as in

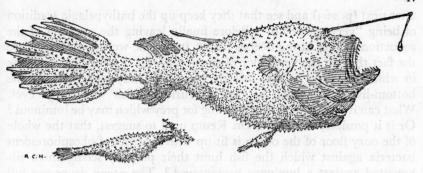

FIG. 83

The female angler fish *Ceratias holbölli* with the small parasitic male attached
($\times \frac{4}{10}$) and enlarged view of male. Drawn from specimens in the British Museum
(Natural History).

cave-animals, variations with poor eyes (or quite blind) will not be
at a disadvantage and so will not be eliminated by selection on that
count. Later on Hjort draws attention to the fact that the luminous
organs tend to diminish in size also with depth and he associates the
decrease in eyes with this as well. He writes (p. 685):

"From what has been said we see that a remarkable coincidence exists between
the development of light-organs and eyes in pelagic fishes. The Scopelidae,
Sternoptychidae, and Stomiatidae, which live above 500 metres, possess well-
developed light-organs and eyes, while from 500 metres downwards light-organs
and eyes both decrease in size."

There is a widespread misconception that the deeper fish are better
equipped with luminous organs. Actually the most highly developed
photophores are to be found on fish in the region of 500 metres and
particularly, as Hjort also pointed out, on those which make extensive vertical migrations. The production of light is certainly one of the
most typical characteristics of bathypelagic fish, crustacea and cephalopods, but it is so prominent a feature of the open sea as a whole that
it seems proper that it should have a chapter to itself. This will be
the next one, so the further discussion of the luminous organs of fish
(and there is much more to be said) will only be deferred for a few
pages, after we have dealt with marine phorphorescence in general.

Another matter which might logically have found a place in this
chapter is the occurrence of deep-water squids, for they certainly
provide a most interesting feature of this bathypelagic life. But here
again we have a subject worthy of special treatment. These large
molluscs are the only invertebrates which have approached the power,
speed and organization of fish. We shall come to them in the chapter

after next (p. 263) and see that they keep up the bathypelagic tradition of being "very queer fish". Before finally leaving the question of the reduction of eyes in the deeper layers, it may be well to mention here the fact that on the very floor of the ocean—far down below the zone in which the deep pelagic fish are blind—we may come across some bottom-living fish, such as the Macruridae, with very large eyes. What can it mean? Are they hunting for prey which may be luminous? Or is it possible, as the late Dr. Kemp used to suggest, that the whole of the oozy floor of the ocean is lit up with a carpet of phosphorescent bacteria against which the fish hunt their prey by seeing them silhouetted against a luminous background? The ocean deeps are full of problems yet to be solved.

CHAPTER 13
PHOSPHORESCENCE AND PHOTOPHORES

PHOSPHORESCENCE is certainly one of the most striking phenomena of the sea, and one that puzzled our sailors, and indeed scientists, for very many years before they knew what it really was. Tachard, an ecclesiastic, expressed the opinion in 1686 that the ocean absorbed the light of the sun by day and emitted it again at night,[1] and our great chemist, Robert Boyle, in the same period believed that the light of the sea was caused by friction, either between the waves and the atmosphere or by the waves striking an object like a ship. Benjamin Franklin at first thought it must be an electrical phenomenon taking place between the particles of water and those of salt; later, however, he gave up this view as the result of an experiment he made in 1750 in which he found that sea water ceased to sparkle after being kept for a time. In the same year the two Venetian naturalists, Professor Vianelli and Dr. Grixellini discovered that phosphorescence in the Adriatic was due to the organism *Noctiluca;* and a little later Spallanzani observed the luminous jelly-fish *Pelagia* in the Mediterranean, although in fact he was only confirming what was already known to Pliny. Then followed the discovery of light-producing marine animals by many naturalists.

[1] Taken from Phipson (1862).

Through the life-long studies of Professor Newton Harvey of Princeton University, which are brought together in his two more recent books (1940, 1952), we now know a good deal about how the light is made and have a wonderful account of the many different kinds of this living luminosity. There are some for whom a knowledge of the chemistry of the process seems to satisfy all their curiosity; to the naturalist, however, who is interested to know what function it may serve in terms of advantage to the organism, these flashing lights are almost as great a mystery as ever. In a few cases we can, of course, see at once a likely use for them: as in the last chapter when we were considering the luminous lures used by so many deep water fish to draw unsuspecting prey towards their rapacious mouths. Other fish, which we shall soon refer to, have been seen to shine a beam of light forwards like a searchlight; these we can also understand. But what of all the photophores—those beautifully designed light-producing organs which stud the undersides of so many fish, squids, prawns and other crustacea?

These photophores are wonderfully made, with a lens in front of the light-source and a concave reflector behind; they are as good a parallel to the bull's-eye lantern as the camera is to the eye. They have been evolved *quite independently* in many different groups, and on each occasion have produced a very similar type of 'lamp': clearly they must serve some very vital function. Variations progressing little by little towards the building of an efficient lantern, would never have been selected towards that end unless at each step they conferred some advantage on their lucky owners in competition with those not so well equipped. There is, perhaps, only one thing more remarkable than their presence upon the bodies of so many different animals: the fact that, in spite of many suggestions, it has been so difficult to find a really convincing explanation of the existence of the great majority of photophores, or indeed of the even more generally held property of giving off flashes of light without having any definite organ for the purpose. One very curious feature must be taken into account; on many of the animals with such organs, the lights point away from the direction in which the eyes are looking, so they can hardly serve the purpose of helping their owner to see more easily in the dark; often the eyes point upwards and the photophores downward, as we see in *Argyropelecus* (Plate 22. p. 245) and in some of the prawns.

Before going any further, however, I must give chapter and verse for the observations I have just mentioned regarding fish which definitely do project a 'spotlight' forwards for the capture of prey in the dark. It is one of the most remarkable accounts of bioluminescence that I know of. It was written in 1931 in the biological log of the R.R.S. *William Scoresby* by the late Mr. E. R. Gunther who was killed in the war; there it lay, apparently overlooked for nearly twenty years, until it was fortunately referred to in a *Discovery Report* by Mr. Robert Clarke (1950) dealing with a deep water angler-fish. He (Clarke) is discussing the evidence for deep-water fish coming up to the surface at night. He cites the occurrence of an eel-shaped stomiatoid fish, with brilliant luminous organs and silvery scales, feeding on krill[1] near the surface; it was seen from the deck of the *William Scoresby* as she lay alongside pack-ice to the east of the South Sandwich Islands. He first quotes the record of the incident from Dr. Mackintoch's Scientific Report of the voyage (unpublished). He then gives the more detailed notes of the feeding activities of the fish found, written by Mr. Gunther, in the Biological Deck Log; these are so exciting that I quote them in full:

"From a pair of luminous organs in the orbital region, the fish (which was 9–12 inches in length) emitted a beam, of varying intensity, of strong blue light which shone directly forwards for a distance of about two feet. The fish had the habit of lurking at a depth of 2–6 feet below the surface, poised at an angle of about 35–40° from the horizontal—this gave the beam an upward tilt: occasionally the fish swam round and with a quick action snapped at the cloud of krill above it.

"In its manner of lurking and of snapping prey it resembled the freshwater pike. From the anal region was seen to trail a length of brown substance which, it was supposed, might have been either genital or faecal product."

Apart from those which have definite luminous organs there are marine representatives of almost every major group of the animal kingdom which can produce flashes of light. In our survey of the main members of the plankton in the earlier chapters, we have mentioned in passing those which are known to be luminous. We will now briefly bring them together into one general discussion of the subject. Some of the very smallest of organisms—some of the bacteria—are brilliantly luminous. We must all, I should think, at some time in our lives, have been surprised, on going into the larder late at night, to see a fish, perhaps a large cod, which we thought was fresh, shining all over like the popular conception of a ghost. This luminescence on dead fish, or sometimes on meat, can be shown to be caused by bacteria; it is possible to infect plates of culture medium from such tainted food, and so establish little luminous centres of growth which gradually increase in size in typical bacterial fashion.

There are some marine animals, for example the tropical East Indian fish *Photoblepharon*, which cultivate their luminous bacteria on

---

[1] Euphausian shrimp-like crustacea.

special areas of their surface and so make their photophores by symbiosis: they give nourishment to the bacteria in exchange for light.
This fish has a luminous patch on each side of its head below its eye;
and it most probably serves as a lure, for Professor Harvey records
(1940) that native fishermen cut out these patches and use them as
bait. He says that the luminosity lasts for seven or eight hours. *Photoblepharon* has a fold of black pigmented skin, like an eyelid, which it
can draw across the lightened area as a blind to cut off the light. In
another fish, *Anomalops*, a similar bacterial photophore is made like a
little culture-plate on the back of a hinged shutter; it can be thrown
open to shine out or swung inwards into a recess to be completely obscured.

Some of the general phosphorescence of the sea may possibly be
caused at times by bacteria, but it is usually due to vast numbers of
little flagellates. The Dinoflagellates, such as *Ceratium* and *Peridinium*
and the aberrant globular form *Noctiluca*, give rise to the most brilliant
displays of this general lighting up of the sea. If you like fireworks it
is always an entertaining experience to take a rowboat out on a dark
night when some of these little flagellates are really abundant, as they
often are in August and September. Every time the oar touches the sea
there is a splash of flame, and as it is drawn through the water it leaves
a trail of fire behind it—as does the boat itself. Let Charles Darwin
give his account of such a night when on his famous voyage of the
Beagle; it is an entry in his journal under the date of 6 December 1833:

"While sailing a little south of the Plata on one very dark night, the sea presented a wonderful and most beautiful spectacle. There was a fresh breeze, and
every part of the surface, which during the day is seen as foam, now glowed with
a pale light. The vessel drove before her bows two billows of liquid phosphorus,
and in her wake she was followed by a milky train. As far as the eye reached,
the crest of every wave was bright, and the sky above the horizon, from the reflected glare of these livid flames, was not so utterly obscure as over the vault
of the heavens."

There is, of course, no need to voyage across the world to see such
displays—sometimes they may be equally brilliant in our own seas.
I have already described (p. 48) how I once saw every fish in a small
shoal outlined in 'fire' and it is not at all rare to see, especially in late
summer, every wave breaking on the beach with a flash of pale greenish
light. The little flagellates flash with light whenever they are violently
agitated. Mr. George Atkinson of Lowestoft recently told me of an
interesting occurrence during the first world war. A zeppelin dropped
some bombs which exploded in the sea a mile or two from land; after
each explosion there was a flash of phosphorescence through the sea
along the shore on which he was standing.

Among the coelenterates many of the small hydroid medusae are said to be luminous, and among the larger jelly-fish there is a very striking example in *Pelagia noctiluca* already referred to on p. 131.

It is the comb-jellies—the Ctenophora—which give us some of the most spectacular displays of brilliant flashing light in our waters. One such example I have mentioned in Chapter 8 (p. 137). They are nearly all capable of emitting sudden vivid flashes. The sea is often full of very small young specimens, each of which may give off quite a bright flash. They are excellent animals to use for demonstrations of spontaneous luminescence. A plankton sample containing these animals can nearly always be relied upon to give a good show—but we must remember that they do not perform at all until they have been in the dark for almost twenty minutes. If you intend to show your friends a good display you must keep your sample of plankton completely covered with light-proof cloth, or in a light-proof cupboard, for this length of time before bringing it out for exhibition in the darkened room.

As a young student I once had an amusing demonstration of this inhibitory effect of light. I had gone over to Brightlingsea to hunt at low tide in the thick Essex mud for the rare and curious worm-like animal *Priapulus*. It was nearly dark before I had found any and it was too late to return to Oxford that night, so I put up at a very old inn where I slept in a four-poster bed in an oak-panelled room. After a strenuous day digging in the mud I retired early and soon dropped to sleep after blowing out my candle. Later in the night I was awakened by some reveller coming noisily to bed in the room next door. I opened my eyes and blinked them with astonishment, for a number of little blue lights were bobbing about in the darkness just over the end of my bed. It was as if there were a lot of little goblins dancing up and down in the air. Before coming to bed I had of course celebrated the finding of *Priapulus*—but only with a pint of bitter; clearly there must be some more objective explanation! I struck a match and lit the candle. I now saw that, level with the end of my bed, was the top of the chimney-piece on which I had placed a row of large glass jars

---

Plate 23. GREY AND SEMI-SILVER FISH FROM THE MIDDLE OCEAN
1. *Cyclothone braueri* (×4)
2. *Vinciguerria attenuata* (×3)
3. *Chauliodus sloanei* (×2)
4. *Valenciennelus tripunctulatus* (×3)
              Drawn on board the R.R.S. *Discovery II*

Plate 23

1

2

3

4

5

6

A.C.H.

Plate 24

lled with sea-water, with a little mud at the bottom of each containing ny precious animals. Getting up and switching on the electric light examined them closely and then saw that the water was full of very oung ctenophores—*Pleurobrachia*, I think—actively swimming up and lown. They had certainly not been flashing when I first turned out he light and got into bed; nor were there any flashes when I settled nto bed for the second time—or rather not at once. I was now well wake, and it was some time before I could get off to sleep again; efore I did so, after about twenty minutes in the dark, the little blue devils' began their dance again.

The light given off by medusae and ctenophores is not produced y well-defined photophores. Of the medusa *Pelagia* it is the whole uter surface of the umbrella which appears to be luminous on stimu-ation, and in an earlier chapter I have already described how it gives ff masses of luminous slime when handled. The ctenophores never eem to give off any slime, and their light appears to be generated in ows of cells along the course of the eight meridional canals which underlie the bands of beating comb-plates. In an aquarium on deck t is not difficult to discern the outline of the animal when it flashes, or the bands of light impart a glow to its whole body.

Passing higher up the animal kingdom we come to the polychaete vorms, a few of which are luminous. Some of the bottom-living ones uch as *Chaetopterus* and the polynoid worms have well defined luminous areas, and the planktonic *Tomopteris* shows an approach to the forma-ion of segmental photophores arranged one on each side of its swim-ning-paddles or parapodia. So like in appearance are these organs o the better developed photophores of higher forms that they were irst described as eyes; they are said to be formed from the segmental xcretory (nephridial) funnels, whose cells have taken over light-roduction and then become surrounded by a globe of other trans-arent cells, apparently to serve as a lens. Each such organ must be under nervous control, for close against it is a special nerve-ending.

Plate 24. BLACK FISH FROM THE GREAT OCEAN DEPTHS

. *Parabrotula plagiophthalmus* × 2

and 3. Young (×4) and mature (×2) specimens of *Lampanyetus pusillus* (a deep-water myctophid)

. *Gonostoma bathyphilum* × 2

. *Saccopharynx johnsoni* × 2

. The angler fish *Melanocetus johnsoni* × 1

Drawn on board the R.R.S. *Discovery II*, except the last, which is from a preserved *Discovery* specimen

Here perhaps we see a step towards the evolution of a 'bull's eye lantern' organ: something originally serving quite another function—in this case excretion—produces by chance some luminous by-product which happens to give the animal some advantage; in the course of time a variety may occur in which adjacent cells become more translucent and so act as a lens—again this variety tends to survive better than others. So gradually, step by step, with varieties linking the cell group more and more closely with nerve endings, we begin to approach a lantern which may be stimulated—switched on and off—at will.

Coming to the great class of the Crustacea we find a number of sub-groups having photogenic members. Many ostracods and copepods, and at least one mysid, are known to flash brightly by the production of light-giving secretions from special glands; and true photophores of elaborate pattern have been evolved in three of the higher malacostracan groups: the euphausiids and two separate divisions of the decapod prawns: the carids and the penaeids. It is the secretions from the ostracod *Cypridina* which have been used so extensively particularly by Professor Newton Harvey, for his study of the chemistry of the light-producing reaction. It is quite outside the scope of this book to go into the biochemistry or physiology of this process in any detail, and for this Professor Harvey's books *Living Light* (1940) and *Bioluminescence* (1952) should be consulted.

From earlier work on the firefly and the glow-worm it had long been known that this animal light is the most efficient form of light production known; it is the coldest light, i.e. that produced with the minimum amount of energy lost as heat. In most forms of illumination more than half of the energy is dissipated as heat; in the glow-worm and the ostracod only about 1% is so lost. If only man could light his cities by the same chemical means there would be an enormous saving of energy! The process is thought to be brought about in the following steps. First, if a substance called luciferin and oxygen are brought together in the presence of an enzyme called luciferase which acts as a catalyst, then the luciferin and oxygen become oxyluciferin and water. This is not a simple combining of the oxygen with the luciferin; it has been shown to be a reduction process: the luciferin becomes oxyluciferin by giving up some hydrogen which then combines with the oxygen to form water. We do not know the exact chemical nature of either luciferin or luciferase, but the latter is certainly an enzyme and is supposed also to be most likely a protein. If we represent luciferin as 'LH_2,' and luciferase as 'A', we could express this first step by a chemical equation as follows:

$LH_2 + \frac{1}{2}O_2$ (oxygen) in presence of $A = L' + H_2O$ (water)

The little stroke (') against the L (oxyluciferin) represents a bit of extra energy (a quantum of energy) excited in L by the action of the catalytic enzyme A.

Now the second step is thought to be a jump back of the quantum of energy from the molecule of oxyluciferin L to the molecule of luciferase A, which now in turn becomes excited by this little turn of extra energy: $L' + A = L + A'$

Now as a final step the molecule of luciferase gives up the extra energy in the form of light.

It is interesting to note that this process is reversible. In darkness and an alkaline medium it goes in the direction indicated above with the liberation of light; in daylight, however, and in an acid medium it absorbs light energy and goes in the opposite direction.

Before passing to those animals which have well-developed photophores—the higher crustaceans, cephalopods and fish—a brief reference should be made to the brilliantly luminous *Pyrosoma*. Although it is a warm-water animal which never comes near our islands, it should be mentioned because it must have been noticed by many readers who have passed through the tropical seas. It is a pelagic tunicate—one of the primitive chordate animals allied to *Doliolum* and *Salpa* referred to on p. 150; a vast number of small individuals, which have been produced by budding, remain together to form a floating colony shaped like an elongated thimble. One species forms colonies several feet in length, but the common kind does not usually exceed about 9 inches. Each little member of the colony glows with a bright blue-green light when agitated. Sometimes they occur in enormous numbers near the surface at night.

I will quote from my journal (*Great Waters*, Collins, 1967) written on the voyage of the old *Discovery*: "On going to the stern the full magnificence of the spectacle was seen. There were thousands and thousands of these bright glowing forms being thrown up to the surface by the churning of the ship's propeller. Outside the track of the ship there was only an occasional flash, but stretching away in our wake for hundreds of yards, at times it seemed for a full half-mile, was a wide band of brightgreeny-blue phosphorescence."

H. N. Moseley, in *Notes by a Naturalist*, records that a *Pyrosoma*, four feet in length and ten inches in diameter, was caught by the *Challenger*; he then goes on to say: "I wrote my name with my finger on the surface of the giant *Pyrosoma*, as it lay on deck in a tub at night, and the name came out in a few seconds in letters of fire.

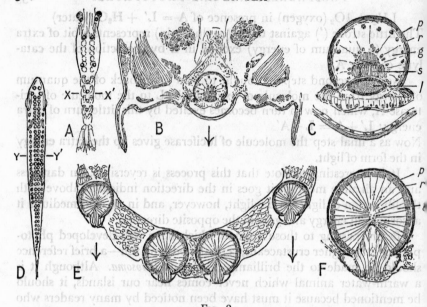

FIG. 84

Photophores of a euphausian crustacean *Meganyctiphanes norvegia* and of a
gonostomatid fish *Vinciguerria attenuata* compared; note how in each case they
are pointing vertically downwards. A, view of *Meganyctiphanes* from below
showing pattern of photophores; B, section through the underside along line
XX′ in A; C, section of photophore enlarged; D, underside of *Vinciguerria*;
E, section through the bodywall along the line YY′; F, section of one of the
photophores enlarged. *g*, glandcells; *l*, lens; *p*, pigment layer; *r*, reflecting
layer; *s*, striated body. Drawn from sections specially cut for me by Miss B.
Jordan in my department at Oxford.

Now let us look very briefly at some crustacean photophores. In
the euphausiids they are placed on the basal joints of the second and
seventh pairs of thoracic limbs, and in a median position on the underside of the first four segments of the abdomen, i.e. between the limbs.
These all point downwards and have a complex structure with reflector, lens and nerve-supply, as shown in Fig. 84 above. Two others of
rather different design are situated one on each eye-stalk.

The deepwater prawns have photophores in many genera; again
they usually point downwards, although on some they are drawn out
into long lines on the sides of the body as well. Again we see a
remarkable parallel in evolution between the penaeid and carid
prawns, this time in the positions occupied by their photophores.
So similar is their arrangement on the body that one might expect

them to have a common evolutionary origin far back in the ancestral stock before the two groups diverged. But no—the beautiful work by Professor Dennell (1940) on the material collected by the *Discovery* expeditions has shown that the microscopic structure of the photophores in the two groups is radically different. What is so remarkable is that, in addition to there being different types of *ectodermal* organs, there are, in *both groups*, curious internal photophores shining downwards through the roofing of their gill chambers. In each group these internal organs are produced by a modification of the cells of the liver tubules—but once again in each group they are of different design, showing evidence of separate but convergent evolution.

Before leaving the higher crustacea, mention should be made of the well authenticated and remarkable power that a number of deep-sea prawns possess of producing luminous clouds in the water. A very good account is given by Colonel A. Alcock (1902) in *A Naturalist in Indian Seas*: a fascinating record of his four years with the Royal Indian Marine Survey ship *Investigator*. He describes how "three large species of luminous deep-sea crustaceans were brought on board alive. . . . Far the most brilliant of them was *Heterocarpus alphousi*, both sexes of which poured out, apparently from the orifices of the "green glands" at the base of the antennae, copious clouds of a ghostly blue light of sufficient intensity to illuminate a bucket of seawater so that all its contents were visible in the clearest detail." Beebe (1935) frequently saw the discharge of luminous clouds by deep-sea crustacea through the window of his bathysphere. "The luminous discharge", he writes, "of large, *Acanthephyra*-like shrimps was a very different sort of illumination. Whereas the photophore-like organs of euphausiids may serve chiefly as recognition marks, or other non-defensive capacities, the luminous matter was obviously discharged only when a shrimp was startled, as it bumped against the bathysphere window." The luminous cloud in the darkness no doubt serves the purpose of baffling a pursuer and covering the line of retreat like a smoke screen or the ink cloud of an octopus or squid; indeed deep-water squids have now been observed to emit such a luminous "smoke screen" instead of one of ink.

Professor Dennell has more recently published (1955) the very interesting observations he made during a special visit to Bermuda to study the luminescence of a number of different bathypelagic prawns which may be obtained there. He was able to see it in eight species including one (*Hymenodora gracilis*) which had not hitherto been thought to be luminous. Most of the organs called photophores found on prawns had been so-called by inference from their

TOS—S

structure and hitherto very few had actually been seen to emit light; this new evidence of Dennell's is therefore very welcome. He was never able to stimulate them to flash, either by mechanical, chemical or electric means, and this made experimental work very difficult; on all the occasions when he saw them light up, they did so quite spontaneously. He found, however, that irritation nearly always caused the ejection of a luminous cloud from the prawn *Systellaspis debilis* and that this was accompanied by a rapid backwards escape action performed by a vigorous forward flick of the expanded tail-fan.

In the last chapter we left out almost all reference to the deep water squids, for they and other cephalopods are to have the next one all to themselves. But we must make some reference to them here, for some have well-developed luminous organs. It is remarkable how similar these are to some of those of fish—yet another example of adaptive convergence. Here again, as a rule, these organs are distributed over the body on the underside. Plate 21 (p. 244) shows the beautiful deep-water squid *Thaumatolampas diadema* with its photophores of different colours and with lenses like pearls; I drew it from a specimen as soon as it had come up in the net on the 1925 *Discovery* expedition. In the same plate are two drawings of the remarkable little *Calliteuthis reversa* which I was lucky enough to see alive on a recent cruise of the *Discovery II*; while its photophores are distributed widely over its surface, most of them are again on the underside.

The typical arrangement of the light-organs on bathypelagic fish are illustrated in Plates 22 and 23 relating to the last chapter; we see once again how they are nearly always on the lower surface and pointing downwards. In Fig. 84 we compare sections of photophores of a luminous fish with those of a euphausian, and their general similarity will at once be seen. Perhaps after all this is not so surprising; all efficient lanterns must have, in addition to the light source, a lens, reflector and a control to turn it on and off, so that a number of different lines of evolution are likely in the long run to finish up with much the same design. What now of their function? Can we hazard a guess?

We can be satisfied that some photophores in special positions must be used as lures to attract prey or as spotlights pointing forwards; and Beebe and Pyl (1944) have given evidence for the brilliant tail lights on some fish having, when suddenly flashed on, a blinding effect on the eyes of possible predators giving chase. Such explanations, however, cannot account for the great majority of light organs spread over the lower aspects of the body. Can they be recognition marks to enable the two sexes of the same species to find each other in the dark—and to prevent them from running after members of closely allied kinds? The rows of lights may perhaps be specific in courtship display between the opposite sexes, or threat postures between rival males, rather as the red breast of the robin or the red belly of the stickleback are used. Marshall, in his recent book (1954) gives some interesting instances of sexual differences in photophore pattern in various species of lantern-fish. It is said that some of the squids with photophores can actually change the colour of their lights, say from white to red, by drawing a little fold of pigmented skin over them as a blind. It has been suggested, too, that they might be used as a signal to advertise to members of the opposite sex the particular state of maturity they are in. Just as there are systems of lights on the front of locomotives to indicate the nature of the train—two white lights on the buffer-beam for an express, one white under the chimney for a local passenger, or a white one on the observer's right of the buffer-beam for a goods train—so possibly squids or fish might use such signs to announce that they are ready to mate, have just mated, or are immature. Perhaps the lights are so often on the animal's underside, pointing away from its eyes, in order that their owner's eyes may be shaded by the body from their glare. Marshall (1954) gives evidence that in some cases this may indeed be so.

Dennell, in his recent paper which I have just referred to, considers afresh the evidence for a view which had been put forward by several authors that large and well-developed eyes among the different kinds of deep-water prawns are usually correlated with the presence of photophores, and poorly developed eyes with their absence; he finds however that there is no real foundation for this supposition. "This conclusion" he says "does not support the view that the function of luminous organs in decapod crustacea involves vision and discrimination on the part of the animal, and seems to weaken the attractive suggestion that the photophores constitute a pattern of recognition marks facilitating swarming and mating." In spite of this, however, he thinks there is some evidence that swarming does take place in the luminous species; while he is writing of prawns, it is certainly true that the euphausiaceans, which are all very luminous, also usually occur in swarms. He quotes Harvey as saying that the luminous paint on watch dials has a brightness of 0.01 to 0.02 millilamberts and says that from his observations he judges that of the prawn photophores to be of about the same order. The ostracod *Cypridina* has a brilliance of 2.16 millilamberts. After discussing researches on the light sensitivity of the arthropod eye and on the penetration of light in sea water, he makes calculations which, most surprisingly, he says "suggest that the green component of the light of decapod photophores may well be mutually perceived at distances of the order of 100 metres."

While some of these suggestions regarding recognition signs and mating signals may well be correct in some instances—and we have the terrestrial examples of the firefly and the glow-worm to lend them plausibility—I find it difficult to believe they can explain such a widespread development of these organs in so many different groups. Apart from actual photophores, the production of light has been evolved independently in so many different groups. Among pelagic animals it is nearly as widespread as is the phenomenon of vertical migration. Is it possible that the two are in some way linked together ? Hjort writes the following passage in *The Depths of the Ocean* (p. 702):

"It was also very interesting to note the remarkable coincidence between the vertical migrations of the fishes and the development of their light-organs . . . (he now refers to a figure in the form of a table showing these vertical occurrences of five specimens of black fish) . . . Evidently we have here a type of deep-sea fishes, living in deep water, but with the power of migrating towards the surface. These forms have retained their well-developed light-organs, which in other black fishes of the deep sea must be considered as extremely reduced, perhaps even quite rudimentary, organs. A perfect analogy is found in the decapod crustacea. The deepest living species have no light-organs and make no vertical migrations. Light-organs, or organs which are believed to produce light, are found only in species living between 150 and 500 metres with a maximum distribution at about 500 metres. These species have been found much higher up in the water during the night than during the day."

Actually the numbers in his tables are much too small to be considered statistically significant, but the suggestion is worth looking into more fully. Can it be mere coincidence that we find photophores so well developed not only on the fish and decapod crustacea, but in the euphausiids, which are known to make most extensive vertical migrations.

In my first edition I then went on to outline some speculations I had made on the subject and described how I had done experiments in Loch Fyne to test one of these hypotheses and so proved it wrong. After suggesting another possible explanation I wrote as follows:

"I fear this spate of speculation has got out of hand; I just want to show how much of a mystery both bioluminescence and vertical migration still are. A few wild ideas may do no harm if they stimulate experiments to disprove them, and perhaps eventually reveal the unexpected truth. The solution to a problem is so often only just round the corner, and when found, looks so very different from what we imagined it would be."

Since then Dr. James Fraser in his excellent book *Nature Adrift* (1962) and Dr. William D. Clarke in an article in *Nature* (vol. 198, pp. 1244-7, 1963) have both, quite independently, provided what I believe must be the answer, not to a possible connection with vertical migration, but as to why so many deepwater animals have luminous organs on their undersides.

I will now quote a discussion of their idea which I give in my more recent book *Great Waters* (1967).

"In these very dimly lit regions the various fish and other animals will hunt by looking upwards to see their prey silhouetted darkly against the blue background of light coming from above. We do in fact see so many of the eyes of deep-sea fish pointing upwards. Now the luminous organs are remarkably evenly spread over the underside of nearly all the animals that have them pointing downwards; they suggest that the lights tend to destroy the dark silhouette appearance and so make their possessors invisible when viewed from below. It might be thought at first that these discrete spots of light would make them more conspicuous; Dr. Clark points out,

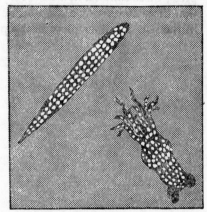

FIG. 84A

Sketches to illustrate Fraser and Clarke's explanation of the function of the luminous organs on the undersides of so many deepwater fish, squid and crustacea. *Left*, a fish and a squid without such organs silhouetted against the faint blue light coming from above when seen from below; *right*, the same but with a typical arrangement of light organs. Seen at a distance, and *out of focus*, the latter are much less conspicuous than those without lights.

however, that at a distance from the predator's eye they are most likely out of focus so that their appearance will be a blurred one giving just the right effect. I show a diagrammatic sketch of such a fish and squid with and without their photophores alight; if you now look at the drawing with a lens out of focus you will see that the ones with their lights on do indeed disappear whereas those which are not shining stand out in sharp contrast to the background. It is a beautifully simple explanation; indeed one which was just round the corner. How foolish one feels not to have looked in the right direction!

"Whilst I believe that his hypothesis correctly explains the widespread occurrence of these downwardly pointing photophores I think in some cases they almost certainly must be used for other purposes as well. Dr. R. H. Kay and I have found that the euphausiacean *Meganyctiphanes norvegica* will at times spontaneously give off bright flashes from its photophores when kept for many hours in darkness."

I go back to the original ending of the chapter which was as follows:

Now, since this chapter was finished and as I send the book finally to press, there comes a new discovery; Captain Brett Hilder (1955) reports that the switching on of ship-borne radar can stimulate phosphorescence in the sea. Another piece of the puzzle is found, but where does it fit?

CHAPTER 14

SQUIDS, CUTTLEFISH AND KIN

FEW MARINE ANIMALS excite our interest and curiosity so much as do the members of the class Cephalopoda, which in our waters are the octopuses, cuttlefish and squids. While this is only one class out of the five making up that great phylum of the Mollusca, its members are so different from those of the other classes—indeed from all other marine invertebrates in their greater powers of locomotion and nervous organization—that it seems proper to give them a chapter to themselves.

It is difficult at first to believe that these animals might reasonably be described as glorified snails, yet they have most surely been evolved from primitive snail-like ancestors. Glorified they certainly are. They are almost as independent of the ocean currents as are the fish. In short bursts of speed they are probably the fastest of all aquatic animals, and, while some are very small, a few of the squids reach a very large size, measuring over all, body and longest tentacles included, perhaps 50 feet in length. They have developed wonderful eyes on the same plan as our own—eyes which in principle are really more efficient than ours, because the retinal cells are facing towards the light rays and have their ganglionic cells behind them. With us vertebrates, by an accident of evolution, it is the reverse: the ganglion cells are on the wrong side of the light-sensitive cells—the side facing the lens— and so in the way of the light-rays forming the image. At any one moment we only see clearly over a very confined area in the middle of the field of vision because, at the corresponding place on the retina, there is a small area (the macula) cleared of ganglion cells to make this possible. It is likely that the cuttlefish has a wider field of clearer vision than we have. The two kinds of eye are compared in Fig. 85 (p. 264). They have brains which allow of a considerable amount of association of different nerve-centres, and so of learning from experience. Their remarkable powers of quick colour-change are unique in the animal kingdom. The proverbial chameleon will not bear comparison in this respect, for it changes but slowly: the cuttlefish and squids

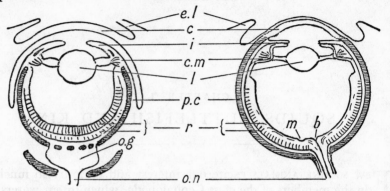

FIG. 85

A remarkable evolutionary parallel: *left*, an eye of a cephalopod (*Sepia*) and *right* that of a vertebrate (man), compared in simplified diagramatic section; the eyes are facing the top of the page. *c*, cornea; *cm*, ciliary muscle; *el*, eyelid; *i*, iris; *l*, lens; *m*, macula; *o.g.*, optic ganglion; *o.n*, optic nerve; *p.c*, protective capsule; *r*, retina.

change their hues in a flash; and not only that—they can alter the pattern of their colour-schemes to suit the occasion in a variety of ways. Further, when this trick of quick-change disguise fails to baffle their enemies, they escape behind an ink-cloud as effective as a smoke-screen thrown out by a warship in action.

The cephalopods are so unlike any other animals we are familiar with, that we shall not understand their way of life until we have some idea of the general plan upon which they are built. Let us very briefly follow in outline their evolution from the primitive snail-like creature which must have been their remote progenitor. Such a simple general-ized snail is shown in Fig. 86A (p. 266); it most likely had a conical cap-like shell on its back—something like a limpet—and crawled by waves of muscular contraction passing backwards along the under-side of the body, which in the modern snail is called 'the foot'. The alimentary canal led from the mouth in front, through a stomach supplied with digestive glands, to the intestine and rectum which opened at the anus into a spacious so-called 'mantle cavity'—a kind of vestibule overhung by the mantle and edge of the shell above and opening widely behind to the outside world. A pair of feathery gills (the ctenidia) projected into this cavity, and here, too, opened the kidney-ducts which, in such a primitive form, also served to discharge the genital products. We deduce all this from our knowledge of the most primitive molluscs of today.

The earliest creatures which can be called cephalopods had external chambered shells to which new and larger chambers were added as they grew. The animals in fact always lived in the last chambers they had made and the former ones—now pushed upwards by the new ones—were filled with gas to act as buoyancy chambers. This was a special adaptation to a floating pelagic life, one which launched the cephalopods upon an entirely new and nautical career. From some such simple beginnings as indicated in Fig. 86B, which is like the early palaeozoic *Piloceras*, must have arisen a more elaborate form which in turn gave rise to two main lines of descent. In one of these the many-chambered shell became coiled, no doubt giving a better balanced float, and produced forms like the *Nautilus* of the tropical seas of today—a survival of a much more flourishing stock of cephalopods which once abounded in the palaeozoic seas of some two hundred million years ago. Allied to these, and also with coiled chambered shells, were the Ammonites, which flourished in the later mesozoic age but became entirely extinct in the Cretaceous; they are among the commonest of fossils and are often thought by country folk to be the petrified remains of coiled-up snakes—'serpent stones' they are sometimes known as. Some hundred million years ago these ammonites were among the most characteristic animals of our open sea.

With these nautiloids and ammonites the chambered shell remained external. In the other main evolutionary line the shell became enclosed within the body, as in the belemnites (Fig. 86E) which, except for this, was very similar to the straight early form of nautiloid *Orthoceras* (86C). These belemnites also flourished in the mesozoic age but became extinct rather later than the ammonites, lasting until the end of the Eocene; they too are very common fossils, looking like stone bullets, and used to be thought by the uninformed to be "thunderbolts" which had fallen like meterorites from the sky. In these forms the chambered shell remained straight, forming a long torpedo-shaped animal; and no doubt the thickened so-called 'guard' at the end of the shell counter-balanced the slight increase of weight of the animal's body at the other end. In some other lines of evolution it curled over at the tip. Occasionally on the west coast of Ireland or Scotland a little curved chambered shell (Fig. 87 p. 268) may be picked up on the beach—the shell of *Spirula*[1]. Like the *Nautilus*, *Spirula* is another lone surviving member from a once flourishing race; it also only lives in tropical seas. When it dies, its little shell floats and may be carried in the Gulf Stream all

[1] Dr. W. J. Rees, of the British Museum, tells me it has also been found on the coasts of Cornwall and Dorset.

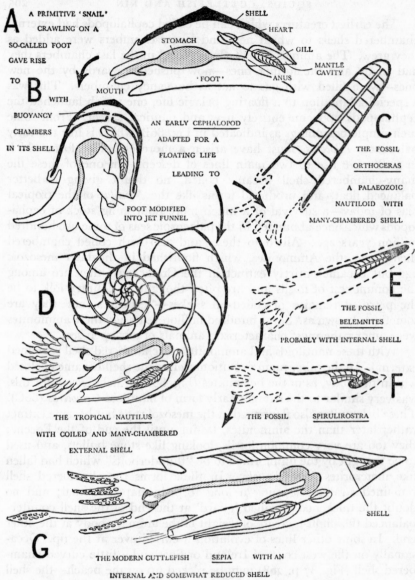

A A PRIMITIVE 'SNAIL'
CRAWLING ON A
SO-CALLED FOOT
GAVE RISE
TO

SHELL

HEART

STOMACH

GILL

MANTLE
CAVITY

'FOOT'

ANUS

MOUTH

B WITH
BUOYANCY
CHAMBERS
IN ITS SHELL

AN EARLY CEPHALOPOD
ADAPTED FOR A
FLOATING LIFE
LEADING TO

FOOT MODIFIED
INTO JET FUNNEL

C THE FOSSIL
ORTHOCERAS
A PALAEOZOIC
NAUTILOID WITH
MANY-CHAMBERED SHELL

D

E THE FOSSIL
BELEMNITES
PROBABLY WITH INTERNAL SHELL

THE TROPICAL NAUTILUS
WITH COILED AND MANY-CHAMBERED
EXTERNAL SHELL

THE FOSSIL SPIRULIROSTRA **F**

SHELL

G
THE MODERN CUTTLEFISH SEPIA WITH AN
INTERNAL AND SOMEWHAT REDUCED SHELL

FIG. 86

Diagrams illustrating the likely course of evolution of the modern
cuttlefish from an early snail-like ancestor; the animals are drawn in
simplified section. A is based on our knowledge of the most primitive
living molluscs; B is hypothetical; C, E and F are known mainly from
their fossil shells.

the way from the West Indies to be cast upon our shores like the South American seeds we spoke of in Chapter 2 (p. 21). In the modern squids and cuttlefish shell, still acting as a buoyancy chamber, is reduced very much in structure: the horny so-called 'pen' of the former and the calcareous 'cuttle-bone' of the latter (Fig. 86G); in the octopus, which has given up a pelagic floating life, it is reduced to but a few calcareous grains.

So much for the modifications to the shell. The next most striking feature of the cephalopod is the development of the long prehensile 'arms' or tentacles which are provided with suckers for the capture of prey. These arms used to be thought to have arisen from divisions of the old basal 'foot' which was supposed to have grown forward to surround the mouth—hence the name: head-footed, from *kephale* a head, and *pous, podos* a foot; it is now more generally considered that the part which represents the original foot is the 'siphon', the funnel-like organ giving these animals their remarkable jet-propulsion. The arms are now regarded as outgrowths from the head itself. The mouth, at the centre of the radiating arms, is provided with powerful horny jaws, somewhat like the beak of a parrot; in handling a living squid it is advisable to keep your fingers away from these, as they can give you a very nasty bite.

In *Nautilus* the siphon is made up of two halves, one derived from each side, which come together in the middle line to form the funnel; in the other modern forms—the octopus, cuttlefish and squid—the two parts are fused early in development into one single organ. By the relaxation of muscles, and by the elasticity of its walls, the mantle-cavity expands and fills with water; when the muscles of the mantle wall suddenly contract, the water is forcibly driven out through the siphon and the animal shoots through the water in the opposite direction at great speed—jet-propelled. The siphon normally points forwards, but this does not mean that the animal can only shoot backwards, as many students think from only seeing a stiff dead specimen; the tip of the siphon in life can be curved round this way and that, and if pointed backwards can shoot the animal forwards just as easily.

The modern cephalopods—apart from that very distinct and ancient survival *Nautilus*—belong to the order Dibranchia (having two gills); this contains two sub-orders: the Octopoda, with eight arms or tentacles, like the octopus, and the Decapoda, with ten arms or tentacles, like the cuttlefish and squids. Here we shall not be very much concerned with the octopods, for their representatives, except for a few bathypelagic forms, have in the main forsaken the swimming habit and taken to crawling and lurking among the rocks along the sea-bottom close against the shore; they belong to the fauna of the coasts rather than to the open sea. Nevertheless they must receive some mention because the very presence of the common octopus *Octopus vulgaris* on our coasts is so much the work of the open sea. We have two kinds of octopus: the lesser octopus, *Eledone cirrosa,* which rarely measures more than thirty inches across the widest span of tentacles, and is to be found right round our coasts including the very north of Scotland; and the larger *O. vulgaris* which may reach a span of up to ten feet and is confined to the south coast of England. The two kinds are very easily distinguished, in that the arms of the former have only a single row of suckers whereas those of the latter are provided with

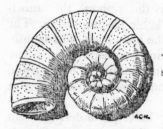

Fig. 87

The floating shell of the tropical *Spirula* which may sometimes be carried to our western shores (×2).

a double set. *O. vulgaris* is really a native of the warmer waters of the Mediterranean, the African coast, the Atlantic islands and coasts of central America; the English Channel is on the northern limit of its range. As a rule it is a comparative rarity on our southern coast but occasionally, as in 1900 and in 1950 and '51, it may appear in very large numbers and wreak havoc among our crab and lobster fisheries, for large crustacea are its favourite food. Dr. W. J. Rees of the British Museum (Natural History) has recently shown (1950a) that it is most probable that it never, or only very occasionally, breeds on our coast, and that our population is recruited year after year by young larvae which have hatched out from eggs laid on the Brittany coast and have been drifted to our shores by the currents up the Channel. The eggs, in small oval capsules, are laid in large numbers, attached like grapes by short stalks to long strands which hang down, many together, to festoon the sides of little cavities between or under rocks, usually just below tide-level; each capsule is rather smaller than a grain of rice. It was hitherto thought that the larval octopus only spent a very few days of life afloat before settling down, but Dr. Rees has shown from

FIG. 88
Larvae of *Octopus vulgaris* from the plankton of the English Channel, ×8, drawn from specimens obtained by Dr. W. J.Rees.

ollections of plankton made in the English Channel during 1948, '49, nd '50 that larvae up to half-an-inch in length may still be drifting n the currents; and these must be several weeks old, for they are atched at only about an eighth of an inch in length. Sketches of hese larvae, drawn from preserved specimens, are shown in Fig. 88 bove. The currents up the Channel pass along the Brittany coast o be deflected northwards past the Channel Islands into mid-Channel nd then the waters may divide, some curving back westward towards)evon and some passing up towards the Sussex coast; thus the larvae nay be dropped at various points along our southern shores, sometimes nore to westward, sometimes more to eastward according to the set f the current. As a rule the sea on our side of the Channel is too cold n winter for many to survive, but occasionally it may be several degrees varmer than usual; it is following such exceptionally mild winters hat we get our plagues of octopus.

As I shall presently be referring to the giant squid, let me say, to void any confusion, that all the stories of octopuses of enormous size ave no foundation in fact. An octopus with a tentacle span of ten feet s exceptionally big.[1] Dr. Rees in an article in *The Times* of 12 August 950 on the habits of the octopus, at the time of its great abundance, vrote as follows:

"It may not be out of place to mention here that few creatures are regarded with such repugnance and even awe by the general public. This is probably due to the influence of Denys de Montford and Victor Hugo. The former, an unscrupulous rascal at one time in the employ of the Paris Museum, "created" a gigantic *poulpe* by which he managed to sink several English men-of-war; England being then at war with France. Although Victor Hugo was undoubtedly familiar with the habits of the octopus, his *pieuvre* in *Les Travalleurs de la Mer* was a fantastic creature far removed from reality. Subsequent writers have drawn on these sources and have often improved on them."

Passing now to the decapods—those with two longer tentacles in ddition to eight other sucker-bearing arms—let us begin with that naster of colour-change, the cuttle-fish *Sepia officinalis*. This also is

[1] G. C. Robson in his *Monograph of the Recent Cephalopoda* (1929–32) describes)*ctopus apollyon* of the North Pacific as reaching a span of 28 feet, but its body is only he size of a quart pot.

not really a pelagic animal; it lives close against the sea-bed in relatively shallow water, but not only against the coasts, for at times it may be caught in trawls over the continental shelf at considerable distances from land. Dr. A. C. Stephen of the Royal Scottish Museum, Edinburgh, has recorded (1944) a specimen trawled at 80 miles N.E. of Aberdeen. Its body is broader and flatter, and its tentacles shorter, than those of the more slender torpedo-shaped squids of truly pelagic habit. Round the margins of its squat body is a thin fin whose undulations propel the animal slowly either forwards or backwards; on its underside, projecting out of the mantle-cavity, is the siphon, which in sudden danger may squirt out a jet of water to send its owner shooting out of harm's way in a flash. Like the flatfish such as plaice or turbot, *Sepia* prefers to live over a sandy bottom and like them too, when settling down to rest, uses its marginal fins for throwing up little clouds of sand, which fall over its edges and so obliterate its outline; also like the flatfish, it combines this trick with a wonderful system of camouflage on its upper surface.

Sepia hunts for shrimps by blowing small jets of water, apparently at random, at the sand in which they hide. The shrimps are so well camouflaged that, even when they are exposed to view, they are not as a rule noticed at once by the cuttle; it is only when they move to throw more sand on their backs that they are instantly seized and eaten. This interesting piece of natural history is retold by N. Tinbergen (1951) from a personal communication by J. Verwey. D. P. Wilson has taken some wonderful photographs of *Sepia* catching prawns, showing the long arms extended in a flash for capturing the prey and then contracted to bring it to the mouth; three of these are reproduced in Plate XXIII (p. 288).

How numerous cuttlefish are it is difficult to know; like squids they are probably able to escape very easily from any slow-going trawl, and it is only the odd one or two which get so trapped. Occasionally some exceptionally adverse condition, such as an unusually cold spell in winter, causes a heavy death-roll among them; then we may get some idea of their importance in the ecology of our waters by the number of dead washed up on the beach. *The Times* of 27 December 1950 records such an exceptional mortality as follows:

"Thousands of dead cuttlefish and great quantities of their oval white bones have been washed ashore on the east coast, chiefly at Lowestoft. Many were washed up close to the fisheries laboratory of the Ministry of Agriculture and Fisheries, where they were identified as unusually fine specimens of *Sepia officinalis*. In one area of Lowestoft beach, 17 yards by eight yards, an observer counted 81 cuttlefish in various stages of disintegration. Some retained their characteristic

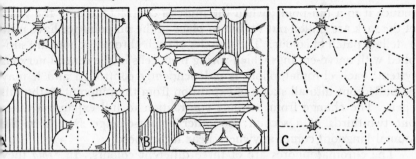

FIG. 89

Colour change in the squid. Diagrammatic views of the same small area of skin highly magnified to show the little star-like pigment sacs (red, blue and yellow) in different states of expansion and contraction. Vertical shading=red, horizontal=blue, unshaded=yellow. In A this part of the animal would appear red because only the red sacs are expanded; in B it would be bluish purple because the blue sacs, and to a lesser extent the red, are expanded; in C it is white, as all the colour sacs are contracted to reveal the dazzling white reflecting layer below. The radiating dotted lines represent the very fine muscle fibres which on contraction pull out the pigment sac as if they were spreading out a coloured handkerchief. Compare with Plate 20 (p. 241).

dull red and brown colouring, but the 10 tentacles were in every case missing. It is believed that the cuttlefish may have been unable to accomodate themselves to the chilling of the sea during the easterly gales before Christmas."

Dr. Stephen (1944) records a similar occurrence in the northern North Sea:

"This species would seem to be an occasional wanderer to northern waters in years when there is a specially strong incursion of Atlantic water. The invasion in 1922 and 1923 seems to have been quite remarkable. Not only were live animals taken in the Scottish area, but pens [cuttle-bones] seem to have been cast ashore in vast numbers, especially in Shetland, where they were described as forming a white band along the tide-marks. Although apparently not recorded previously from the Faroes, large numbers were washed ashore from the middle of February to the end of May 1923."

Let us now consider the cephalopods' remarkable powers of colour-change, as exemplified by *Sepia* in which it has been so fully studied by Dr. William Holmes (1940); similar mechanisms for quick change also occur in the octopus and the squids. The secret of this mechanism lies in the nature of the tiny spots of pigment which cover the entire surface of the animal; these little bodies which carry the colours are appropriately called chromatophores. Each chromatophore is a little transparent elastic sack filled with a coloured substance and having radially attached to its circumference a series of thin muscle-strands. When these muscle-fibres are relaxed the elastic walls of the sack reduce it to a pin-point; when the muscles contract strongly the walls of the sack are pulled out so that in place of the pinpoint we have now a stellate patch of colour. The muscles of each chromatophore are supplied with nerve-endings which spring from cells in the central nervous system; each chromatophore is independently controlled and it may be pulled out into all stages of expansion from the pinpoint to the full open. The change from extreme contraction to full expansion takes but two-thirds of a second; no other colour-change system in the animal kingdom comes anywhere near it for speed.

The chromatophores are of three different colours. In *Sepia* they are arranged in three layers below the skin: those of the outer layer contain bright yellow pigment, those of the middle layer orange-red, and below these are ones which are dark brown and sometimes almost black. Below these expanding and contracting spots of colour is a layer of immobile cells which appear a brilliant white, for they reflect all the light like snow-crystals. If all the chromatophores were contracted the animal would appear a dazzling white; if the yellow alone are expanded it appears yellow; if the red alone are expanded it appears red, and if the dark brown and black are expanded the animal will of course be dark in hue. All manner of combinations of colour are possible, and by the contraction of some chromatophores here and the expansion of others there, all manner of patterns may be produced and changed at will. Alternate waves of contraction and expansion may run over the body to produce moving pictures of ripples—or more like the moving shadows on the sand caused by ripples on the surface above.

The cephalopods have mastered the process of 'colour-printing.' Anyone who is not familiar with the four-colour half-tone process of colour printing should look at one of the colour-plates in this book with a lens. In an ordinary black-and-white half-tone reproduction of a photograph the picture is of course produced by a vast number of tiny black dots which range in size from ones that are so small as to be almost invisible points, to ones which are so thick that they all but touch the sides of the adjacent dots of similar size. Areas occupied by the former appear white and by the latter black; by varying the size of the dots all gradations of intermediate tones may be produced. In four-colour printing there are four sets of dots—one set for each of the three primary colours, red, yellow and blue, and one for black. Where, for instance, a yellow patch of colour appears in the picture, the blue and the red dots will be invisibly small, but the yellow ones as wide as possible; where green appears the yellow and blue dots will be well developed but the red invisible and so on: all manner of shades of

Plate XXI. An oceanic squid *Sthenoteuthis (Ommastrephes) caroli* stranded at Looe, Cornwall, in November 1940. Dorsal view, with a metre rule alongside.

(*Douglas Wilson*)

Plate XXII. A giant squid *Architeuthis*, stranded at Bay of Nigg, Aberdeen, on 30th November, 1949.

green may be reproduced by varying the relative sizes of the yellow and blue dots—or even bringing in some of the red dots in the brown-greens. This is exactly what the cephalopod does; but it goes further, it presents a moving picture by continually altering the emphasis of its colours at will, sa shown in Fig. 89 (p. 271).

Some of the squids such as *Sthenoteuthis* (*Ommastrephis*) have the three primary colours of red, yellow and blue, instead of the red, yellow and black of *Sepia*. Plate 19 (p. 240) shows a reproduction of colour-sketches made on board the R.R.S. *Discovery II* of living speci-mens of *Sthenoteuthis pteropus* caught on our voyage to the Azores; at one moment the animal may be a silver-grey and at the next blushing orange, crimson or purple. This species and the very closely allied *S. caroli*, which will be further discussed on p. 281, are both sometimes found stranded on the shores of Britain.[1] In Plate 20 (p. 241) we see details of four small areas of the surface enlarged six times to show the relative states of expansion of the red, yellow and blue chromato-phores which produce the general colour-effect shown in the sketch of the whole animal. These were drawn from a specimen just after it had died; in life the chromatophores are continually altering their size so that different colour-effects sweep over the animal in rapid succession.

But let us return to *Sepia* and the observations of Dr. Holmes. Figs. 90 *a–f* (p. 274) show six distinct phases of the colours and pattern-changes evoked under different conditions, the drawings *c* and *e* being made from two of his photographs and the rest being reproductions of his original drawings. *a* represents the colour-pattern as seen in actively swimming *Sepia*, in one of the largest aquarium tanks at the Plymouth laboratory with a variety of back-grounds. The back is dark brown, due to the dark chromatophores being expanded and the others contracted, while at the sides there are irregular stripes of expanded and contracted dark chromatophores; this pattern tends to break up the outline of the animal against the variety of backgrounds in which it is moving. In addition to this dis-ruptive camouflage the animal is beautifully counter-shaded, being dark on the top surface and light on the underside, where all the croma-tophores are contracted: this counteracts the natural daylight effect of high light on top and shadow below. *b* shows a "light mottle" pattern assumed when *Sepia* settles down in the sand and, as already explained, throws up sand with its fins to obliterate its outline. The irregular triangular markings round its edge tend to break up its outline

[1] See footnote on p. 281.

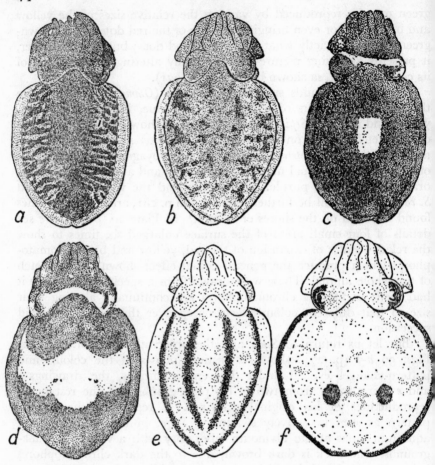

FIG. 90

Colour changes in the cuttlefish *Sepia*. *a-f* different patterns possible in the
same individual under different circumstances as explained in the text.
After Holmes (1940).

and the general colour of the rest is a sandy brown just like its background. When the animal is kept in a very dark-coloured or black tank,
its whole upper surface become very dark. Now if a white piece of porcelain is placed near the animal, thus producing a strong contrast
of black and white in its background, then a remarkably square
patch of white appears in the middle of the back by the contraction
of all the chromatophores in this area which allow the light-reflecting

area to become visible as in *c*. This pattern, and also that with the very white stripe shown in *d*, are frequently produced in cuttlefish swimming in a well-lit outdoor aquarium having a strong contrasting background of dark and light stones and shells on the bottom. Still more striking are the pattern- and colour-changes when the animal is much disturbed; to describe these phases I will quote Dr. Holmes' actual words:

"Strong optic, static, or tactile stimulation of *Sepia*, such as is the result of a violent movement in its visual field or of touching it with the hand, is always followed by a series of colour-changes of remarkable rapidity and completeness, involving the whole body of the animal; and in which a succession of patterns is displayed, each one enduring only for a few seconds.

"The first result of the stimulation is usually the appearance of the 'black spot' pattern, consisting of two black spots which arise in an invariable position on the dorsal surface of the mantle, one a little to each side of the middle line. . . .

"Slightly stronger stimulation, or a repetition of the same adequate stimulus, causes a rapid and total paling of the rest of the animal, and the accentuation of the black spots themselves by the still further expansion of the dark chromatophores which produce them. Thus for a moment the two black spots stand out most vividly on the background of the iridescent white animal. The total pallor is never maintained for more than a few seconds, but its transitoriness makes it the more striking. Often this is followed by the animal's contracting its mantle violently and shooting away by the action of its siphon, and this movement is accompanied by a colour-change which makes it amazingly difficult to follow, even to the human eye. For while at one moment one's eyes are fixed on a white animal with two black spots, at the next it seems to have disappeared, for its rapid movement is accompanied by total darkening of its body, by full expansion of all chromatophores. This complete colour-change, most deceptive to a human observer, must be even more effective in nature in deluding predators. . . .

"If further irritated, the animal may respond by a total paling of the whole of its body, and upon this background may appear longitudinal black stripes, at the base of the fins and along the middle of the back (*e*). These lines flicker vividly over the pallid back, and then suddenly disappear, to be followed perhaps by a reappearance of the black spots, another total darkening, or a brief reappearance of the zebra pattern. All this time the animal darts about rapidly, as if to avoid the irritation, and its final action when it cannot do so is to eject a cloud of ink. Then at once it becomes motionless, and hides below the black cloud which it has produced, and its colour can be observed no more.

"Sometimes, when *Sepia* is resting in a corner, or behind stones, it does not move away when irritated strongly, but instead shows a strange reaction involving the flattening of its body and the appearance of a striking colour-pattern. The whole animal broadens itself and decreases in thickness, so that it seems to shrink against the sea floor. The change in the shape of its head which results from the flattening brings its eyes into such a position that they are directed upwards rather than in the normal lateral direction (*f*), and they are made still more conspicuous by the raising up of the iris which normally covers most of the eye. At the same time, a black line of fully expanded chromatophores appears along the outer edges of the fins, and striking black rings are similarly formed around the sides and lower edges of its eyes. This remarkable colour-pattern is always accentuated by the total paling of the rest of the body, and this combines with the black markings and the condition of the eye to give the animal the appearance of 'threatening' the disturber. It seems not unreasonable to suppose that those features of this pattern that make the animal seem so 'fierce' to human observers have a special protective function in nature."

The colour patterns of *Sepia* are not only used for concealment; they play an important part in a courtship behaviour which shows a remarkable parallel to that of certain sexually dimorphic fish. I refer to a study by L. Tinbergen which has been briefly summarized in English in his brother's book *The Study of Instinct* (1951) in the following words:

"A male *Sepia* in mating condition assumes a strongly variegated pattern of alternating white and dark purple bars, and displays the most conspicuous part, the broad flattened, lateral surface of the fourth arm, towards other individuals. Reactions of males and females to this display differ essentially: a male returns the display, a female in mating condition keeps quiet and allows the male to copulate. Experiments show that the male's nuptial colours, and especially the colour and display of the arm, released fighting in other males, while all models coloured and 'behaving' like females were treated like females."

Before leaving the cuttlefish we should mention the observations on its learning capacity made by Dr. F. K. Sanders and Professor J. Z. Young (1940) in the course of an investigation of the functions of the different parts of its nervous system. Here we will only be concerned with the natural behaviour of the intact animal. By attaching a prawn to a thread they were able to draw it out of sight of the *Sepia* round a corner, and to show that the cuttlefish is able to realise where the prey has gone to and so follow it round; in this way quite a hunt could be set up. If they placed a sheet of glass between a cuttlefish and a prawn, the cuttle would at first make repeated attempts to seize the prawn but presently give up and if again presented in a few hours' time with a prawn behind glass it would feel the glass and then give up trying to get the prawn at once. It learnt by experience; it remembered that it was unable to get the prawn if glass was present. Further they introduced a sheet of glass with a white disc on it; now *Sepia* learnt to associate the presence of the white disc with failure to reach the prawn, and so, after a bit, when shown the white disc with the prawn, it no longer first felt if the glass was there, but made no attempt whatever to get the prawn. The higher mollusc, like a higher vertebrate, can learn to associate one event with another and remember the significance of the association. While dealing with behaviour I should mention that it is in the nervous system of the cephalopods,

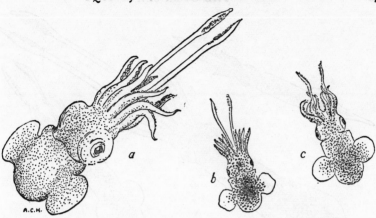

FIG. 91

Smaller relations of the cuttlefish from the seas round Britain: *a, Rossia macrosoma; b, Sepiola atlantica; c, Sepietta oweniana. a* and *b* drawn from specimens in the British Museum and *c,* from one at the Fisheries Laboratory, Aberdeen; all shown natural size.

particularly that of the squid, *Loligo,* that Professor Young has demonstrated the existence and the importance of the remarkable giant nerve-fibres which transmit impulses at great speed. Such giant fibres are found in animals with very quick reactions.

It is impossible to mention all the British cephalopods, but I will here just note three other species, which like *Sepia,* are commonly found close against the bottom. They belong to three closely related genera, all with much rounder bodies (almost spherical in fact) and with rather rounded fins which are large in proportion to the body. The three are shown together for comparison in Fig. 91 above. *Sepiola atlantica* is a very small animal, rarely more than an inch in length. It is often very abundant just below low-tide mark along sandy beaches. Mr. J. A. Stevenson (1935) says it may be caught in large numbers in shrimp-nets along the Yorkshire coast in summer, but may also be taken by trawlers up to 40 miles from the coast. Dr. E. S. Russell (1921) records 256 specimens being taken in a single trawl haul in the Firth of Forth. *Sepietta oweniana* is a little larger and reaches about 3 inches in length; it is very much an inshore species and is commoner on the coasts of Scotland than of England. *Rossia macrosoma* is larger still and is said to reach a length of 10 inches, but specimens more than 5 inches are not often found; it is rare on the English coast and on the east coast of Scotland south of the Moray Firth, but on the north and west

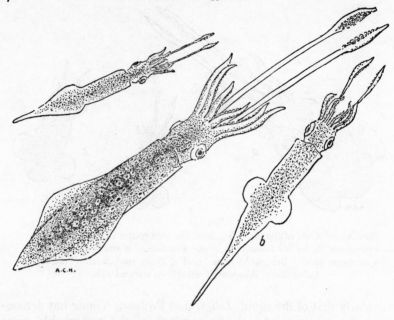

FIG. 92

Some common squids from British seas: *a, Loligo forberii* ($\times \frac{1}{3}$); *b* and
c, male and female respectively of *Alloteuthis subulata* ($\times \frac{1}{2}$).

coasts of Scotland it is very common. *Rossia* has its mantle free all
round, whereas the other two have it fused to the back of the
head.

Now let us pass to the more pelagic squids. How important they
are in the ecology of our seas is very difficult to say, for it is almost
certain that at present we have a very inadequate idea of their true
abundance. They can swim so fast by their jet propulsion, and swim
both backwards and forwards, that they very probably avoid most
of our nets, or if once inside, may easily get out again. On the *Discovery*
expedition to the Antarctic in 1925–27, when we spent much time
studying the general biology of the whaling grounds round the island
of South Georgia, we used almost every kind of net and trawl, but
caught only very few squid; yet the stomach of every elephant-seal
we examined was full of their remains. The elephant-seals occur there
in huge numbers; to keep that stock nourished the squids must be
enormously abundant and occupy a very important place in the ecology
of those waters.

The animal to which most of our fishermen give the name squid is *Loligo forbesi*, which occurs all round our coasts. It is shown opposite in Fig. 92; it may reach a length of up to 2½ feet, but 8 to 12 inches is a more usual size. Stevenson (*loc. cit.*) says "it is probably the most abundant cephalopod frequenting the coastal waters of Yorkshire during the summer and autumn. It does not appear to be caught close inshore at any time by the fishermen, but at times large numbers are taken in the nets of trawlers fishing from 7 to 15 miles off shore." Miss Anne Massy (1928) describes it as distributed all round the coasts of Ireland in about 5–65 fathoms; an idea of its abundance, she says, may be gained from the fact that the Irish fisheries research ship *Helga* trawled 1070 specimens in 40 hauls. The smaller and more slender squid *Alloteuthis subulata* appears to be still more abundant than *Loligo*, and is shown for comparison in the same Fig. 92; it rarely exceeds 6 inches in length. To quote Stevenson again "it is probably the most plentiful cephalopod found off the coast of Yorkshire during the winter and early spring. It is particularly abundant at this time in the deeper off-shore waters (40 fathoms and down), and is occasionally taken in very large numbers in the trawl-nets. The fishermen inform us that these small squids have a habit of clinging to the exterior of their nets, sometimes in such large numbers that the nets appear quite white with them." Miss Massy (*loc. cit.*) describes it as being taken in very large numbers off Ireland, particularly in October; for example over 500 specimens were taken in 10 trawls off Clogher Head. Dr. Anna Bidder (1950) records that "*Loligo*'s natural food is fish, crustacea and smaller squid, all of which have been found in the stomach of captured animals," and she shows the same for *Alloteuthis*. Watching *Loligo* feeding in an aquarium she found that a fish is always "seized behind the head and held obliquely with the tail uppermost and is so carried until the head is bitten off and dropped. . . . The trunk of the fish is then held horizontally in the arms, in line with the body . . . (and then) . . . the squid bites through the fish from head to tail by a series of transverse bites . . ."

Many readers of the wonderful *Kon-Tiki* narrative were surprised to learn of the flying squids which sometimes left the water in shoals and 'flew' like flying-fish. Another good recent account is given by members of the *Pequena* expedition from Cape Town to Tristan da Cunha in 1948. I quote the following from a special number of the *South African Shipping News* devoted to the expedition:

"On Monday, 2nd February, 1948, on the outward voyage to Tristan da Cunha, during heavy rain and squall, the *Pequena* was eased down owing to heavy swell with a W.N.W. wind force 7, just over 900 miles from Cape Town.

"What was thought to be a flying fish was seen to take off out of a trough of a deep swell 40 to 50 yards away abeam on the port side, and came in at an angle at great speed, overtaking the ship with its wing-like pectoral fin all a-quiver. It landed on the fore well-deck.

"When collected, everybody aboard, including the marine biologists, were amazed to find that it was a squid about 8 in. long and pale green in colour. No-one had ever heard of a flying squid before.

"Later on in the voyage other squids were observed to jet-propel themselves out of the water, stern first, using the tail fins as a plane for gliding, covering quite considerable distances, and not unlike flying fish in appearance."

A few years ago Dr. Rees (1949), on examining the cephalopod collections brought back by a number of oceanographic expeditions, noticed that specimens of the hooked squid *Onychoteuthis banksi* were never caught in nets, but were all picked up on the decks of ships whilst they were on passage from one observation station to another. This made him look at the older collections in the museum and here again in the majority of cases he found the same thing, indicating, as he says "that the species has the strange habit of 'flying' or leaping out of the water in common with *Sthenoteuthis bartrami* (Lesueur)." He then quotes Professor A. E. Verrill as writing in 1880 that this latter "is an exceedingly active species, swimming with great velocity, and not rarely leaping so high out of the water as to fall on the decks of vessels". On this account it has been called the "flying squid" by sailors.

Flying squids have thus been known for a long time. There appear to be at least two which have this habit. *Onychoteuthis banksi* is distributed throughout the oceans of the world and occasionally comes into British waters; in addition to specimens taken by the Irish fishery research ship *Helga* S.W. of Ireland, one was stranded at Banff on the Moray Firth in 1853, and quite recently, in 1939, several were stranded at Hastings.[1] It is said to be one of the principal foods of the tunny and allied fish caught by the fishermen of the Breton ports. On returning from the Azores on our recent *Discovery* cruise we saw from time to time small squids, often several together, leap through the surface of one wave and disappear into the next and occasionally make longer leaps; I think these were most likely *O. banksi*; no doubt, as with the flying fish, this is their means of escape from fast pursuing enemies.

The squid *Ommatostrephes sagittatus* is stranded on our coasts in small numbers in most years, but occasionally it comes into our waters in enormous numbers. In the summers of 1930 and '31 great shoals

[1] J. M. Baines (1939).

invaded the Scottish herring grounds, in 1930 principally the Shetland grounds, but in 1931 extending south to those off Peterhead. The fishermen believed that these 'ink-fish', as they call them, caused a sudden break-up of the fishery; whether they actually had an effect on the herring shoals is not known, but they were said to have done considerable damage to the herring while in the drift nets. There were many of these squids stranded on the east coast of Scotland in January and February 1920 and again in February 1937; also in February 1937, between Lincolnshire and the north of Scotland, Dr. F. C. Fraser of the British Museum recorded (1937) a series of strandings of the Common Dolphin and found that many of them bore the marks made by the horny sucker-rings of some squid which closely resembled those of *O. sagittatus*. Often on the jaws of toothed whales, particularly those of the sperm whale, are to be seen such circular scars left by the sharp horny denticles which have armed the suckers on the arms of squids struggling desperately in the course of being eaten. I will refer again (p. 288) to the importance of these animals in the diet of cetaceans when I have described some of those of more substantial size.

We now come to the squids of the genus *Sthenoteuthis*[1] which are sometimes classed as giant squids although they never reach the huge size of those real giants, the species of the related genus *Architeuthis*. Two species of *Sthenoteuthis*, *S. caroli* and *S. pteropus* are from time to time stranded on our coasts. Fig. 93 (p. 282), which is compiled from the charts published by Dr. Rees (1950), shows the positions of these strandings together with those of *Architeuthis*. We see that there are two points on the coast where such strandings occur more frequently than at any other places: in the region of the Firth of Forth and on the Yorkshire coast. Clearly these animals come into the North Sea in the main Atlantic influx from the north, see Fig. 8 on page 27, and then probably find conditions during the winter in the North Sea too severe, so that they die and are washed ashore where two swirls sweep round to the coast; nearly all the strandings at these two localities occurred from

---

[1] There is considerable confusion in the literature over this name. Rees in his 1950 paper to which I have referred, says in a footnote that he has followed Winckworth in referring the species he is discussing "to *Ommastrephes* although most authors have recorded the species under the name *Sthenoteuthis caroli*." Here I propose to use the latter name because at present it is that most widely used in the literature and also on account of the name *Ommastrephes* being further confused with *Ommatostrephes*. Also note that Rees himself in his paper of the year before (1949) quoted on p. 280 refers to the allied species as *Sthenoteuthis bartrami*. More recently still (1955) he expresses his belief that *pteropus* and *bartrami* may prove to be one species—and he tells me that he is now inclined to regard *caroli* as but a fully developed female of the same kind.

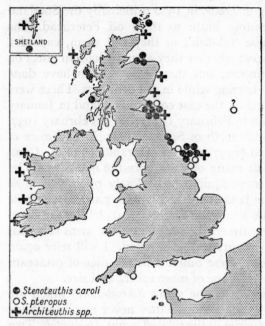

SHETLAND

● Stenoteuthis caroli
○ S. pteropus
✛ Architeuthis spp.

FIG. 93
Chart showing the strandings
and other records of large
squids in British waters, taken
with kind permission from Rees
(1950) with one more recent
addition. Note that *Stenoteuthis* = *Ommastrephes* (the
name used by Rees); I have
used the former name for the
reasons given in the footnote
on p. 281.

December to March. Normally both species only appear to be stranded
—or at any rate are only noticed—every four or five years, but the
five years 1927 to '31 saw an exceptional invasion of *S. caroli*. The
measurements of the total length of the five specimens taken on the
Yorkshire coast from 1925 to 1931 are given by Steenson (*loc. cit.*)
as follows: 7 ft. 1½ in.; 5 ft. 7 in.; about 4½ ft.; 3 ft. 9 in. and 5 ft. 10 in.
After giving the particulars of these specimens he makes the following
remarks which show indeed that there are likely to be many more
squids stranded than reported:

"The above are all the definite records we have of *Sthenoteuthis caroli* in this
district. From time to time, however, we hear of exceptionally large squids being
found upon the shore and cut up for bait before we can examine them; these are
very probably referrable to this species, since of the larger forms this seems to be
the most abundant on the Yorkshire coast. Mr. Clarke furnishes some interesting
notes upon the occurrence of some of these large squids:

"Giant squid, unidentified, 'caught on rod and line by an angler from the east
pier at Scarborough on 20th May 1929. It was measured—4½ feet in extreme
length, and was cut up for bait. Another, about 5 feet in length, was stranded
alive on the south sands in the second week in June, 1929. It was returned to the
water and swam away.' Again Mr. Clarke writes (13th February, 1933):

'. . . Last week two more [squids] came ashore, one 6 feet over all, the other
with a body, exclusive of tentacles, of 3 feet (estimated lengths), but I was not

told anything about them in either case until they had been cut up, so I have no idea what species they were.'

"When it is considered what a small area of the coastline of Yorkshire has hitherto been investigated with regard to these strandings, it would be quite safe to assume that these giant squids are not nearly so rare as they have been thought to be."

A photograph of a specimen of *S. caroli* which was stranded at Looe in Cornwall is shown in Plate XXI (p. 272).

On the recent *Discovery* cruise to the Azores, which I have already referred to, we made attempts to capture large squids by big baited traps, like enormous lobster-pots, lowered to various depths; by towing baited hooks at many depths; by baited hooks on vertical paternoster lines when the ship was stopped; and by shining powerful lights over the stern to attract them. The traps failed to catch any; so did the baited lures, although we had evidence that many squids had attacked the bait; not only was the bait taken on many occasions, but several times suckers from the tentacles of quite large squids were left behind on the hooks—their owners had evidently wrenched their arms away, tearing out the suckers to escape. However, at the lights, using rod and line, with model myctophids or prawns as bait, we were successful. All the specimens we secured were *Sthenoteuthis* (*Ommastrephes*) *pteropus* (Plate 19, p. 240). They advanced into the circle of light like torpedoes, sometimes one at a time, sometimes two or three together. Their power of leaping was well illustrated when we placed a small very active individual about 6 inches long in an aquarium tank some 2½ feet long, 1½ wide and about 1½ deep. The aquarium was of plate-glass held together by welded angle-iron, so that round the top edge of each side there was an overhanging flange of iron on the inside. In a matter of half a minute or less the squid had made repeated attempts to strike out in different directions, hitting all of the walls in turn; then it shot upwards in each direction, up the sides so that it came up against the angle iron edge and fell back; finally, going to the middle of the tank, it suddenly shot upwards like a rocket, cleared the edge of the tank and hit the far wall of the deck laboratory. *S. pteropus,* like the closely allied *S. bartrami,* is certainly a 'flying squid'; in addition it has remarkable powers of exploring and summing up a position in a very short time, and then taking the only way out. No wonder we do not catch them in our large and slow-going tow-nets. I have already, on page 273, referred to its colour-changes and illustrated them in Plates 19 and 20. It is difficult to explain the significance of its hues: crimson, orange, purple or white, or a mixture of them. No doubt sudden changes in colour, as it darts this way and that, will make it more difficult for a pursuer to follow. Its principal enemies however, if whales and dolphins, may, like most mammals, be colour-blind, though not, of course, shade-blind.

We come now to the real giants of the genus *Architeuthis*. When they are described as being up to 50 feet in length we must remember that the greater part of that length is taken up by the pair of very long thin tentacles; the eight other arms are much shorter and the body itself is usually only about a quarter of the total length. The fullest and best account of these animals we have is still the remarkable study made by Professor A. E. Verrill and published in 1882 as an appendix to the *Report of the United States Commission of Fish and Fisheries* for 1879. It is largely based on the many strandings in the Newfoundland area and

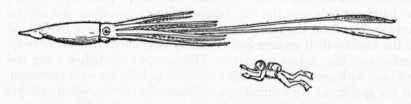

FIG. 94

A giant squid, *Architeuthis princeps*, of average size, and a frogman drawn to the same scale. The proportions of the squid are based on those of the 40 foot specimen figured by Verrill (1882).

owes much to the activities of a local amateur naturalist, the Rev. M. Harvey, who took measurements and preserved as much of the specimens as possible; without his interest and help they would certainly have been lost. It is fitting that his name should be attached to one of the animals he secured. Verrill gives measurements from the best authenticated records up to his time and refers them to two species: *Architeuthis harveyi* and *A. princeps*. The accompanying table gives the total length (from tip of longest tentacle to tail end of body) and the length of the body where known (i.e. from the base of the arms to the tail end) of ten specimens; it should be remembered that there may ' ˑ considerable differences in the state of extension or contraction of the long tentacles in different examples, also that some of the measurements were made from fresh specimens and others after the parts had been preserved. In Fig. 94 above an average *Architeuthis princeps* is compared in size to a man.

A. harveyi		A. princeps	
Total length	Length of body	Total length	Length of body
52 ft.	10 ft.	55 ft.	20 ft.
45 ft.	?	52 ft.	15 ft.
40 ft.	?	44 ft.	11 ft.
31 ft. 10 in.	7 ft. 8 in.	40 ft.	9½ ft.
		38 ft.	9 ft.
		37½ ft.	11 ft. 8 in.

While more specimens have been stranded in the Newfoundland region than anywhere else, the British Isles come next in number of records. Fig. 95 (p. 287) shows the positions of strandings or authentic observations in the North Atlantic. Some of the early Scandinavian records go back to the sixteenth and seventeenth centuries, and it is not perhaps unlikely that the stories of the dreaded Kraken in Norse legend are founded on *Architeuthis*. Mr. A. G. More, Assistant Naturalist in the Museum of the Royal Dublin Society, describes in the *Zoologist* of August 1875 a species of *Architeuthis* captured off Boffin Island, Connemara. He writes as follows:

"The history lately given in a newspaper, the 'Galway Express,'—which has also been published in the 'Zoologist' for June (S.S.4502),—is no myth, and the great size of the animal is sufficiently proved both by letters received from Sergeant O'Connor, of the Royal Irish Constabulary stationed in Boffin Island, and by the portions of this great squid which he has sent up to Dublin. Though imperfect, both tentacles and arms are represented, and the huge beak, about five inches across, is now to be seen in the Museum of the Royal Dublin Society.

"The animal was killed on the 25th of April; and, as the men who attacked it were in a small boat, they could only bring ashore the head and some of the arms—viz., the tentacles and two of the short arms. The head and eyes were unfortunately destroyed, but Sergeant O'Connor managed to rescue, and has transmitted to us, the greater part of both tentacles, one short arm, and the beak. He measured the tentacles when fresh as reaching to the length of thirty feet, and the portions of them which we have received—shrunk and distorted as they now are—still measure fourteen and seventeen feet, when the pieces are put together."

Earlier in the same volume (July 1875) he reproduces an account published in 1673 entitled:

"A true Relation of a strange Monster that lately was by a Storm driven ashore at *Dingle-I-cosh* in the county of *Kerry* in Ireland, with letters testimonial under the hands of credible Eye-witnesses."

From among the several letters printed describing the event I will select one which leaves no doubt that the 'monster' was an *Architeuthis:*

"Letter No. 2, from Thomas Hooke (Dublin) to Mr. John Wickins (London). December 23, 1673—

'Loving Friend, I send you this onely pursuant to my former of the Fish, which I now confirm to be as I gave you the first Account with this addition of certainty, that knowing the man by name James Steward, and hearing two or

three nights since of his being at a Printers neer our house to get the Lord Lieuten-
ants Order Printed, which he gave him for exposing what he hath of the fish to
view, I sent, desiring to speak with him, and he came, having then the Picture
with him of the Fish, and he gave me himself the full account of it, viz.

'That in the month of October last, I think about the 15th day he was alone
riding by the sea-side, at Dingle-I-cosh and saw a great thing in the Sea, which
drew his eye towards it, and it came just to him; when he discerned the horns it
began to look frightfully, he said he was sometimes afraid to look on it, and when
he durst look on it, it was the most splendid sight that ever he saw; The Horns
were so bespangled with those Crowns, as he calls them; they shewed he saith
like Pearls or precious Stones; the Horns it could move and weild about the Head
as a Snail doth, all the ten; the two long ones it mostly bore forwards, the other
eight mov'd too and fro every way; When it came to shore its fore parts
rested on the shore, and they lay; He got help after awhile, and when he saw
it stirred not to fright them, he got ropes and put them about the hinder parts, and
began to draw it on shore, and saw it stir'd not to hurt them, they grew bold, and
went to pull with their hands on the Horns, but these Crowns so bit them, that
they were forced to quit their hold: the crowns had teeth under every one of them,
and had a power to fasten on anything that touched them; they moved the Horns
with handspikes, and so being evening they left it on the shore, and came in the
morning and found it dead. The two long Horns are about 11 foot, the other 9;
the other 8 Horns, about 6 and 8 foot long a piece, and as thick as a man's arm
every one of them. He hath brought up to Dublin but two short Horns of the
Crowned ones, and the little Head, being not able to bring the rest the way is so long.

'The certainty is attested by many at the place, and is no doubt a very certain
truth, the mantle was all red on the out-side, which for the colour sake he kept a
piece of it, it was five inches thick, and white under; when they cut the Fish it had
not a drop of blood, nor scale, nor fin, my man took a draught of the Picture which I
have here enclosed, he said it was as big as any horse as ever he saw, it had no leggs.'
                                   'Your loving friend, THOMAS HOOKE.' "

The horns are, of course, the tentacles and the crowns bespangling them
the suckers, which with their toothed edges do look just like little crowns

   There are two other Irish records of *Architeuthis*, one from Kilkel,
Co. Clare, in 1918, and another whose remains were found in the
stomach of a sperm whale at the Belmullet Whaling Station by Mr.
J. E. Hamilton in 1914; the latter was probably *A. harveyi*. The most
complete description of a British specimen is that given by the late
Mr. G. C. Robson (1933) of the British Museum in describing a new
species which was stranded at Scarborough in January 1933 and
obtained by that energetic naturalist the late Mr. W. J. Clarke, who
has played the same important part in reporting and securing squids
on the Yorkshire coast as the Rev. M. Harvey played in Newfoundland
last century. The new squid differed in a number of respects from both
A. harveyi and *A. princeps* and Robson appropriately named it *A. clarkei*.
The 'hands' or ends of the long tentacles had unfortunately been cut
off by a souvenir-hunter before Mr. Clarke could obtain the specimen;

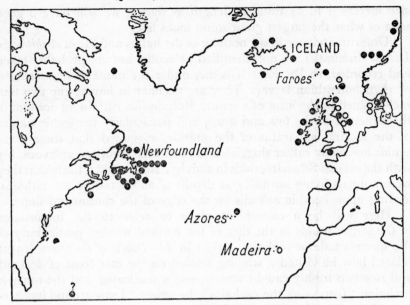

FIG. 95

Records of the giant squid (*Architeuthis*) in the north Atlantic Ocean taken from the map by Robson (1933) with more recent additions (open circles) kindly provided by Dr. W. J. Rees.

allowing 2 feet for their probable length, the total length of the speci-men would be 19 feet and that of the body about 5 ft. 5 inches. There are seven records from Scotland from the following points: between Hillswick and Scalloway, Shetland in 1860 or '61; Dunbar in November 1917; one taken at sea near the Bell Rock off Angus in November 1937; one stranded on North Uist, Outer Hebrides in February 1920; one at Caithness in 1922; one at the Bay of Nigg near Aberdeen in November 1949 and one at Carnoustie in January 1952. The identity of any of these specimens is by no means certain. Some of the earlier ones seem most likely to have been *A. harveyi*, but the Nigg specimen shows much agreement with *A. clarkei* although it differs from it in a few respects; it was approximately the same size as the Scarborough specimen, having a total length of 19 feet 3 inches and a body-length of 4 feet 9 inches, but Dr. B. B. Rae who describes it (1950) considers that it may be a somewhat younger individual and that this may account for the differences found. Plate XXII (p. 273), shows a photograph of it taken from Dr. Rae's paper; if we imagine its size increased to $2\frac{1}{2}$ times its length as shown, we will have a good idea of what the largest giant squid looks like.

Unfortunately we know nothing of the habits and life of *Architeuthis*. Larval *Architeuthis* are not identified by name, but are probably taken and recorded, so Dr. Rees tells me, under the name of *Rhynchoteuthis* or rhynchoteuthian larvae. They are peculiar in having the two tentacles joined in the form of a spout. Robson (*loc. cit.*), in an account of the structure of the fins and arms, and particularly the feeble nature of the locking apparatus of the mantle, suggested that these giant squids are really rather sluggish animals which compare unfavourably with the smaller Sthenoteuthids in activity. He thought it likely that they were animals living normally at depths of about 100 or 200 fathoms, feeding upon benthic animals on the edge of the continental slope.

Here will be a convenient place to refer to the importance of the larger squids in the diet of the toothed whales, particularly of the sperm whale or cachalot. Hjort in *The Depths of the Ocean* (1912) related how he visited a whaling station on the east coast of Iceland and saw two freshly caught whales, one a nordcaper and the other a cachalot. "Inspecting the cachalot", he writes, "I saw around its enormous jaws several long parallel stripes, consisting, as closer scrutiny revealed, of great numbers of circular scars or wounds about 27 mm (one inch) in diameter. . . . It occurred to me that these scars must have been left by the suckers of a giant squid, and following up this idea I found in the whale's mouth a piece of squid tentacle 17 cm (6¾ inches) in maximum diameter. In the stomach of the whale many squid-beaks of various sizes were found, the largest measuring 9 cms (3½ inches) in length, besides some fish bones; and the men who had shot the whale told me that in its death-flurry it disgorged the arm of a squid 6 metres long." The real pioneer, however, in the study of the food of the sperm whales was the Prince of Monaco who took up whale hunting for the express purpose of adding to our knowledge of these giant squids in the eighteen-nineties. Mr. J. Y. Buchanan (of the *Challenger*), who had accompanied him on many of his expeditions, describes in his *Accounts Rendered* (1919) how, while visiting the Azores in 1895 a sperm whale was mortally wounded by the local whalers and charged under the Prince's yacht. As it came up to the surface to die on the other side of the ship it spewed out the remains of a large squid which the Prince secured to find that it was a species new to science. It was this which fired him with the enthusiasm to equip his yacht for whale-hunting and so to give us almost the only information we have about these elusive monsters except that from the specimens already discussed.

Plate XXIII. The Cuttlefish *Sepia* catching a prawn (*top*) *a*. *Sepia* approaching a prawn with outstretched arms and protruded tentacles.
(*centre*) *b*. The prawn is successfully seized. The head of the Sepia is craned forward on the neck, and the arms are parted to each side ready to receive the prey when it is drawn back on to them.
(*bottom*) *c*. Sepia is settling down to eat the prawn.
All exposures by synchronised flashlights at 1-100 sec.　　　　　　　(*Douglas Wilson*)

Plate XXIV

While there can be no doubt that these giant squids are from time to time eaten by the sperm whale, it would be a mistake to imagine that they form its principal food, at any rate in the Azores area. Mr. Robert Clarke of the National Institute of Oceanography, who has recently been investigating there, tells me in a personal communication, prior to publication, that he examined the stomach contents of 39 whales at Fayal in 1949. He finds that squids are certainly their staple diet, but the largest he measured out of a total of 112 was, excluding the long arms, not more than 8 feet in length; the average length he found to be just over 3 feet. The two commonest species taken were *Histioteuthis bonelliana* (59 per cent by numbers) and *Cucioteuthis unguiculatus* (39 per cent). Occasionally large fishes are also eaten.

In the two preceding chapters dealing with bathypelagic life and phosphorescence I postponed dealing with some of the deep-water and luminous cephalopods until I had reached this chapter devoted to the group. It has already turned out longer than I intended; so I must cut down all written reference to the many deep-water forms, some beautiful and some grotesque, to a minimum; I illustrate a group of different types in Fig. 96 drawn from specimens in the British Museum which have been caught in the Atlantic. I would just call attention to the remarkable *Chiroteuthis veranyi* with its long, extremely slender, tentacles, and to *Histioteuthis bonelliana* which, like *Calliteuthis reversa* in Plate 21, is studded all over with complex photophores. I was lucky to see *Calliteuthis* brought up alive on the *Discovery II* in 1954. It is remarkable not only on account of its vast numbers of opalescent light organs but because the right eye is so different from the left; this is a character of the genus. The right eye is the smaller and more sunken and is surrounded by a ring of photophores which are absent from the enlarged left eye. *Thaumatolampas* (*Lycoteuthis*) *diadema* has luminous organs of different colours, some white, some red and some blue; the colour drawing in Plate 21 (p. 233), was made from a freshly-caught specimen on the 1925-27 *Discovery* expedition; it is one of the most beautiful animals I have ever seen. I think it will be a fitting end to the chapter if I quote a description of it given by Professor Chun who

---

Plate XXIV. Plankton collecting gear. (*top left*) *a.* The continuous plankton recorder, with internal mechanism taken out and placed alongside. (*top right*) *b.* The small plankton indicator for use on fishing craft. (*bottom left*) *c.* The vertical Hensen net; the coarse netting on the outside protects the net of fine silk gauze within (*Fisheries Laboratory, Lowestoft*). (*bottom right*) *d*, Large deep water tow-net, 4½ metres in diameter, used in the R.R.S. *Discovery*.

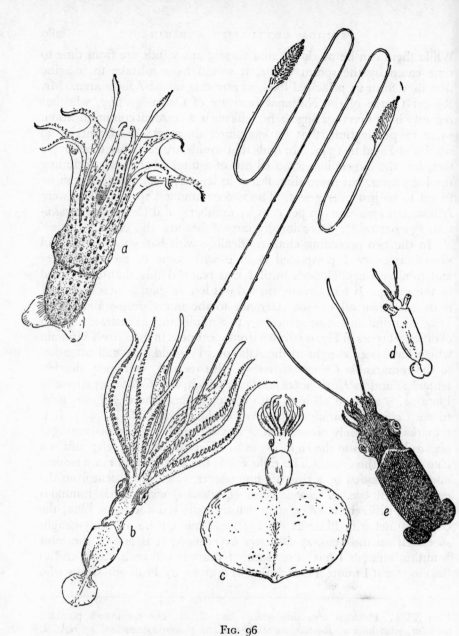

FIG. 96

Some deepwater squids of the Atlantic. *a, Histioteuthis bonelliana*, taken at
1,500 metres depth off the west of Scotland ×½; *b, Chiroteuthis (Loligopsis)
veranyi*, ×⅓; *c, Octopodoteuthis sicula*, ×½; *d, Taonidium sp.* ×½; *e,
Benthoteuthis megalops*, ×½. All drawn from specimens in the British Museum
except '*a*' which was drawn at the Fisheries Laboratory of the Scottish
Home Department at Aberdeen.

discovered it on the *Valdivia* expedition and gave it its name; the following passage is a translation given by Dr. W. E. Hoyle in his Presidential Address to Section D (Zoology) of the British Association at Leicester in 1907:

"Among all the marvels of coloration which the animals of the deep sea exhibited to us nothing can be even distantly compared with the hues of these organs. One would think that the body was adorned with a diadem of brilliant gems. The middle organs of the eyes shone with ultramarine-blue, the lateral ones with a pearly sheen. Those towards the front of the lower surface of the body gave out a ruby-red light, while those behind were snow-white or pearly, except the median one, which was sky-blue. It was indeed a glorious spectacle."

CHAPTER 15

PLANKTON AND THE FISHERIES

IN THE PREFACE I have explained how I originally intended to include the fish and fisheries of the sea in this one volume. Such a book, if kept to the size of the others in the series, would have meant a very inadequate treatment of the whole subject. It was decided therefore, that the fish (except the deep pelagic species already dealt with), the whales and the life of the sea-bed should all become the theme of another volume; a sequel which will appear before very long. However, in order to round off our present study, we must briefly relate the planktonic world to these other kinds of life. Can our knowledge of this drifting community, which has been won by so much research, help us in our fishery problems?

The fish of the open sea may be divided into two main categories— pelagic and demersal, according to their habits. The former are those like the herring, mackerel, sprat or pilchard which swim up into the upper layers of water to feed directly on the plankton; they are usually caught by nets suspended from floats at the surface. The demersal fish are those frequenting the sea-bed, such as cod, haddock, plaice and skate, which, with many more, are caught in the trawl; these prey upon the bottom-living animals which are in turn fed by the rain of dead and dying material, falling from the pelagic world above. These trawl-caught fish, however, have a much closer link with the plankton in early life; their baby fry live for a time in the upper layers of the water, not only feeding on the plankton, but actually being part of it, drifting with the currents of the sea.

In Chapter 5, p. 78, we saw how markedly the plankton may vary, both in quantity and kind, and in Chapter 2, p. 13, how the waters carrying this mixed and uneven population are constantly on the move. While the main current-systems round our islands are well known, they, too, are continually varying in detail. The inflow of Atlantic water into the North Sea from the north may be stronger in some years than others, likewise that from the Channel through the

Straits of Dover; the resulting eddies and swirls as they flow into this very irregular basin (Fig. 8, p. 27) are continually being modified. At one time there may be a rich zone of herring food slowly on the move; in another year, in the same season, the plankton may be quite different. The study of these changes has always held a particular attraction for me: surely no-one can doubt that here must lie important keys to a better understanding of success or failure in the fisheries. In this chapter I will describe some of the attempts at hunting for these keys in this world we cannot see.

Pelagic fish such as the herring, because they feed on the plankton, are more likely to be affected by it than others; so let us begin with them. Curiously enough, the first evidence of a planktonic influence on the herring was not concerned with the actual presence of their food, but with that of certain other elements which they appeared to avoid. F. G. Pearcey (1885) made a voyage in one of the old herring sailing luggers from Leith round Shetland, taking a tow-net with him. He came across many areas of dense phytoplankton (mainly the diatoms *Rhizosolenia* and *Thalassiosira*) and in these hardly any herring were caught; in the clear regions, on the other hand, the catches were high. At Lowestoft, where I was appointed Assistant Naturalist at the Fishery Laboratory in 1921, I began to study the feeding habits of the herring and its relations to the plankton in general. It was a poor autumn fishing that year. Going out on a drifter with a tow-net, I had just the same experience as Pearcey: the water was thick with phytoplankton, also *Rhizosolenia*, so that the meshes of my net were clogged with it. The skipper called this water "weedy water" or "Dutchman's baccy juice", because the nets came up slimy and brown, and he said they were unlikely to catch fish in it.

I then started experimenting with a little instrument I called a plankton indicator, a metal torpedo-shaped device, hollow and open at each end to allow the water to flow through it. It was simple and sturdily built to stand rough treatment on a fishing boat; when first designed it was taken down by a weighted tow-rope, but later it was fitted with the little diving planes shown in Plate XXIV (p. 289). Before it was thrown out a gauze disc was placed across its water tunnel to sieve out the plankton as it was towed along. Fishermen were asked to use it to sample the sea where they fished, and then to wrap up the plankton-coated disc in a piece of calico provided, and drop it into a tin of preservative fluid (formalin). Later, when they had hauled their nets, they filled in a printed label giving the date, position and the number of herring caught. The idea was to get a series of

plankton samples together with records of the catch of fish, so as to
have definite evidence as to whether the fisherman's belief in the poor
catches to be expected in "weedy water" was based on fact or not.
At that time I had not yet come across Pearcey's paper. The device was
first tried in the Lowestoft fishery of the autumn of 1922 and spring
of 1923. Only fourteen records were obtained, but the results seemed
to be very striking (Hardy, 1926). Seven of the discs were coloured
a distinct green with phytoplankton, either due to *Phaeocystis* (p. 49)
in the autumn, or, on one occasion, to the diatom *Chaetoceros* (p. 44)
in the spring, the other seven were quite blank or colourless with
only very few diatoms. The herring catches corresponding to the
green discs ranged from ½ to 6 crans with an average of 3—a cran
being a measure of about a thousand fish according to their size;
the catches corresponding to the other discs ranged from 15 to 45
crans, with an average of 25. The odds against such a clear-cut
result being due to chance are very high. It appeared to be a strong
confirmation of Pearcey's original observations; as it turns out, however,
this particular case may well have another explanation which I shall
come to in a moment.

Having left the North Sea for a time, to go south with the *Discovery*,
it was not until some years later that I was able to start a similar set of
experiments; they clearly showed the phytoplankton effect, but not to
such an extent as in 1922 (Hardy, Lucas, Henderson and Fraser, 1936;
Part IV). In each of the four years of the tests, 1931 to 1934, the
average catch corresponding to the clear discs was higher than that
for the green discs by a considerable amount: 330, 191, 145 and 69
per cent. respectively; in the first three of these years the results were
almost all from the summer fisheries.[1]

Dr. D. H. Cushing, of the Fishery Laboratory at Lowestoft, has
now kindly shown me the draft of a paper which he will shortly publish
(1956) giving evidence that there is no effect of actual exclusion of

---

[1] Most of the observations were taken in the summer Scottish fisheries or off
Shields, where the catches are not so high as at East Anglia. In these investigations
the fisheries were divided into different areas and into half-monthly periods. In
the summer of 1931 25 green discs were taken, for which the average herring catch
was 1¼ crans; in the same areas and periods the number of clear discs was 93 with
an average catch of 5¾ crans. Similar figures for 1932 were 23 green discs with an
average catch of 1⅔ crans, and 103 clear discs with an average of 4¼ crans; in 1933
there were only 7 green discs giving an average of 1⅔ crans as against 52 clear discs
giving an average of 4 crans. In 1934 a more intensive experiment was made in the
East Anglian autumn fishery, using a much smaller instrument; here the catches
were much higher; 42 green discs were obtained with an average catch of 21½
crans, whereas 108 clear discs gave an average of 36¼ crans.

herring by the plants at the time of the autumn fishery. He does not deny that the presence of dense phytoplankton in the *summer* fishery indicates the likelihood of a poor catch of herring as found by Pearcey and in our experiments of 1931, 1932 and 1933; he believes, however, that this effect is only found in the summer when the herring are feeding. The rich phytoplankton areas are likely to be those with little herring food in them, because this food consists of animals which in turn feed on the plants; i.e. if the animals had been abundant they would keep the plants down—this is the grazing effect we have discussed on page 215. In support of his view he presents the important evidence of 62 records of phytoplankton of varying density and corresponding herring catches, all taken within a period of 18 hours

FIG. 97

Examples of dense autumnal patches of plant plankton in the southern North sea: *left*, *Phaeocystis* in November, 1927 (Savage and Hardy, 1935); *right*, *a* and *c* *Rhizosolenia styliformis* and *b* *Biddulphia sinensis* in October 1933 (Savage and Wimpenny, 1936). The English and Dutch coasts are shown in outline.

in the autumn fishery of 1948, and these do not confirm the exclusion effect. This seems conclusive and I consequently think he must be right in attributing the apparently significant correlation of my 1922 results to what is technically known as a serial effect.[1]

After the "weedy water" had been found stretching across the fishing grounds in the poor season of 1921, we decided on another line of work: a series of annual cruises to chart the phytoplankton zones

[1] In the autumn my twelve results were spread in time between October 13th and November 24th. In the first half of that period these mainly gave low phytoplankton and high herring catches, and in the second half the reverse; thus, the results might be simply due to a good early fishery and a poor late one which were independent of the appearance of the heavy phytoplankton in November. In the following spring I got two more indicator records from a drifter, one of high and one of low phytoplankton, with 1½ and 24 cran catches respectively; these fitted so well with the former results that I unfortunately discounted the possible serial effect.

and see if we could relate them to the fortunes of the fisheries. Beginning in 1922, they were continued for many seasons and showed that the size and position of the patches could vary enormously from year to year (Savage and Hardy, 1935; Savage and Wimpenny, 1936). Two examples are given in Fig. 97 (p. 295). It was thought likely that, when these concentrations occurred near the main shoaling grounds, the herring might be delayed in their arrival or perhaps deflected to other areas. Dr. Cushing, however, in his forthcoming paper, doubts the validity of these conclusions, partly on account of his phytoplankton and herring observations just referred to, and partly because there is now more knowledge available regarding the path taken by the herring in approaching the fishing grounds. Whether the fish in the autumn are deflected by such patches or not, the evidence of the indicator experiments makes it clear that "weedy water" on the herring grounds in the summer is likely to have an effect on the *drift-net* fishery because smaller catches are nearly always made in it. This is further supported by the observations of the Russian fishery workers in the Barents Sea, who have carried out intensive tests with plankton indicators similar to mine: they are described by Manteufel (1941)[1] and I will give a slightly abridged quotation from a translation of his summary of "practical instructions for reconnoitring herring."

"Strongly 'weedy water' acts negatively upon the shoaling of the herring. Therefore in May and June slimy, bright green or brown-green plankton indicator discs as well as green mists in samples collected by Hensen nets, with a rather sharp smell, characterise waters unfavourable for the fishing of herring. If it is necessary to fish herring in these waters one has to use drift-nets on long ropes below the zone of algae. . . . In July discs and samples of plankton may also indicate 'weedy water' (summer weedy) which is unfavourable for herring".

Savage and Wimpenny, in their study of the phytoplankton of 1933 and '34, showed how the herring appeared to mass against the edge of the concentrations. This behaviour has also been reported by the Russians. I quote again from the same translation:

"In May and June the shoaling of the herring takes place near the boundaries of the shrinking zone of 'weedy water'. . . . The presence of small numbers of herring within the regions of the boundaries of 'weedy water' often indicate the presence of large shoals of herring 5 to 10 miles westward or south westwards (against the current)".

Now let us turn to the feeding relationship. The gills of the herring are provided with fine comb-like structures called gill-rakers, which prevent the escape of the plankton through the gill-slits at the back of the mouth. About sixty of these little comb-teeth go to an inch of gill—

[1] I have elsewhere given some extensive extracts from a translation of this paper. (Hardy, 1951).

the same, as it happens, as the number of meshes to the inch in a Hensen plankton net, or in the gauze of an indicator disc. It is often thought that the herring feeds by simply swimming forward with its mouth open so that the water flows out continuously through the gill-rakers which sieve out the plankton like a tow-net; it appears, however, that the herring snaps at the little plankton animals individually, like a swift or swallow feeding on the tiny insects in the air, and that the gill-rakers are there to prevent the small creatures from passing out of the gill openings and being lost. If its favourite food, the copepod *Calanus*, is in the plankton in large numbers, it will feed on this and little else, neglecting the smaller copepods and many other animals which may be present (Cheng, 1942). Some very interesting observations were made by Commander G. C. Damant (1921) when diving for salvage. While he was suspended below the keel of his ship, he saw mackerel feeding on plankton animals which were in the shadow of the hull and so made conspicuous in dark silhouette against the sunlit waters beyond; they were picking them up one by one by rapid snapping motions. I have also seen young herring in an aquarium snapping at individual plankton animals. As a matter of fact, herring are not entirely plankton feeders, for in the spring, in the southern North Sea, they take young sand-eels, up to four inches in length, in large numbers. I have found eighteen (averaging 2½ inches in length) in one herring stomach, packed tightly like sardines in a tin, and as many as 53 much smaller ones in another. Plankton, however, is certainly their main food.

Many naturalists[1] in the past have believed that the movements of the herring are governed by the abundance or scarcity of their plank-tonic food, but until recently this has been difficult to prove. In 1923, in addition to using the plankton indicator for the phytoplankton tests, I had used it in the summer Shields fishery to see if it could demonstrate a positive correlation with their food. It failed, but I then realised that, while efficient in the sampling of little plants, it was by no means so well adapted to the capture of the animal forms: very often the weight was insufficient to take it more than a little way below the surface. It was then that the small diving planes were added and a new series of experiments made from 1930 to '33. They were on quite a large scale as we were fortunate in receiving the willing help of the skippers of no less than thirty-two drifters. The same system of providing them with labels and tins of formalin was adopted, and they were asked to tow the instrument for a mile just before they reached

[1] For a general review of these earlier observations see Hardy (1924*b*).

the position at which they intended to fish. Over 1,400 records of catches were obtained from the Scottish and English fisheries, each with a corresponding sample of plankton ready for analysis (Hardy, Henderson, Lucan and Fraser, *loc. cit.*).

We divided the fisheries into regions separated by degrees of latitude and dealt with the samples from each in half-monthly periods. These we examined to see if there was any correlation between the numbers of herring caught and those of the copepod *Calanus*, which they prefer for food, found at the same place. I must ask the reader to forgive a little explanation of our procedure, which is necessary for an understanding of the results. We wrote down the numbers of *Calanus* found in each sample and arranged them in ascending order from the lowest to the highest: we then put them down in two equal columns, that on the left with the *lower Calanus* numbers, and that on the right with the *higher* ones. Next, in columns against the *Calanus* figures, we wrote down the corresponding quantities of herring caught in crans; now we can total up the crans of herring taken in the poorer *Calanus* water, and compare them with those from the richer water. An example will make it clear; the figures in the accompanying table are from the eastern Scottish fishery, latitude 59°—60° N, for the second half of July, 1932.

In poorer *Calanus* water		In richer *Calanus* water	
Calanus numbers	Herring (in crans)	*Calanus* numbers	Herring (in crans)
1	0	120	36
1	24	125	47
20	20	165	2
25	4¾	180	9½
40	0	204	61
42	4½	245	37½
68	¾	280	63
88	72	330	59½
96	30	332	10¾
101	1	480	19
105	0	1420	0
TOTALS 587	157	3881	345¼
AVERAGES 53	14¼	353	31½

We see that while some good catches were made in the poorer *Calanus* water, and some poor catches were made in the richer water, the total catch in the richer water was more than double that in the poorer: an average of 31½ as against 14¼ crans. This example is illustrated in the diagram shown in Fig. 98. For the Scottish fisheries, from Shetland southwards, the average catches in the richer *Calanus*

HERRING
IN CRANS

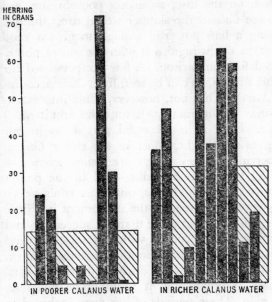

FIG. 98
Histograms comparing the
individual catches of herring
(black columns) and the
average catch (shaded areas)
taken in poorer and richer
Calanus water off the north-
east of Scotland in the period
July 16—August 1, 1932.

IN POORER CALANUS WATER   IN RICHER CALANUS WATER

water exceeded those in the poorer water for fourteen out of eighteen
such periods of experiment.

Could the indicator be used commercially? We wanted to find
out what would be the advantage to a skipper if he actually used it to
guide him to the richer grounds. In our example we see that the total
catch of herring during the period was $502\frac{1}{4}$ crans (i.e. 157 plus $345\frac{1}{4}$);
that is what he caught by fishing at random. If he used the indicator
to test the water as he steamed out from port—taking samples every
few miles and coming back to fish where he got the most *Calanus*—
he should each time have come into the richer water instead of only
sometimes. Now if all the catches had been in this water, instead of
only half, he should have caught $345\frac{1}{4}$ crans twice over, that is $690\frac{1}{2}$,
instead of the $502\frac{1}{4}$ crans he got by random fishing. That is a gain of
$188\frac{1}{4}$ crans or $37\frac{1}{2}$ per cent. This, of course, is only an estimated gain,
and reliance must not be placed upon just one such example; if,
however, we take the whole series of eighteen periods, including the
four in which the catch of herrings was actually less in the richer water,
we get an average estimated gain for the whole of the eastern Scottish
fisheries of $24\frac{1}{2}$ per cent. In the Shields fisheries the results were not
so good, but still gave an average estimated gain of $12\frac{1}{2}$ per cent.

At last it seemed that planktology was about to yield practical

300 THE OPEN SEA

results. A green indication on the disc, signifying poor fishing, can be seen at a glance. To see *Calanus* the skipper would drop the disc into a little frame carrying a lens powerful enough to give a good magnification; he soon learns to distinguish it with the aid of photographs which are provided for comparison. A few skippers used the indicator with considerable success, as can be seen from their accounts published in the *Fishing News*[1]; it has not, however, come into extensive use for several reasons. Unfortunately, it must be admitted, it is not sufficiently reliable. Even in a successful period, as in our example, it frequently gives false indications; in the richer *Calanus* water there were five herring catches which are below average— including no herring at all for the richest disc—and in the poorer water there were two above average. That is, out of 22 trials, seven indications were wrong—or one in three. If the instrument is a little trouble to use, the failures will stand out more than the successes—and the skippers feel they might have had success without its use anyway! But more serious there may be whole half-monthly periods—four in the Scottish and five in the Shields fishery during our three years' experiments—when, if used commercially, it would have given an average loss instead of a gain. These periods of loss were included in the total estimated gains for the Scottish and Shields fisheries given above; over reasonably long periods of use the losses are offset by the greater gains—nevertheless, they are too discouraging. Now science has given the fisherman a more direct method of locating his quarry; the echo-sounding gear which has already been described (p. 235) in relation to the deep scattering layer. While it is being extensively used in detecting shoals of cod and similar fish swimming a little way from the bottom, it is of even greater importance in showing the presence of shoals of pelagic fish such as the herring, sprat and pilchard; few drifter skippers now shoot their nets without first getting an indication of fish on their echo-meter.

The periods of negative results shown by the indicator may sometimes be due to a shortage of *Calanus* when the herring may turn after other food (Cheng, 1941), or sometimes, perhaps, to ripe fish moving towards the spawning grounds, rather than looking for richer plankton. If the herring congregate on the denser concentrations of *Calanus* we may first expect a positive indication; then, as they graze down the population until its density becomes less than that in the surrounding areas, we may expect negative results until the herring move on to richer grounds again. This is what happened in

[1] Quoted in Hardy, Lucas, Henderson and Fraser (1936), p. 172.

the Shields fishery in the latter half of the 1931 season. (Lucas, 1936)[1]. The same thing has later been demonstrated by Manteufel in the Russian fishery. More intensive research on the indicator correlations is at present being carried out in the Fraserborough area by the staff of the Oceanographical Laboratory at Edinburgh, and we may expect further information on these problems before very long. While the indicator has its drawbacks,[2] I still think, as more is known about when to use it, that it may be a valuable adjunct to a herring boat; indeed, it would be this, if for no other reason than that at times, in spring and summer, it would save a loss by warning the fishermen to move out of thick phytoplankton water, which often may not be detected until the slimy nets have been hauled. As a matter of fact I find that the indicator is not really a novel invention at all; quite recently, Mr. G. McPherson, of the Scottish Home Department's Fishery Laboratory at Aberdeen, has shown me the following note written by one of their Collectors of Fishery Statistics, Mr. David Buchan, of Peterhead, who can remember the old days of the fishery at the turn of the century:

"I have seen them in the sail boats take a square piece of sacking, tie it at the corners and tow it like a small drogue to catch the small marine growth. If it was the small gritty material they rubbed it between their fingers and said 'That's good feeding matter for herrings.' If, as sometimes happened, it was the slimy vegetable material left in the drogue which had a green smell they said 'Herrings dinna like to rise in that water.' I have noted the truth in this ever since."

If positive indications sought for by an individual boat frequently fail, there can be no doubt about the value of a general survey of the plankton over the herring grounds for the information of the fleet as a whole. Let us now turn to some recent work in this direction.

The Fishery Laboratory at Lowestoft was, I believe, the first to issue plankton reports for the guidance of herring fishermen when it began before the war making announcements over the radio as to the distribution of phytoplankton patches in relation to the East Anglian autumn fishery. Manteufel, in the paper already referred to, has described the success—and partial failure—of Russian attempts to supply their fleet with the latest plankton information. I quote again from the translation:

"During these months (May and June) the herring can be successfully found only on condition that all the fishing boats are armed with apparatus for collecting

[1] Part III in Hardy, Lucas, Henderson and Fraser (1936).
[2] The indicator has, incidentally, proved useful as a research instrument for it may be used at much higher speeds than a tow-net; various modifications of it are described by Glover (1953). It could be very handy for a naturalist yachtsman to sample the plankton at intervals without slowing down his craft.

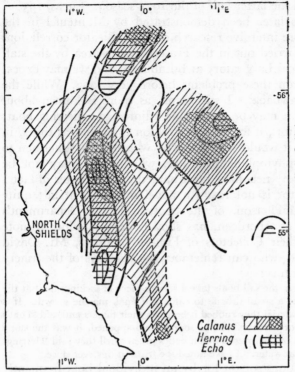

Fig. 99
A chart showing the correspondence between the concentrations of herring as revealed by an echo survey (thick line contours) and patches of the copepod *Calanus* (shaded contour areas), from Cushing (1953).

plankton (plankton indicators, Hensen's nets). It is necessary to collect together in good time the data obtained and to give operative scientific advice to the herring fleet. An experiment of giving such advice was very successfully fulfilled in May, 1939, when a detailed survey of hydrology and of plankton was made by the crew of s.s. *Persey* in the south-western part of the Barents Sea. This survey, supplemented by data obtained from fishing drifters, gave a clear picture of the distribution of plankton, and thus constant advice could be given to the reconnoitring fleet. Helped by this information, the herring fleet found several large shoals of herring and fished them successfully. However, as at that time some of the instructions were not sufficiently studied and as the drifters did not always use the advice received, a large number of drifts were made in zones of plankton unfavourable for the shoaling of herring."

Dr. Cushing (1952, 1953), of the Lowestoft Laboratory, has made a series of cruises charting the plankton, and at the same time using the echo-sounding gear to make a survey of the distribution of the herring shoals. He shows how, in a series of cruises, the herring were seen to aggregate in increasing numbers about patches of *Calanus*. Fig. 99, above, is a reproduction of one of his charts. A recent Ministry of Agriculture and Fisheries leaflet, *Sea Fisheries Research Notes*, 1954, records how the research ship *Sir Lancelot* again followed a *Calanus* patch and "at the same time as she was keeping in touch with the plankton . . . watched for the arrival of the herring. When dense echo-traces showed up ENE from the Tyne she was able to pass the word to North Shields, and the fishing, which had hung fire, then began."

Similar relationships between the plankton and our next most important pelagic fish, the mackerel, are being established. Mackerel are plankton-feeders for about half the year—through the spring to midsummer; from the end of June onwards they feed largely upon small fish, and it is then that we can catch them so easily on a line with a bright spinner or even with a small white feather on the hook. Nearly fifty years ago G. E. Bullen (1908) began plankton studies in the important western mackerel fishery which is centred on Newlyn in the spring. He recorded that the fishermen agreed that the largest catches are usually made in what they call 'yellow water'; and this he showed to be so coloured by the presence of large numbers of the two copepods *Calanus* and *Pseudocalanus*. They also told him that it was useless to shoot their nets in what they called 'stinking water'; this, they said, was "so dense that a man looking over the side of a sailing drifter cannot see down to the keel," and it had "a distinctly noxious smell . . . similar to that of decaying seaweed." This must usually have been very dense phytoplankton, although in the only sample Bullen got for examination he could not detect any odour and found only a moderate preponderance of diatoms over zooplankton. G. A. Steven (1949), who again investigated the fishery, gives ample confirmation that the highest mackerel catches are taken in the 'yellow water' and that the colour is due to the same copepods as recorded by Bullen; he did not himself come across examples of stinking water, but says "there appears to be no doubt that when it does occur poor fishing must be expected in it."

Turning for a moment from fish to whales, all the larger species, except the sperm whale, are toothless and feed on plankton. It is indeed remarkable that the very largest of animals—far larger than the greatest dinosaurs of the past—should be nourished on plankton; it speaks volumes for the high nutritive value of this diet. The blue whale, which measures about one hundred feet in length when fully grown, is 23 feet long at birth and over 60 feet when two years old. The whale has a mouth with an enormous capacity, and takes in a huge gulp of sea with its hosts of small crustacea; then, by raising the

floor of its mouth and expanding its gigantic tongue, it forces the water
out sideways through those remarkable sieves of horny fibres
which hangs down from the inner margins of the upper jaw like a
vast internal moustache. The plankton deposit is them wiped off by
the tongue and swallowed. Hjort and Ruud (1929) have shown how
in the north the blue and fin whales congregate in regions of great
swarms of krill, as the Norwegians call the prawn-like euphausiacians
(p. 171) on which they feed; they also show that the somewhat smaller
sei whales, which have finer horny filters for capturing copepods, tend
to collect in the regions of greatest *Calanus* production. During the
Discovery investigations in the Antarctic, the whaling gunners at South
Georgia and the South Shetlands kept log books showing the positions
of all the whales harpooned, so that the spread of the different species
could be charted month by month. When the abundance of their food,
the southern krill *Euphausia superba*, was plotted on the same maps—
from the plankton surveys made by the R.R.S. *Discovery*—a remarkably
exact correspondence was shown between the concentrations of whales
and the regions of greatest krill production. (Hardy and Gunther, 1935).

Returning to our fish, it is worth noting, before leaving the pelagic
species, that the largest of these are also plankton-feeders: the great
whale-sharks of the southern seas and the thirty-foot basking sharks
which visit the west coasts of Ireland and Scotland in large numbers
in the summer months. They have enormous gill-openings which are
equipped with fine networks of interlocking combs. These are gill-
rakers, which are developed to a great length, but have been, of course,
evolved quite independently of those of the herring, for the two groups,
within the class of fish, are as widely separated as they could be in their
past racial history. Unlike the herring, the basking sharks do in fact
catch their plankton as a tow-net does. When feeding they swim at only
about two knots, with the mouth kept continually open, so that the
water flows in a steady stream out through the gills, leaving the plankton behind on the gill-raker netting. Like good planktologists, they
know that it is useless to push such a net through the water any faster;
as we noted in Chapter 3 (p. 39) it just will not filter more quickly.
When *Calanus* is abundant the basking sharks are a common sight
round the Isle of Arran and up the lower reaches of Loch Fyne; they
cruise slowly along, just below the surface, so that the large triangular
fins in the middle of their backs stick up out of the water like so many
little sails.

The majority of demersal fish, as we have already noted, feed
upon the animals of the sea-floor, although the cod, in addition,

preys extensively upon the herring, which spends much of the day near the bottom. The growth of the bottom fauna, which depends for its food on the life above, must vary considerably from year to year. We have just seen examples of the great phytoplankton masses which are sometimes produced in the autumn, but in other years they may be much smaller. If they do erupt in their full glory in two consecutive years, they rarely do so in just the same place. The quantity of animals available as fish-food on the sea-bed must surely reflect such changes as these in the supply of plant-life above. Quite apart from their normal migrations to and from their spawning areas, the bottom-living fish shift their main concentrations to different parts of the sea in different years. This may be seen from a detailed study of the monthly trawling statistics obtained by the Ministry of Agriculture and Fisheries and plotted on squared maps of the North Sea. It is not unlikely that their movements are in part linked with earlier planktonic changes in the waters above.

An important part of fishery investigations is concerned with the fluctuations in the relative strength of the different age-groups in the fish populations. In one stock, for example, we may find many more four- and six-year-olds, than, say, five-year-olds. This brings us to the other link which the trawl-caught fish have with the plankton; one in their early life when as small fry they form part of the pelagic community. It is almost certain that a shortage of five-year-old fish means that five years ago the conditions in the plankton were less favourable for the young of this particular species; it has been clearly shown that usually it is not due to any shortage of adults supplying the spawn. Not only may the quantities of suitable planktonic food for the fry vary greatly in different years, but the time of its main production may sometimes be earlier and sometimes later than usual. It was Hjort, I believe, who first suggested that these fluctuations in the year-classes were due to shortages of food at the right time in early development; in later years the Danish biologists have given strong evidence to support him. Johansen (1927) showed that during the years 1904 to '26 the numbers of young plaice varied with the temperature and salinity of the water; in hard winters there is a greater outflow of cold and less saline water from the Baltic, and, to quote his words, "the tiny plaice larvae fail to obtain sufficient nourishment, and die of starvation *en masse*." Poulsen (1944) was able to go further: from the years 1923 to '39 he could demonstrate a definite correlation between the numbers of young cod which he took in his young fish trawl in April and May, and the quantity of plankton present in the previous three months. He further showed a correspondence between these young fish fluctuations and those of the older fish later on. Lack of food, however, is not the only hazard of their pelagic life. In Chapters 6 and 7 we have seen how small medusae, comb-jellies and arrow-worms all levy a heavy toll on the tiny fry, as well as competing with them for the supply of small crustacea; in some years these voracious enemies are present in much greater numbers than in others, and Russell (1935) has given us an example of their probable effect on the young fish populations in the Plymouth area in 1929.

Certainly, if less obviously, the plankton affects the fisheries for bottom-living fish no less than those for the surface-swimming kinds. There is, however, yet another way in which plankton studies may help in solving fishery problems. The movements of fish may be affected directly by changes in the water itself. Not only does the sea vary in temperature and saltness, but in much more subtle chemical ways. These different waters, as already noted in Chapter 2 (p. 29) may often be more readily recognised by the characteristic kinds of plankton animals they carry. If changing water conditions are affecting a fishery, a knowledge of the plankton may give us the key to the trouble, as it has done in the Western Channel fisheries (p. 30). R. S. Glover, in his study of the eastern Scottish herring fisheries has recently (1955) shown that over the years 1947 to '53 there has been a southerly movement of the main centres of fishing; at the same time he records increasing quantities and earlier appearances of a number of planktonic animals, such as the pteropod *Clione*, which indicate a progressively larger or earlier Atlantic inflow. It looks, as he suggests, as if we may put two and two together.

Those who wish to make a fuller study of plankton research in relation to the fisheries should read two excellent reviews of the subject which give many more examples and references to the literature: one by Russell (1952) and the other by Lucas, which is now in the press. The latter, which I have kindly been shown in proof, forms two chapters in an important new book *Sea Fisheries and their Investigation* edited by Michael Graham, which should be published in 1956.

The changes of fortune in the fisheries, apart from those brought about by man's own folly in overfishing, must have their causes in the natural world; the relation between cause and effect, however, will only be understood when we have an unbroken record of the changes in the sea month by month over a very wide area for very many years. To solve our problems we must have a service of information from the sea like the meteorological observations from the air which now enable us to link the ups and downs in agriculture with variations in the sunshine, rain and wind. Such a service, moreover, might not only help us to trace cause and effect; it might allow us to infer the probable course of events to come. It is now possible to forecast the weather a day or two in advance; since ocean currents move so much more slowly than the streams of air, it is likely that, given comparable information, we will in time be able to forecast events in the sea a week or two ahead—long enough to be of value in guiding fishing boats to more economic fishing. A broad scale recording of the principal variations in the plankton may be of greater service than one showing only those of a physical kind; the plankton, as we have just seen, is the all-important background to the fisheries and at the same time indicates the major changes in the movement and conditions of the sea. Is such a survey possible? Perhaps it may not be out of place if I finish the volume with a brief account of an attempt which is now being made to supply just such a service; it is one with which I have been specially concerned.

Trying to make out the actual *changing* nature of the plankton is indeed very like a study of the weather. We are concerned with the development and movement, not of centres of high and low pressure, but of centres of high and low animal and plant production. It would be impossible for us to trace the paths of pressure in the atmosphere, and so forecast the weather, if we had only one or two meteorological stations; equally we can never hope to study the continuous progression of the plankton over very wide areas with just one or two ships. We must have many points of simultaneous observation—or better still, lines of observation. Would it be possible to get the help of a number of commercial ships, running regularly on different routes across the sea, to act as moving plankton observatories which would automatically sample the plankton as they went along? It seemed worth attempting.

I have already described in Chapter 5 (p. 76) how, after being so impressed with the patchy nature of the plankton, I devised the first plankton-recorder for use on the *Discovery* Antarctic expedition of 1925–27, to give a continuous line of sampling through the sea. When later I became professor at the new University of Hull (University College, as it was then) I thought it might be possible to re-design it, as a relatively compact machine, for towing behind the ships which run from Hull across the North Sea to various ports on the other side. It is this revised model which I illustrate in Fig. 19 (p. 77). Although

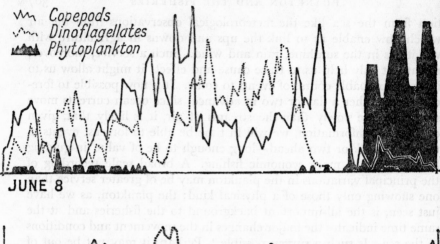

JUNE 8

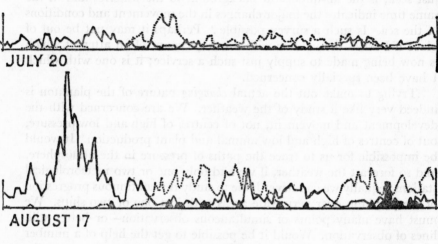

JULY 20

AUGUST 17

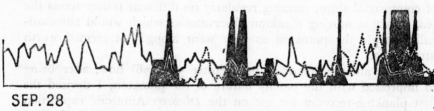

SEP. 28

FIG. 100

Examples of graphs made from the analysis of the plankton collected by the continuous recorders as they are towed across the North Sea; they show the fluctuations in three types of organism on 180 miles of the Hull-Hamburgh run during June, July, August and September, 1932. (The first four such records to be taken in the North Sea.)

originally invented to get over the falsity of taking isolated observations, it turned out to be well adapted as an automatic sampling machine for use on commercial ships. It is set ready for working in the laboratory before going on board. When the ship passes some pre-arranged point, on leaving our coast, the device is lowered into the water and run out astern on a standard length of cable; on reaching a similar point on the other side it is wound in again, on a special winch provided, and kept for return to the laboratory. It requires no other attention; as soon as it enters the water it begins to work as the propeller driving the mechanism is set in motion by the rush of water past it. In the survey I am about to describe the recorders have always been run at a standard depth of 10 metres (approximately 35 feet); it will be remembered that they have the property of remaining at a constant depth in spite of a considerable variation in the speed of the ship.

We began in 1932 with a five years' trial over the southern half of the North Sea, on lines radiating from Hull to the Skagerak, to Hamburg and to Rotterdam. From this small beginning to the greatly extended survey now being undertaken, we have been dependent upon the generous co-operation of a number of different shipping companies and their officers, and upon financial support from the Development Commission; without either, this venture would have been impossible. Let me also gratefully acknowledge the support given to the initiation of this research by the University of Hull and by a private benefactor who still insists on remaining anonymous.

Month by month we obtained continuous records of the varying plankton on these three lines, which we could graph and, for this limited area, chart like weather maps. Just as was expected from the results of tow-net surveys, we saw patches of high production and areas of paucity of the different organisms spread across the sea; but now, as is not possible with a single survey, we could see these patterns of distribution change as the year advanced, and compare the course of events in one year with another (Lucas, 1940; Rae and Fraser, 1941; Henderson and Marshall, 1944). Fig. 100, opposite, shows four consecutive monthly records across the middle of the North Sea in 1932—actually the first four such records to be made. In one month large concentration of copepods may be close against the coast; at a the same time next year the peak of abundance on the same line may be far out in the middle of the run—while perhaps, in its place against the coast may be a belt of plant plankton. Comparing the records for different years we see striking contrasts in the populations of many of

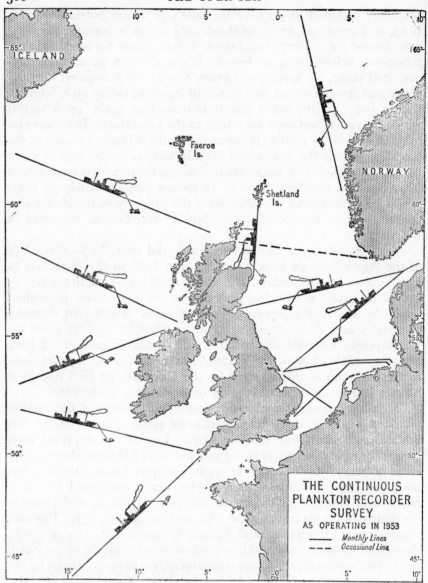

FIG. 101
The plankton recorder survey as operating at the present time (1956).

the principal species; we see, for example, that the little copepod *Pseudocalanus*, which is so important as the food of the very young herring-fry, may in some years be three or four times as numerous as in others. In one year a particular species may reach its peak of abundance much earlier than in another; and some may have two periods of high production in some years but only one in others. In 1936 a new line from London to Esbjerg in Denmark was started to cut across the others and give more detail. It was now clear that in principle the method worked; it *was* possible, without the use of special research ships, to chart the plankton over wide areas and study its changing distribution with the passage of time.

In 1938 the survey was extended over the whole of the North Sea by a laboratory opened at Leith to operate lines from there to Shetland, to the Skagerak and to Hamburg; ships sailing from Glasgow also ran recorders from the north of Scotland to Bergen and again to Hamburg. The last-mentioned line passed down the middle of the area and cut across other lines to give us confirmation of results at several points. In the spring of the following year we began a line from the Pentland Firth across 400 miles of the Atlantic towards Iceland. Then came the war, but fortunately not before this extended survey had given good promise for the future (Lucas, 1941, '42; Rae and Rees, 1947; Marshall, 1948).

As the monthly charts of 1938 were made, it became possible to mark the edge of the Atlantic inflow by the front of a wave of oceanic animals advancing down the east of Britain and spreading out over the North Sea from August to the end of the year; *Sagitta elegans*, the pteropods *Limacina retroversa* and *Clione limacina*, and the copepod *Metridia lucens* were the principal indicator species. Variations in the timing and extent of this invasion must have a profound effect upon the North Sea as a whole. Clearly it was worth following up. I had now left Hull, but Dr. C. E. Lucas, who had begun the work with me in 1932, drew the team together again in 1946, and directed the resumption of the survey till he himself left in 1948 to become Director of the Scottish Fishery Laboratory at Aberdeen. For some years after the war, shipping lines, particularly those to Germany, were much curtailed so that we could not get so full a service in the North Sea as we had before. This loss in intensity, however, was offset by an immense gain in extent; the new weather-ships, two British, two Dutch and one Norwegian, have joined in the campaign, and now tow recorders when going out to take up their positions in the Atlantic and Arctic. These, with the renewed Icelandic run, now give us a

series of lines radiating west and north from our more immediately
surrounding waters: it is thus possible to follow movements of plankton
in the Atlantic Current, which may eventually be linked up with
later plankton changes in the North Sea (see for example, Glover,
1952). Fig. 101 (p. 310) shows a map of the present-day recorder
lines. The survey, now under the direction of Mr. K. M. Rae,
has outgrown the phase of being a pioneering University research;
it has become a long-term project with its centre of gravity much
more in the north. It has now a new home of its own—the Edinburgh
Oceanographic Laboratory, and has become part of the organisation
of the Scottish Marine Biological Association.

It would be out of place here to go into this work in detail. We
may note, however, that in addition to the more typical plankton,
the fish eggs and young are recorded, (Stubbings, 1951; Henderson,
1954), as well as the larval stages of many bottom-living invertebrates

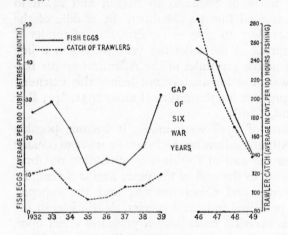

FIG. 102
A graph showing the cor-
respondence between the
number of fish eggs taken by
the plankton recorder in the
southern North Sea (con-
tinuous line) and the fish
landed at British ports by
trawlers working in the same
area (broken line) from
1932-39 and 1946-49. From
Henderson (1954).

which are so important as the food of the trawl-caught fish, (Rees 1951,
1952). Each of these lines of study may prove to be of more than
academic interest. The results of this survey have now filled three
volumes of the *Hull Bulletins of Marine Ecology*.[1] Let me draw attention
to one point. In Fig. 102, above, I reproduce a graph from Dr.
Henderson's 1954 paper on the eggs and young of fish; it compares
the numbers of eggs taken by the recorder in the southern North Sea
in the years 1932 to '49 (except the war years) and the quantities of
fish landed by trawlers from the same area. The agreement between

[1] Since the removal of the Headquarters to Edinburgh the series is being continued
as the *Bulletins of Marine Ecology*.

the two is surely remarkable. It is, of course, really what we should expect; the number of eggs spawned should vary in proportion to the number of adult fish on the grounds; what takes us a little by surprise is that the agreement should be so good—it indeed gives us a measure of the reliability of our method as a means of recording some, at any rate, of the events in the sea. It also gives support to Victor Hensen's original contention that the charting of fish-egg abundance could be used to estimate the strength of the stocks of adult fish; note well that it refers to the stocks that spawned the eggs, not those to which the eggs will give rise—we have seen how greatly the survival of the young may vary in different years.

There are exciting possibilities showing themselves in the recorder results, but it is too early yet to be counting our chicks; I will, however, give just one hint as an example of the sort of possibility I mean. Fig. 103 (p. 314), which is taken from Glover's recent paper (1955), shows on the left the typical monthly progression, as recorded in our survey, of the plankton species indicating the flow of mixed Atlantic water from the north; and on the right, the positions and dates of the main spawnings of the herring in the North Sea. The two series of events show the same north-to-south sequence over the same area. Are they related, as Mr. Rae has tentatively suggested? Does an earlier or later influx of water from the north affect the times of arrival of the shoals coming into spawn and so also the times of the fisheries? If it did, then here could be an excellent means of forecasting. We have just seen (p. 306) Mr. Glover's evidence for the possible Atlantic influence on the eastern Scottish fisheries; this lends some colour to what at the moment is no more than a tantalizing possibility.

In stressing the use of the recorder in this survey I do not wish to give the impression that I believe it has replaced the tow-net and other types of sampler for all kinds of plankton study; there are many classes of investigation for which it is quite unsuitable. In the periodic survey, too, it has its limitations. At present we use it at one standard depth, so that for many animals which have a marked diurnal vertical migration we must carefully distinguish those parts of a record taken during the day from those taken at night. It would be better if we could afford to run two or three at different depths (as has been done recently by American workers for charting fish eggs and larvae) or, over deep water, cause a single machine, perhaps with movable diving planes, to undulate between the surface and deeper water for every mile of tow.[1] No doubt before long the old recorder will be

---

[1] Experiments in this direction have been made but not yet with complete success.

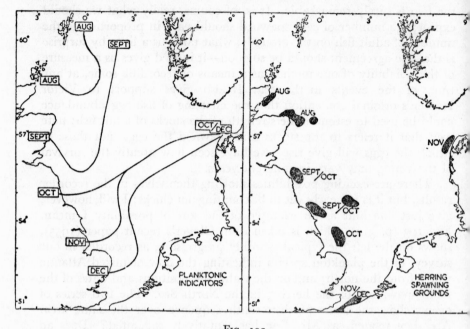

FIG. 103

Left, the typical month by month southern limit of plankton species indicating a flow of mixed Atlantic water from the north, as shown by the continuous plankton recorder survey. *Right*, the position and dates of the more important herring spawning grounds. From Glover (1955).

obsolete, as it is superseded by a new invention. I venture to believe however, that any future forecasting of the distribution of fish must be made with an up-to-date knowledge of the changes in the sea over a wide area, and that this may best be gained by some rapid method of recording the varying distribution of its plankton. In saying this, I must also make it clear that I do not for a moment wish to imply that such a survey will cut out the necessity for the continued efforts of the ecologists on the fisheries research ships; rather, it will be a supplementary background to their work. Their great contribution to marine science will be the subject of our next volume, *The Open Sea, Part II*, which will deal principally with the fish and fisheries.

It must take many years yet before we have enough knowledge of this changing plankton to establish what may be considered the *normal* state of distribution for any one time of year; only when this has been done, of course, can we see in which particular ways the different years have been *abnormal*. Not until then can we safely think of trying to predict the effect of such changes on a fishery. We must not be discouraged at having to go slowly; we must always remember what a big place the sea is, and how obscure. The first attempts will no doubt be poor ones, but were not the first weather maps poor things when compared with the forecasts of today which are of such service to marine and aerial navigation?

While I have sought in this last chapter to show how our plankton studies may be of service to the fisheries, I must not end on a utilitarian note. We are the new naturalists, looking at the sea with the ecological outlook, exploring it for our pleasure and hoping perhaps to add some new facts to its natural history. Let me end by reiterating the hope expressed in the opening chapter, that my book may be a lure to entice some naturalists, who have hitherto confined their attentions to the land, to enjoy the marine habitat as well; and I shall be still more pleased if I may have attracted new recruits to natural history from among the very many who spend their holidays by the sea or afloat on cruising craft. It is the astonishing variety of life that makes the sea such a fascinating hunting ground. Get a tow-net, dredge and quite a simple microscope, and a new world is yours—a world of endless surprises.

The delights of dredging have only just been touched upon in the present volume; an account of them will be *continued in our next.*

ADDENDUM

Regarding my statement on p. 14 of the first edition that Maury wrote the first text book of oceanography, Dr. J. N. Carruthers of the National Institute of Oceanography, has kindly drawn my attention to Count Marsilli's *Histoire physique de la mer*, published at Amsterdam in 1785. I had not been aware of this book.

BIBLIOGRAPHY

THIS IS NOT a bibliography in the sense of being in any way a full catalogue of the books and papers dealing with the natural history of the open sea; that, of course, would be quite impossible in a volume of this size. It is nothing more than a list of references to the books and more important research papers which have been mentioned in the text; it cannot pretend to give the original source of discovery of every bit of natural history recorded.

References to many important papers are omitted if their main findings have been brought together in a later more comprehensive work which has been quoted; for example the earlier papers on the ecology of *Calanus* by Drs. Marshall and Orr are not included because they are so well summarized in their authors' recent book *The Biology of a Marine Copepod* (1955).

From the further lists of references within the books and papers included, and so on yet to others, the student will be able to follow the trail of any particular line which is attracting him. All the abbreviations used are those given in the *World List of Scientific Periodicals*.

AGASSIZ, A. (1833). Exploration of the surface fauna of the Gulf Stream III: The Porpitidae and Velellidae. *Mem. Mus. comp. Zool. Harvard, 8*, No. 2.

ALCOCK, A. (1902). *A Naturalist in Indian Seas*. London, John Murray.

ALLEN, E. J. (1914). On the Culture of the Plankton Diatom *Thalassiosira gravida* Cleve, in Artificial Sea-water. *J. Mar. biol. Ass. U.K., 10*, 417–439.—(1919). A contribution to the Quantitative Study of Plankton. *Ibid., 12*, 1–8.

ALLEN, E. J. and NELSON, E. W. (1910). On the Artificial Culture of Marine Plankton Organisms. *Ibid., 8*, 421–474.

BAINBRIDGE, R. (1952). Underwater observations on the swimming of marine zooplankton. *Ibid., 31*, 107–12.—(1953). Studies on the interrelationships of zooplankton and phytoplankton. *Ibid., 32*, 385–447.

BAINES, J. M. (1939). A rare cephalopod at Hastings. *Hastings Nat., 5 b*, 266–268.

BAKER, J. R. (1948). The Status of the Protozoa. *Nature, 161*, 548 and 587.

BEEBE, C. W. (1935). *Half Mile Down*. London, J. Lane.

BEEBE, C. W. and VANDER PYL, M. (1944). Eastern Pacific Expeditions of the New York Zoological Society, XXXIII. Pacific Myctophidae (Fishes). *Zoologica, N.Y., 29*, 59–95.

DE BEER, G. R. (1940). *Embryos and Ancestors*. Oxford, Clarendon Press.—(1954). The Evolution of Metazoa, in *Evolution as a Process* edited by Julian Huxley, A. C. Hardy and E. B. Ford. London, Allen and Unwin.

BERRILL, N. J. (1930). On the Occurrence and Habits of the Siphonophore *Stephanomia bijuga* (Delle Chiaje). *J. Mar. biol. Ass. U.K., 16*, 753–55.

BERRY, S. S. (1920). Light Production in Cephalopods. *Biol. Bull. Wood's Hole, 38*, 141–195.

318 THE OPEN SEA

BIDDER, A. M. (1950). The Digestive Mechanism of the European Squids . . . *Quart. J. micro. Sci.*, *91*, 1–43.

BOWDEN, K. F. (1953). Physical Geography of the Irish Sea. *A Scientific Survey of Merseyside.* Liverpool, University Press.

BRANDT, K. and APSTEIN, C. (ed.) (1901–28). *Nordisches Plankton* (many authors), Kiel and Leipzig, Lipsius and Tischer.

BROWNE, E. T. (1898). On Keeping Medusae alive in an Aquarium. *J. Mar. biol. Ass. U.K.*, *5*, 176–180.

BUCHANAN, J. Y. (1919). *Accounts Rendered.* London.

BULLEN, G. E. (1908). Plankton Studies in Relation to the Western Mackerel Fishery. *J. Mar. biol. Ass. U.K.*, *8*, 269–302.

BURFIELD, S. T. (1927). Sagitta. *L.M.B.C. Memoirs, 28.*

CALMAN, W. T. (1909). Crustacea. In Lankester's *Treatise on Zoology*, Part VII, fasc. III. London, A. and C. Black.—(1911). *The Life of Crustacea*, London, Methuen.

CANNON, H. G. (1940). On the Anatomy of *Gigantocypris mulleri.* '*Discovery*' *Rep.*, *19*, 185–244.

CARRUTHERS, J. N. (1926). A New Current Measuring Instrument for Purposes of Fishery Research. *J. Cons. int. Explor. Mer.*, *1*, 127–139.—(1928). New Drift Bottles for the Investigation of Currents in connection with Fisheries Research. *Ibid.*, *3*, 194–205.—(1930). The Water Movements in the Straits of Dover. *Ibid.*, *5*, 167–91.—(1954). Sea Fish and Residual Sewage Effluents. *Intelligence Digest Supplement*, October issue.

CHENG, C. (1941). Ecological Relations between the Herring and the Plankton off the North-East Coast of England. *Hull Bull. Mar. Ecol.*, *1*, 239–54.

CLARKE, G. L. (1932). Quantitative Aspects of the Change of Phototropic Sign in *Daphnia. J. Exp. Biol.*, *9*, 180–211.—(1934). Further Observations on the Diurnal Migration of Copepods in the Gulf of Maine. *Biol. Bull., Wood's Hole*, *67*, 432–55.

CLARKE, R. (1950). The bathypelagic angler fish *Ceratias holbölli* Kröyer. '*Discovery*' *Rep.*, *26*, 1–32.

COOPER, L. H. N. (1952). The physical and chemical oceanography of the waters bathing the continental slope of the Celtic Sea. *J. Mar. biol. Ass. U.K.*, *30*, 465–510.

COUSTEAU, J. -Y. (1954). To the Depths of the Sea by Bathyscaphe. *Nat. Geogr. Mag.*, 106, 67–79. Washington.

CRAIG, R. E. and BAXTER, G. I. (1952). Observations in the Sea on the Reaction of Ultra-violet Light on certain Sound Scatterers. *J. Mar. biol. Ass. U.K.*, *31*, 223–27.

CROFTS, D., (1955). Muscle morphogenesis in primitive gastropods and its relation to torsion. *Proc. zool. Soc. London*, *125*, 711–50.

CUSHING, D. H. (1951). The Vertical Migration of Planktonic Crustacea. *Biol. Rev.*, *26*, 158–92.—(1952). Echo-Surveys of Fish. *J. Cons. int. Explor. Mer.*, 18, 45–60.—(1953). Studies on Plankton Populations. *Ibid.*, *19*, 3–22.

DAMANT, G. C. C. (1921). Illumination of Plankton. *Nature*, *108*, 42.

DARWIN, C. (1845). *Journal of Researches . . during the voyage of H.M.S. Beagle round the world* . . . London, J. Murray (Home and Colonial Library).

DEACON, G. E. R. (1933). A general account of the Hydrology of the Southern

Ocean. '*Discovery*' *Rep.*, 7, 171–238.—(1937). The Hydrology of the Southern Ocean. *Ibid.*, 15, 1–124.

DENNELL, R. (1940). On the Structure of some Photophores of some Decapod Crustacea. *Ibid.*, 20, 307–82.—(1955). Observations on the Luminescence of bathypelagic Crustacea Decapoda of the Bermuda Area. *J. Linn. Soc. Zool.* 42, 393-406.

DOBELL, C. C., (1911). The Principles of Protistology. *Archiv. Protistenk. Jena.* 23, 269–310.

DROOP, M. R. (1954). Cobalamin requirement in Chrysophycae. *Nature,* 174, 520. —(1955). A pelagic marine Diatom requiring Cobalamin. *J. Mar. biol. Ass. U.K.,* 34, 229–231.

EYDEN, D. (1923). Specific gravity as a factor in the vertical distribution of the plankton. *Proc. Camb. phil. Soc. biol. Sci.*, 1, 49–55.

FLEMING, R. H. (1939). The Control of Diatom Populations by Grazing. *J. Cons. int. Explor. Mer.,* 14, 3–20.

FOWLER, G. H. and ALLEN, E. J. (1928). *Science of the Sea,* Oxford, Clarendon Press.

FRASER, F. C. (1937). Common Dolphins in the North Sea. *Scot. Nat., Edinb.,* 1937 103–5.

FRASER, J. H. (1949). The distribution of Thaliacea in Scottish waters, 1920–1939. *Sci. Invest. Fish. Div. Scot., 1949 (1).*—(1952a). Hydro-Biological Correlation at the Entrances to the Northern North Sea in 1947. *Rapp. Cons. Explor. Mer.,* 131, 38–43.—(1952b). The Chaetognatha and other zooplankton of the Scottish area and their value as biological indicators of hydrological conditions. *Mar. Res. Scot., 1952* (2).—(1955). The Plankton of the Waters Approaching the British Isles in 1953. *Ibid., 1955 (1).*

GAMBLE, F. W. (1912). *The Animal World.* London, Williams and Norgate.

GARDINER, A. C. (1937). Phosphate production by planktonic animals. *J. Cons. int. Explor. Mer.,* 12, 144–146.

GARSTANG, W. (1928). The Morphology of the Tunicata and its bearings on the Phylogeny of the Chordata. *Quart. J. micr. Sci.,* 72, 52–187.—(1946). The Morphology and Relations of the Siphonophora. *Quart, J. micr. Sci.,* 87, 103–193. —(1951). *Larval Forms and other Zoological Verses.* Oxford. Blackwell.

GLOVER, R. S. (1952). The Euphausiacea of the North-Eastern Atlantic and the North Sea, 1946–1948. *Hull Bull. Mar. Ecol.,* 3, 185–214.—(1953). The Hardy Plankton Indicator and Sampler: a Description of Various Models in Use. *Bull. Mar. Ecol.,* 4, 7–20.—(1955). Science and Herring Fishery. *Advancement of Science,* 11, 426–434.

GRAHAM, M. (1938). Phytoplankton and the Herring, Part III, Distribution of Phosphate in 1934–36. *Fish, Invest. Lond.,* Ser. II, 16, No. 3.

GRAN, H. H. (1912). Pelagic Plant Life. Chapter VI in Murray and Hjort's *The Depths of the Ocean.* London, MacMillan.

GROSS, F., RAYMONT, J. E. G., MARSHALL, S. M. and ORR, A. P. (1944). A Fish-Farming Experiment in a Sea Loch. *Nature,* 153, 483.

GURNEY, R. (1942). *Larvae of the Decapod Crustacea.* London, The Ray Society.

HAFFNER, R. E. (1952). Zoogeography of the Bathypelagic Fish, *Chauliodus. Systematic Zoology,* 1, 113-33.

HARDY, A. C. (1923). Notes on the Atlantic Plankton taken off the East Coast of England in 1921 and 1922. *Circ. Cons. Explor. Mer.,* No. 78.—(1924a). Report on the Possibilities of Aerial Spotting of Fish. *Fish. Invest. Lond.,* Ser. II, 7, No. 5.

—(1924b). The Herring in Relation to its Animate Environment, Part I, The Food and Feeding Habits of the Herring. *Ibid.*, Ser. II, 7, No. 3.—(1925). ——, Part II, Report on Trials with the Plankton Indicator. *Ibid.*, *8*, No. 7.—(1936). The Continuous Plankton Recorder. *'Discovery' Rep.*, *11*, 457–510.—(1939). Ecological Investigations with the Continuous Plankton Recorder: Object, plan and methods. *Hull Bull. Mar. Ecol.*, *1*, 1–57.—(1951). Towards a Programme of Herring Research. *Rapp. Cons. Explor. Mer.*, *128*, 9–18.—(1953). On the origin of the Metazoa. *Quart. J. micr. Sci.*, *94*, 441–43.—(1954). Escape from Specialisation, in *Evolution as a Process*, edited by Julian Huxley, A. C. Hardy and E. B. Ford. London, Allen and Unwin.

HARDY, A. C. and BAINBRIDGE, R. (1951). Effect of pressure on the behaviour of decapod larvae (Crustacea). *Nature*, *167*, 354.—(1954). Experimental Observations on the Vertical Migrations of Plankton Animals. *J. Mar. biol. Ass. U.K.*, *33*, 409–48.

HARDY, A. C. and GUNTHER, E. R. (1935). The Plankton of the South Georgia Whaling Grounds and Adjacent Waters, 1926–27. *'Discovery' Rep.*, *11*, 1–146.

HARDY, A. C., HENDERSON, G. T. D., LUCAS, C. E., and FRASER, J. H. (1936). The Ecological Relations between the Herring and the Plankton Investigated with the Plankton Indicator. *J. Mar. biol. Ass. U.K.*, *21*, 147–291.

HARDY, A. C. and PATON, W. N. (1947). Experiments on the Vertical Migrations of Planktonic Animals. *J. Mar. biol. Ass. U.K.*, *26*, 467–526.

HARRIS, J. E. (1953). Physical factors involved in the vertical migration of plankton. *Quart. J. micr. Sci.*, *94*, 537–50.

HARRIS, J. E. and WOLFE, U. K., (1955). A laboratory study of vertical migration. *Proc. roy. Soc., B.*, *144*, 329–54.

HARVEY, E. N. (1940). *Living Light.* Princetown University Press. (1952). *Bioluminescence.* New York, Academic Press.

HARVEY, H. W. (1928). *Biological Chemistry and Physics of Sea Water.* Cambridge, University Press.—(1934). Measurement of phytoplankton population. *J. Mar. biol. Ass. U.K.*, *19*, 761–73.—(1942). The Production of Life in the Sea. *Biol. Rev.*, *17*, 221–246.—(1945). *Recent Advances in the Chemistry and Biology of Sea Water.* Cambridge, University Press.—(1955). *The Chemistry and Fertility of Sea Water.* Cambridge, University Press.

HARVEY, H. W., COOPER, L. H. N., LEBOUR, M. V., and RUSSELL, F. S., (1935). Plankton Production and its Control. *J. Mar. biol. Ass. U.K.*, *20*, 407–441.

HENDERSON, G. T. D. (1954). The Young Fish and Fish Eggs 1932–39 and 1946–49. *Hull Bull. Mar. Ecol.*, *3*, 215–252.

HENDERSON, G. T. D. and MARSHALL, N. B. (1944). The Zooplankton (other than Copepoda and Young Fish) in the southern North Sea, 1932–37. *Ibid.*, *1*, 255–275.

HENDY, N. J., CUSHING, D. H., and RIPLEY, G. W. (1954). Electron Microscope Studies of Diatoms. *J. R. micr. Soc.*, *74*, 22–32.

HERDMAN, W. A. (1923). *Founders of Oceanography and their work.* London, Arnold.

HILDER, B. (1955). Radar and Phosphorescence at Sea. *Nature*, *176*, 174.

HJORT, J. and RUUD, J. T. (1929). Whaling and Fishing in the North Atlantic. *Rapp. Cons. Explor. Mer.*, *56*, 1–123.

HOLMES, W. (1940). The Colour Changes and Colour Patterns of *Sepia officinialis*, L *Proc. Zool. Soc. Lond.*, *110*, 17–36.

HUNT, O. D. (1952). Occurrence of Pelagia in the River Yealm Estuary, South Devon. *Nature, 169*, 934.

JACKSON, P. (1954). Engineering and Economic Aspects of Marine Plankton Harvesting. *J. Cons. int. Explor. Mer., 20,* 167–174.

JOHANSEN, A. C. (1927). On the fluctuations in the quantity of young fry among plaice and certain other species of fish, and causes of the same. *Rep. Danish Biol. Stat.* 33, 5–16.

JOHNSTONE, J. (1908). *Conditions of Life in the Sea,* Cambridge University Press.

JOHNSTONE, J., SCOTT, A., and CHADWICK, H. C. (1924). *The Marine Plankton.* London, Hodder and Stoughton.

KEMP, S. (1910). Notes on the Photophores of Decapod Crustacea. *Proc. Zool. Soc., 1910,* 639–651.—(1938). Oceanography and the Fluctuations in the Abundance of Marine Animals. *Brit. Assoc. Rep., 1938,* 85–101.

KEMP, S. and HARDY. A. C. (1929). The Discovery Investigations: Objects, Equipment and Methods, Part II. *'Discovery' Rep., 1,* 151–222.

KNIGHT-JONES, E. W. and QASIM, S. Z. (1955). Responses of Some Marine Plankton Animals to Changes in Hydrostatic Pressure. *Nature, 175,* 941.

LEBOUR, M. V. (1922–23). The Food of Plankton Organisms. *J. Mar. biol. Ass. U.K. 12,* 644–677; *13,* 70–92.—(1925). *The Dinoflagellates of Northern Seas.* Marine Biological Association, Plymouth.—(1930). *The Planktonic Diatoms of Northern Seas.* Ray Society, London.

LUCAS, C. E. (1938). Some Aspects of Integration in Plankton Communities. *J. Cons. int. Explor. Mer., 8,* 309–22.—(1940). The Phytoplankton in the southern North Sea, 1932–37. *Hull Bull. Mar. Ecol., 1,* 73–170.—(1941–42). Phytoplankton in the North Sea, Part I—Diatoms, and Part II—Dinoflagellates, Phaeocystis, etc. *Ibid.,* 2, 19–70.—(1947). The ecological effects of external metabolites. *Biol. Rev. 22,* 270–295.—(1956a). External Metabolites in the Sea. *Deep Sea Res.* (In the press).—(1956b). Chapters on Plankton in *Sea Fisheries and their Investigation,* ed. by M. Graham, London. (In the press).

LUCAS, C. E. and RAE, K. M. (1946). The Plankton of the North Sea in relation to its Environment, Part I. The Hydrological Background in the southern North Sea, 1930–37. *Hull Bull. Mar. Ecol. 3,* 1–33.

LUCAS, C. E. and STUBBINGS, H. G. (1948). Size variation in Diatoms and their Ecological Significance. *Hull Bull. Mar. Ecol., 2,* 133–171.

LUMBY, J. (1932). Current Systems of the North Atlantic and the North Sea. *J. Ecol., 20,* 314–25.

MACBRIDE, W. E. (1914). *Text-book of Invertebrate Embryology.* London, MacMillan.

MACKINTOSH, N. A. (1937). The Seasonal Circulation of the Antarctic Macroplankton. *'Discovery' Rep., 16,* 367–412.

MANTEUFEL, B. P. (1941). Plankton and Herring in the Barents Sea. *Trans. Knipovitch polyar. Sci. Inst., 7,* 125.

MARSHALL, N.B. (1948). Zooplankton (other than Copepoda and young Fish) in the North Sea, 1938–39. *Hull Bull. Mar. Ecol., 2,* 173–213.—(1951). Bathypelagic Fishes as Sound Scatterers in the Ocean. *J. Mar. Res., 10,* 1–17.—(1954). *Aspects of Deep Sea Biology.* London, Hutchinson.

MARSHALL, S. M. (1925). On *Proterythropsis vigilans,* n.sp. *Quart. J. micr. Sci., 69,* 177–84.—(1949). On the biology of the small copepods in Loch Striven. *J. Mar. biol. Ass. U.K., 28,* 45–122.

MARSHALL, S. M. and ORR, A. P. (1928). The Photosynthesis of Diatom Cultures
TOS—Y

in the Sea. *J. Mar. biol. Ass. U.K.*, *15*, 321–60.—(1955). *Biology of a Marine Copepod*. Edinburgh, Oliver and Boyd.

MASSY, A. L. (1928). The Cephalopoda of the Irish Coast. *Proc. R. Irish Acad.*, *38*, 25–37.

MAURY, M. F. (1855). *The Physical Geography of the Sea*. London, T. Nelson & Sons.

MINCHIN, E. A. (1922). *An Introduction to the Study of the Protozoa*. Second Impression. London, Arnold.

MORTON, J. E. (1954). The Biology of *Limacina retroversa*. *J. Mar. biol. Ass. U.K.*, *33*, 297–312.

MOSELEY, H. N., (1892). *Notes by a Naturalist*, London, John Murray.

MURRAY, G. and BLACKMAN, V. H., (1898). On the nature of the Coccospheres and Rhabdospheres. *Philos. Trans. London, B.*, *190*, 427–41.

MURRAY, J. (1913). *The Ocean*. London, Williams and Norgate.

MURRAY, J. and HJORT, J. (1912). *The Depths of the Ocean*. London, MacMillan.

NICHOLLS, A. G. (1933). On the Biology of *Calanus finmarchicus III*. Vertical Distribution and Diurnal Migration in the Clyde Sea-Area. *J. Mar. biol. Ass. U.K.*, *19*, 139–164.

ORTON, J. H. (1922). The mode of feeding of the jelly-fish, *Aurelia aurita* on the smaller organisms in the plankton. *Nature*, *110*, 178.

OSTER, R. H. and CLARKE, G. L. (1935). The Penetration of the Red, Green and Violet Components of Daylight into Atlantic Waters. *J. opt. Soc. Amer.*, *25*, 84–91.

PARKE, M., (1949). Studies on Marine Flagellates. *J. Mar. biol. Ass. U.K.*, *28*, 255–86.

PEARCEY, F. G. (1885). Investigations on the Movements and Food of the Herring and additions to the Marine Fauna of the Shetland Islands. *Proc. R. phys. Soc. Edinb.*, *8*, 389–415.

PHIPSON, T. L. (1862). *Phosphorescence*. London, Lovell, Reeve & Co.

PICKEN, L. E. R. (1953). A Note on the Nematocysts of *Corynactis viridis*. *Quart. J. micr. Sci.*, *94*, 203–227.

POOLE, H. H. and ATKINS, W. R. G. (1926). On the Penetration of Light into Sea Water. *J. Mar. biol. Ass. U.K.*, *14*, 177–198.

POULSON, E. M. (1944). On fluctuations in the size of the stock of cod in the waters within the Skaw during recent years. *Rep. Danish Biol. Stat. 46 (1941)*. 1–36.

RAE, B. B. (1950). Description of a Giant Squid Stranded near Aberdeen. *Proc. Malacol. Soc. Lond.*, *28*, 163–67.

RAE, K. M. and FRASER, J. H. (1941). The Copepoda of the Southern North Sea, 1932–37. *Hull Bull. Mar. Ecol.*, *1*, 171–238.

RAE, K. M. and REES, C. B. (1947). The Copepoda in the North Sea, 1938–39. *Ibid.*, *2*, 95–132.

REES, C. B. (1949). The Distribution of *Calanus finmarchicus* (Gunn.) and its two forms in the North Sea, 1938–39. *Hull Bull. Mar. Ecol.*, *2*, 215–275.—(1951). First Report on the Distribution of Lamellibranch Larvae in the North Sea. *Ibid.*, *3*, 105–134.—(1952). The Decapod Larvae in the North Sea, 1947–1949. *Ibid.*, *3*, 157–184.

REES, W. J. (1949). Note on the Hooked Squid, *Onychoteuthis banksi*. *Proc. Malacol. Soc. Lond.*, *28*, 43–45.—(1950a). The Distribution of *Octopus vulgaris* Lamarck in British Waters. *J. Mar. biol. Ass. U.K.*, *29*, 361–378.—(1950b). On a giant squid *Ommastrephes caroli* stranded at Looe, Cornwall. *Bulletin of the British Museum (Natural History)*, *1*, No. 2.

REGAN, C. T. (1925). Dwarfed Males Parasitic on the Females in Oceanic Angler-Fishes (Pediculati Ceratioidea). *Proc. ray. Soc. London B 97*, 386—399.—(1930). A Ceratoid Fish (*Caulophryne polynema, sp. n.*) Female with Male, from off Madeira. *J. Linn. Soc. London., 37*, 191–195.

RILEY, G. A. and BUMPUS, D. F. (1946). Phytoplankton-zooplankton Relationships on the Georges Bank. *J. Mar. Res., 6*, 33–47.

ROBSON, E. A. (1953). Nematocysts of *Corynactis:* the activity of the filament during discharge. *Quart. J. micr. Sci., 94*, 229–35.

ROBSON, G. C. (1933). On *Architeuthis clarkei*, a new species of giant squid, with observations on the genus. *Proc. Zoo. Soc. Lond. 1933*, 681–697.

RUSSELL, E. S. (1922). Report on the Cephalopods collected by the research steamer *Goldseeker* during the years 1903–1908. *Fish. Scot. Sci. Invest., 1921*, No. 3.

RUSSELL, F. S. (1925–34). The vertical distribution of marine macroplankton I–XII. *J. Mar. biol. Ass. U.K., 13–19.*—(1927). The Vertical Distribution of Plankton in the Sea. *Biol. Rev., 2*, 213–262.—(1932–33). On the Biology of *Sagitta*, I–IV. *J. Mar. biol. Ass. U.K., 18*, 131–160, 555–574.—(1935*a*). The seasonal abundance and distribution of the pelagic young of teleostean fishes caught in the ring-trawl in off-shore waters in the Plymouth area, Part II. *J. Mar. biol. Ass. U.K., 20*, 147–179.—(1935*b*). On the Value of Certain Plankton Animals as Indicators of Water Movements in the English Channel and North Sea. *J. Mar. biol. Ass. U.K., 20*, 309–332.—(1936). Observations on the Distribution of Plankton Animal Indicators made on Col. E. T. Peel's yacht "St. George" in the Mouth of the English Channel, July 1935. *Ibid., 20*, 507–22. —(1938–40). On the Nematocysts of Hydromedusae, I, II and III. *Ibid., 23*, 145–165, 347–359; and *24*, 515–523.—(1939). Hydrographical and Biological Conditions in the North Sea as Indicated by Plankton Organisms. *J. Cons. int. Explor. Mer., 14*, 171–192.—(1940). On the seasonal abundance of young fish. VII. The year 1939, January to August. *J. Mar. biol. Ass. U.K.*, 265–270.—(1952). The Relation of Plankton Research to Fisheries Hydrography. *Rapp. Cons. Explor. Mer., 131.* 28–34.—(1954). *The Medusae of the British Isles.* Cambridge, University Press.

RUSSELL, F. S., and KEMP, S., (1932). Pelagic Animals off the South-West Coasts of the British Isles, *Nature, 130*, 664.

RUSSELL, F. S. and YONGE, C. M. (1928). *The Seas*, London, Frederick Warne.

SANDERS, F. K. and YOUNG, J. Z. (1940). Learning and other Functions of the Higher Nervous Centres of *Sepia. J. Neurophysiol., 3*, 501–26.

SARS, G. O. (1895–1928). *An Account of the Crustacea of Norway.* Christiana (Oslo) and Bergen.

SAVAGE, R. E. and HARDY, A. C. (1934). Phytoplankton and the Herring, Part I. *Fish. Invest. Lond., 14*, No. 2.

SAVAGE, R. E. and WIMPENNY, R. S. (1936). Phytoplankton and the Herring, Part II. *Ibid., 15*, No. 1.

SMITH, H. G. (1936). Contribution to the anatomy and physiology of *Cassiopeia frondosa. Pap. Tortugas Lab., 31*, 19–52.

SOUTHWARD, A. J. (1955). Observations on the Ciliary Currents of the jelly-fish *Aurelia aurita* L. *J. Mar. biol. Ass. U.K., 34*, 201–216.

STEPHEN, A. C. (1944). The Cephalopoda of Scottish and Adjacent Waters. *Trans. roy. Soc. Edinb., 61*, 247–270.

STEVEN, G. A. (1949). Contributions to the Biology of the Mackerel, *Scomber scomber*

L., II: A study of the Fishery in the Southwest of England, with special reference to Spawning, Feeding and "Fishermen's Signs". *J. Mar. biol. Ass. U.K.*, *28*, 555–81.

STEVENSON, J. A. (1935). The Cephalopods of the Yorkshire Coast. *J. Conch.*, *20*, 102–116.

STUBBINGS, H. G. (1951). Fish Eggs and Young Fish in the North Sea, 1932–1939. *Hull Bull. Mar. Ecol.*, *2.*, 277–281.

SVERDRUP, H. V., JOHNSON, M. W. and FLEMING, R. H. (1942). *The Oceans: their Physics, Chemistry and General Biology*. New York, Prentice Hall.

TAIT, J. B. (1952). *Hydrography in Relation to Fisheries*. London, Arnold.

TCHERNAVIN, V. V. (1947). Six specimens of Lyomeri in the British Museum. *J. Linn. Soc.*, *41*, 287–350.—(1948). On the Mechanical Working of the Head of Bony Fishes. *Proc. Zool. Soc.*, *118*, 129–143.

THOMSON, C. WYVILLE, (1873). *The Depths of the Sea*. London, MacMillan.

TINBERGEN, N. (1951). *The Study of Instinct*. Oxford. Clarendon Press.

TOTTON, A. K. (1954). Siphonophora of the Indian Ocean. *"Discovery" Rep.*, *27*, 1–162.

VERRILL, A. E. (1882). Report on the Cephalopods of the North-Eastern Coast of America. *Rept. U.S. Comm. Fish. for 1879*, 211–260.

WATERMAN, T. H., NUNNEMACHER, R. F., CHASE, F. A., and CLARKE, G. L. (1939). Diurnal Vertical Migrations of Deep-water Plankton. *Biol. Bull. 76*, 256–79.

WATERMAN, T. H. (1948). Studies on Deep-sea Angler-fishes (Ceratioidea) III. *J. Morph.*, *82*, 81–150.

WERNER, B. (1954). Uber die Fort pflanzung der Anthomeduse *Margelopsis haeckeli* Hartlaub durch Subitan-und Dauereier und die Abhangigkeit ihrer Bildung von aussern Faktoren. *Verh. Deutsch. zool. Ges. Leipzig. 1954*, 124–33.

WIBORG, K. F. (1954). Investigations on zooplankton in coastal and offshore waters of western and north-western Norway. *Rep. Norweg. Fish. Invest.*, 11, (1).

WILLIAMSON, D. I. (1956). Plankton studies in the Irish Sea. *Bull. Mar. Ecol.*, *4*, (in the press).

WILSON, D. P. (1932). On the Mitraria larva of *Owenia fusiformis* Delle Chiaje. *Phil. Trans. B.*, *221*, 231–334.—(1937). The Habits of the Angler-fish *Lophius piscatorius* L., in the Plymouth Aquarium. *J. Mar. biol. Ass. U.K.*, *21*, 477–96. —(1946). A Note on the Capture of Prey by *Sepia officionalis* L. *Ibid.*, *26*, 421–424. —(1947).The Portuguese Man-of-War *Physalia physalia* L., in British and Adjacent Seas, *Ibid.*, *27*, 139–72.—(1948–54). Several papers on the settlement of larvae *J. Mar. biol. Ass. U.K.*, 1948, *27*, 723–60; 1953, *31*, 413–38; *32*, 209–33 and 1954, *33*, 361–80.—(1951). A Biological Difference between Natural Sea Waters. *Ibid.*, *30*, 1–20.—(1952). The influence of the nature of the substratum on the metamorphosis of the larvae of marine animals . . . *Ann. Inst. Oceanogr. Monaco*, *27*, 49–156.

WIMPENNY, R. S. (1936). The Size of Diatoms. *J. Mar. biol. Ass. U.K.*, *21*, 29–60.

WOLLASTON, H. J. B. (1915). Report on the Spawning Grounds of the Plaice in the North Sea. *Fish. Invest. Lond.*, Ser. II, 2, No. 4.—(1923). The Spawning of the Plaice in the Southern Part of the North Sea in 1913-14. *Ibid.*, *5*, No. 2.

WOODCOCK, A. H. (1944). A theory of surface water motion deduced from the wind-induced motion of the *Physalia*. *J. Mar. Res.*, *5*, 196–205.

WOODWARD, B. B. (1913). *The Life of the Mollusca*. London, Methuen.

YONGE, C. M. (1949). *The Sea Shore* (The New Naturalist). London, Collins.

GLOSSARY

THE ZOOLOGICAL names of different groups of animals are given in the index where the first page cited is that on which they are described. Where technical terms are explained in the text, and used more than once, the page references to such explanations are given below; those which are explained in the text and not used again are not included.

Adoral Band: p. 190.
Asexual: reproducing without sex, i.e. by budding or simply dividing in two.
Atrium (Atrial): p. 149.
Barbel: filament hanging from lower jaw of some species of fish.
Bathypelagic: referring to life in the ocean depths, but not on the bottom.
Benthic: referring to life on sea-floor.
Bioluminescence: light produced by animals.
Biramous (Limbs): typical crustacean limbs having two branches, *cf. uniramous* with only one.
Blastostyle: p. 106.
Boreal: of the north.
Calcareous: containing lime.
Carapace: p. 167.
Chitin: horny substance forming the outer cuticle of many animals (especially crustaceans).
Chloroplast: p. 40.
Cilia: p. 90.
Compound Eye: p. 177.
Dactylozooid: p. 114.
Demersal: living near the sea bottom.
Detritus: particles produced by breakdown of animal and plant bodies, or by erosion of rocks or soil.
Dimorphism: when an animal has two forms.
Diurnal: daily.
Ecology: the science of interrelations between animals and plants and their surroundings.
Endostyle: p. 150.
Fission: usually dividing into two (simple reproduction) but also multiple fission.
Flagellum (pl. *Flagella*): p. 46.
Ganglion: a nerve centre (hence ganglionic).
Gastrozooid: p. 114.
Herbivore: animal feeding on plants (hence herbivorous).
Hydroid: a polyp member of the class Hydrozoa (p. 95).
Hydrologist: scientist studying properties of water.

326   THE OPEN SEA

Insectivorous: feeding on insects.
Lasso-cells: p. 135.
Manubrium: p. 96.
Maxillae: p. 157.
Medusoid: of or like a medusa (jellyfish).
Metatroch: p. 189.
Motile: capable of motion.
Nematocyst: p. 92f.
Notochord: flexible rod forming basis of spinal column in primitive or young chordate (vertebrate) animals.
Operculum: p. 187.
Parapodium (pl. *Parapodia*): p. 144.
Pelagic: of the open sea.
Photic Zone: p. 59.
Photosynthesis: p. 3.
Phototropism: a turning or movement towards the light.
Phylum: a major division of the animal kingdom.
Polyp: p. 95.
Preoral Band: p. 182 and 191.
Protoplasm: p. 40.
Prototroch: p. 182.
Statolith (or *Statocyst*): sense organ telling animal its position in relation to gravity.
Siliceous: containing the mineral silicon (or its dioxide: silica).
Striations: fine linear marks or furrows on surface of animal or plant.
Symbiosis: p. 89.
Telotroch: p. 189.
Telson: tailpiece of a crustacean.
Tentaculocyst: p. 124.
Thoracic: belonging to *thorax,* division of body next to the head in crustaceans (and insects).
Uniramous: see *Biramous.*
Uropods: limbs on last segment of crustacean body forming with telson a tail-fan.
Vacuole: minute cavity containing fluid in animal or plant cells.
Vesicle: small bladder, bubble or hollow structure.
Vestigial: referring to part of animal reduced in course of evolution to small remains.

INDEX

Numbers in heavy type refer to pages opposite which photographic or coloured illustrations will be found

ADDENDUM

The remarkable spiny decapod larvae in Plate 16, Fig. 5 (opposite p. 171) has now been identified by Dr. Isobella Gordon, of the British Museum (Natural History), as a very rare form belonging to Stebbing's larval genus *Problemacaris*—surely a most appropriate name.